Topics in Atmospheric and Oceanic Sciences

Editors: Michael Ghil Robert Sadourny Jürgen Sündermann

Multiprocessing in Meteorological Models

Edited by
G.-R. Hoffmann and D. F. Snelling

With 106 Figures

Springer-Verlag
Berlin Heidelberg New York
London Paris Tokyo

Dipl.-Math. Geerd-R. Hoffmann
David F. Snelling

European Centre for Medium-Range Weather Forecasts
Shinfield Park, Reading Berkshire RG2 9AX, United Kingdom

Series Editors:

Dr. Michael Ghil, Courant Institute of Mathematical Sciences,
New York University, 251 Mercer Street, New York, NY 10012 / USA

Dr. Robert Sadourny, Laboratoire de Météorologie Dynamique,
Ecole Normale Supérieure,
24 rue Lhomond, 75231 Paris Cedex 05 / France

Dr. Jürgen Sündermann, Universität Hamburg,
Institut für Meereskunde,
Heimhuder Straße 71, 2000 Hamburg 13 / FRG

ISBN-13:978-3-642-83250-5 e-ISBN-13:978-3-642-83248-2
DOI: 10.1007/978-3-642-83248-2

Softcover reprint of the hardcover 1st edition 1988

2132/3130-543210

Contents

List of Attendees

C. Arnold	ETA SYSTEMS, Inc., 1450 Energy Park Drive, St. Paul, MN 55108, USA
A. Barklie	Floating Point Systems U.K. Ltd., Apex House, London Road, Bracknell/Berks. RG12 2TE, United Kingdom
M. Baylis	Floating Point Systems U.K. Ltd., Apex House, London Road, Bracknell/Berks. RG12 2TE, United Kingdom
R. Bell	IBM UK Ltd., 414 Chiswick High Road, London W4 5TF, United Kingdom
R. Benzi	European Center for Scientific and Engineering Computing (ECSEC), IBM Italia Spa, 00147 Roma, Via Giorgione 159, Italy
T. Bloch	CCVR, Ecole Polytechnique, 91128 Palaiseau Cedex, France
C. Burton	IBM UK Ltd., 414 Chiswick High Road, London W4 5TF, United Kingdom
R. Carruthers	Control Data Ltd., Control Data House, 179/199 Shaftesbury Avenue, London WC2H 8AR, United Kingdom
G. Carver	Department of Meteorology, The University, James Clerk Maxwell Building, King's Buildings, Mayfield Road, Edinburgh EH9 3JZ, United Kingdom
G. Cats	Royal Netherlands Meteorological Institute, P.O. Box 201, 3730 AE de Bilt, The Netherlands
J. Clochard	Direction de la Météorologie Nationale, 77 Rue de Sèvres, 92106 Boulogne Billancourt Cedex, France
T. Davies	Meteorological Office, London Road, Bracknell/Berks. RG12 2SZ, United Kingdom

D. Deaven	NOAA, National Meteorological Center, W/NMC2, WWB, Room 204, Washington, DC 20233, USA
A. Dickinson	Meteorological Office, London Road, Bracknell, Berks. RG12 2SZ, United Kingdom
K. Droegemeier	University of Oklahoma, School of Meteorology, 2oo Felgar Street, Norman, Oklahoma 73019, USA
I. Duff	Harwell Laboratory, Computer Science and Systems Division, U.K. Atomic Energy Authority, Harwell, Oxfordshire OX11 0RA, United Kingdom
R. Dumelow	Meteorological Office, London Road, Bracknell, Berks. RG12 2SZ, United Kingdom
Y. Ernie	Service Météorologique Métropolitain, 2 Avenue Rapp, 75340 PARIS Cedex 07, France
K. Fechter	Deutscher Wetterdienst, Postfach 100465, 6050 OFFENBACH, Fed. Rep. of Germany
J. Fleming	Cray Research (UK) Ltd., Cray House, London Road, Bracknell, RG12 2SY, England
K. Fielding	ETA Systems Inc., c/o CDL, Genesis Centre, Birchwood Science Park, Birchwood, Warrington, Cheshire, WA3 7BH
T. Garcia-Meras	Instituto Nacional de Meteorologia, Apartado 285, Madrid 3, Spain
S. Grønaas	Norwegian Meteorological Institute, Postboks 320, Blindern, 0314 Oslo 3, Norway
B. Groskup	Control Data, MINNEAPOLIS, MN, USA
Q. Guoqing	Beijing Meteorological Centre, 46 Beishiqiaolu, Western Suburb, BEIJING, People's Republic of China

J. Gurd — Department of Computer Science, The University of Manchester, Manchester M13 9PL, United Kingdom

N. Gustafsson — Swedish Meteorological and Hydrological Institute, S-60176 Norrköping, Sweden

M. Hammett — IBM UK Ltd., 414 Chiswick High Road, London W4 5TF, United Kingdom

J. Haugen — Danish Meteorological Institute, 100 Lyngbyvej, 2100 Copenhagen, Denmark

P. Helfrich — Royal Netherlands Meteorological Institute, P.O. Box 201, 3730 AE de Bilt, The Netherlands

M. Herscuilz — CCVR, Ecole Polytechnique, 91128 Palaiseau Cedex, France

R. Hockney — The University of Reading, Dept. of Computer Science, Whiteknights Park, READING RG6 2AX, England

M. Hortal — Instituto Nacional de Meteorologia, Ciudad Universitaria, Apartado 285, MADRID, Spain

C. Hsiung — Cray Research, Inc., Highway 178 N, Development Building, Chippewa Falls, WI 54729, USA

G. Hurst — Control Data Ltd., Control Data House, 179/199 Shaftesbury Avenue, London WC2H 8AR, United Kingdom

M. Ikeda — Fujitsu Ltd., Design Section 4, Computer Engineering Department I, Main Frame Div., Computers Group, 1015 Kamikodanaka, Nakahare-ku, Kawasaki 211, Japan

M. Imbard — Direction de la Météorologie Nationale, 77 Rue de Sèvres, 92106 Boulogne Billancourt Cedex, France

A. Iwaya — EDP Planning Office, NEC Corporation, 5-33-1, Shiba, Minato-ku, TOKYO 108, Japan

S. Järvenoja — Finnish Meteorological Institute, P.O.Box 503, SF-00101 Helsinki 10, Finland

C. Kers — Royal Netherlands Meteorological Institute, P.O. Box 201, 3730 AE de Bilt, The Netherlands

P. Kesel — Science Applications, 205 Montecito Avenue, Monterey, CA. 93940, USA

A. Labajo — Instituto Nacional de Meteorologia, Apartado 285, Madrid 3, Spain

J. Larson — Cray Research, Inc., Highway 178 N, Development Building, CHIPPEWA FALLS, WI 54729, USA

A. Larsson — Swedish Meteorological and Hydrological Institute, S-60176 Norrköping, Sweden

B. Leasure — Kuck & Associates, Inc., 1808 Woodfield Drive, Savoy, IL 61874, USA

A. Lim — IBM, Neighbourhood Road, KINGSTON, NY 12401, USA

E. Linnebur — Los Alamos National Laboratory, P.O. Box 1663, Mail Stop 260, Los Alamos, New Mexico 87545, USA

A. Lund — DTH, Bldg. 305, DK-2800 Lyngby Denmark

B. Machenhauer — Danish Meteorological Institute, 100 Lyngbyvej, 2100 Copenhagen, Denmark

N. Marie — Direction de la Météorologie Nationale, 77 Rue de Sèvres, 92106 Boulogne Billancourt Cedex, France

B. Maxwell — Fleet Numerical Oceanography Center, Monterey, CA 94943-5005, USA

W. Macintyre — National Center f. Atmospheric Research, P.O.Box 3000, BOULDER, Colorado 80307, USA

G. McBride	Cray Research (UK) Ltd., London Road, Bracknell, Berks. RG12 2SY, United Kingdom
O. McBryan	Courant Institute, New York University, 251 Mercer Street, NEW YORK, NY 10012, USA
A. Moon	IBM UK Ltd., 414 Chiswick High Road, London W4 5TF, United Kingdom
E. Olson	Control Data, 4105 North Lexington Avenue, Arden Hills, MN 55126, USA
M. O'Neill	Cray Research (UK) Ltd., London Road, Bracknell, Berks. RG12 2SY, United Kingdom
R. Perrott	Dept. of Computer Science, Queen's University, BELFAST, N. Ireland PS Division, CERN, CH-1211 GENEVA 23, Switzerland
M. Pierrard	Direction de la Météorologie Nationale, 77 Rue de Sèvres, 92106 Boulogne Billancourt Cedex, France
L. Prahm	Danish Meteorological Institute, 100 Lyngbyvej, 2100 Copenhagen, Denmark
M. Rattenborg	Danish Meteorological Institute, 100 Lyngbyvej, 2100 Copenhagen, Denmark
P. Rigsbee	Cray Research, Inc., 1440 Northland Drive, Mendota Heights, Minnesota 55120, USA
I. Ron	Floating Point Systems U.K. Ltd., Apex House, London Road, Bracknell, Berks. RG12 2TE, United Kingdom
T. Rosmond	Department of the Navy, Numerical Modeling Department, Naval Environmental Prediction Research Facility, Monterey, CA 93943-5106, USA
H. Rothmund	ETA Systems Inc., MINNEAPOLIS 55108, USA

E. Rowe	ETA SYSTEMS, Inc., 1450 Energy Park Drive, St. Paul, MN 55108, USA
Y. Sakata	Fujitsu Ltd. Chiyoda-Ku, TOKYO 100, Japan
D. Samond	Cray Research (UK) Ltd., London Road, Bracknell, Berks. RG12 2SY, United Kingdom
I. Schmiedel	Direction de la Météorologie Nationale, 77 Rue de Sèvres, 92106 Boulogne Billancourt Cedex, France
J. Sela	NOAA, National Meteorological Center, W/NMC2,WWB, Room 204, Washington, DC 20233, USA
W. Sell	Max-Planck-Institut f. Meteorologie, Bundesstr. 55, D-2000 Hamburg, F.R. of Germany
P. Sguazzero	European Center f. Scientific and Engineering Computing (ECSEC), IBM Italia spa, 001437 Roma, Via Giorgione 159, Italy
B. Smith	Denelcor Inc., 17000 East Ohio Place, AURORA, Colorado 80017, USA
D. Snelling	Denelcor Inc., 17000 East Ohio Place, AURORA, Colorado 80017, USA
D. Soll	IBM Kingston, Neighbourhood Road, Kingston, NY 12401, USA
A. Staniforth	Canadian Atmospheric Environment Service, 2121, Voie de Service Nord, Porte 508, Route Trans-Canadienne, Dorval, Québec , Canada H9P 1J3
W. Stern	NOAA, Geophysical Fluid Dynamics Laboratory, Princeton University, P.O.Box 308, Princeton, New Jersey 08542, USA
R. Strüfing	Deutscher Wetterdienst, Frankfurter Strasse 135, 6050 Offenbach, F. R. of Germany
M. Suarez	NASA/Goddard Space Flight Center, Code 611/Building 22, GREENBELT, Md. 20771, USA

P. Swarztrauber — National Center for Atmospheric Research, Scientific Computing Division, P.O. Box 3000, Boulder, Colorado 80307, USA

R. Sweet — National Bureau of Standards, Mailstop 713, 325 Broadway, BOULDER, Colorado 80303, USA

D. Tanqueray — Floating Point Systems U.K. Ltd., Apex House, London Road, Bracknell, Berks. RG12 2TE, United Kingdom

K. Uchida — Fujitsu, Computer Engineering 1, 1015 Kamikodanaka Nakahara-ku, KAWASAKI, Japan

A. Ullerstig — Swedish Meteorological and Hydrological Institute, S-60176 Norrköping, Sweden

R. Van Lierde — Institut Royal Météorologique de Belgique, Avenue Circulaire 3, 1180 Bruxelles, Belgium

S. Watanabe — NEC Corporation, 33-1, Shiba 5-Chome, Minato-ku, TOKYO 108, Japan

J. Welsh — NOAA, Geophysical Fluid Dynamics Laboratory, Princeton University, P.O. Box 308, Princeton, New Jersey 08542, USA

A. White — Los Alamos National Laboratory, P.O.Box 1663, Mail Stop 260, Los Alamos, New Mexico 87545, USA

P. White — Meteorological Office, London Road, Bracknell, Berkshire, RG12 2SZ, England

T. Wilde — Department of the Air Force, Headquarters Air Force Global Weather Central (MAC), Offutt Air Force Base, Nebraska 68113-5000, USA

R. Wiley — Meteorological Office, London Road, Bracknell, Berks. RG12 2SZ, United Kingdom

Lennart Bengtsson	E C M W F	Director
Horst Böttger	"	Operations, Met. Operations
Dave Burridge	"	Head Research Dept.
Frédéric Delsol	"	Head Meteorological Division
David Dent	"	Research, Numerical Aspects
Richard Fisker	"	Operations, Systems
Rex Gibson	"	Operations, Met. Applications
Mats Hamrud	"	Research, Numerical Aspects
Janet Haseler	"	Research, Data Assimilation
Claus Hilberg	"	Operations, Systems
Anthony Hollingsworth	"	Head Data Division
Michel Jarraud	"	Research, Numerical Aspects
Norbert Kreitz	"	Operations, User Support
Andrew Lea	"	Operations, User Support
Peter Lönnberg	"	Research, Data Assimilation
Dimitrios Maretis	"	Operations, User Support
Brian Norris	"	Operations, Met. Applications
Pam Prior	"	Operations, User Support
Adrian Simmons	"	Research, Numerical Aspects
David Snelling	"	Operations, Consultant
Daniel Söderman	"	Head Operations Dept.
Gilles Sommeria-Klein	"	Head Model Division
Neil Storer	"	Operations, Systems
Per Undén	"	Research, Data Assimilation

<u>Chairman:</u>

Geerd-R. Hoffmann	E C M W F	Head Computer Division

Introduction: The Case for Multi-Processing in Meteorological Applications

GEERD-R. HOFFMANN

European Centre for Medium-Range Weather Forecasts, Shinfield Park, Reading, Berkshire RG2 9AX, United Kingdom

1. HISTORICAL BACKGROUND

The use of computers for weather forecasting has always been considered a natural application, because, although Richardson had in 1922 already pointed out how numerical processes could be applied to predict the weather (Richardson (1922)), the number of arithmetic operations needed for numerical weather prediction was beyond human capacity . Therefore, as soon as computers became commercially available, meteorological services started to acquire them for weather prediction.

1.1 Use of Uniprocessors

All early computers were uniprocessors. Consequently, the meteorological services wrote uniprocessor applications and went to great length to sequentialise their problems. However, as soon as parallelism became available in uniprocessors they began to exploit these features in order to decrease the elapsed time needed for a specific weather prediction.

1.2 Parallelism in Uniprocessors

According to the model of a computer used by von Neumann, the parallelism in a uniprocessor may be introduced in its central processing unit (CPU), the main memory and/or the input/output (I/O) area. In addition, parallelism may be incorporated by software.

1.2.1 Parallelism in CPU

There are various ways to use parallel architecture in the construction of CPU's. One possibility exploited by the Control Data Corporation, for example in its CDC6600 in 1964, is the duplication of arithmetic functional units. Other possibilities implemented by various manufacturers include the use of pipelines for instruction decoding and arithmetic operations. A special case of the latter are vector-pipelines which are used to apply the same arithmetic operator to a sequence of operands.

1.2.2 Parallelism in Memory

In order to allow memory accesses to happen at the same frequency as the instructions issued by the CPU, memory modules are built independent of each other. Interleaving of memory banks or parallel memories are commonly used.

1.2.3 Parallelism in I/O

Very early in the development of computers it was recognised that the difference in speed between CPU and I/O devices was so great that the I/O had to be completed without the CPU waiting for it, i.e. it became essential to overlap CPU and I/O operations. For this reason, DMA (direct memory addressed) channels with interrupts, or similar solutions, were incorporated in almost all computers. Independent I/O processors were later introduced for the same purpose.

1.2.4 Parallelism In Software

After the initial phase of single-user machines, multiprogramming and time-sharing operating systems were developed. Some of them also contained the notion of "multitasking" as part of their design. Tasks in these operating systems ran sequentially on the same CPU in multiprogamming mode. However, due to the unpredictability of interrupts etc., a programming style had to be applied as if these tasks were truly run in a multiprocessor environment.

1.3 Coupled Uniprocessors

In the second half of the 1960's, uniprocessors were already being combined into coupled systems by allowing up to four processors to have access to the same memory or I/O devices. However, as the synchronisation times of these systems were prohibitively large and since more powerful uniprocessors were available, the meteorological services, with some exceptions (most notably the US Navy, using a dual CDC6500), did not make use of these facilities.

2. CURRENT SITUATION

As the requirement for numerical computations in the time period according to the validity of the intended weather forecast still exceeds the maximum computational speed of today's supercomputers, the most advanced meteorological services today use the fastest vector processors commercially available, like CRAY X-MP, CRAY-2, CYBER 205, HITACHI S/810 and others. Using highly optimised programs adopted to the machine architecture, performances for a numerical weather prediction model may average between 70 and 150 million floating point operations per second (MFLOPS) on a single vector processor. Since these speeds are still insufficient for both higher resolution models and longer forecasting periods, a small number of meteorological services, for instance the European Centre for Medium-Range Weather Forecasts (ECMWF), the Canadian and the French Weather Services, have used the multiprocessing features of the most advanced supercomputers to make more computing power per time unit available for their problems. They have reached average MFLOPS ratings of around 300 for their most sophisticated programs with peak rates approaching or even exceeding one GFLOPS (10^9 floating point operations per second). Quite a number of programming problems had to be overcome before such high performance figures could be achieved.

2.1 Parallelisation Of Programs

Amdahl's Law states that the speed-up S achieved by p processors for a problem which contains a percentage TS of sequential code is

$$S = \frac{1}{TS + \frac{1-TS}{p}}$$

In order to achieve a reasonable speed-up it is essential to eliminate as much sequential code as possible from a program. For example, just 2% sequential code will restrict the speed-up to about 75% of that possible on a 16 processor system. Likewise, an infinite number of processors would only produce a speed up of about a factor of 50 for that particular code. Therefore, techniques to decompose a program into parallel tasks, without generating sequential code, are essential. In addition, the workload should be uniformly distributed over the available processors to avoid unbalanced waiting. Tools to achieve these goals automatically are not yet available.

2.2 Code Granularity

The decomposition of programs has to take into account the overheads introduced by necessary task synchronisation, because otherwise these overheads may result in too high a proportion of sequential code. The granularity thus imposed is machine architecture dependent and consequently has to be modified, if the same program is to run on a different machine. Tuning aids for this problem are still unavailable.

2.3 Program Debugging

Since the order in which the operating system processes the different parallel tasks may vary from run to run, the reproducibility of results is no longer guaranteed and debugging such a program consequently becomes very difficult. New methods of discovering programming errors which inevitably occur have to be investigated.

3. FUTURE REQUIREMENTS

According to Bengtsson (1984), numerical weather prediction (NWP) models will require between 10^{12} and 10^{16} computations to be carried out within 3 to 24 hours. This results in a requirement for a computer to produce up to an average of 10^2 GFLOPS, i.e. the performance of today's fastest computer must be upgraded by a factor of at least 300. Such an increase in performance can only be achieved with currently developing technology if the number of processors being used in parallel is increased dramatically.

4. CONCLUSION

For NWP to achieve its goals in the not too distant future, the problems raised by using massive parallel computers have to be solved, but first, the difficulties already encountered using only a few processors in parallel have to be overcome. The workshops organised by ECMWF provide a forum to air these problems and to focus the attention of the experts on this field of research.

5. ACKNOWLEDGEMENTS

The author would like to thank all participants to the workshops for their contributions, his colleagues at ECMWF for their support and encouragement and last, but not least, Springer Press for making the papers from the workshops more widely available.

6. REFERENCES

Bengtsson, L. (1984): Computer Requirements For Atmospheric Modelling. This report.

Richardson, L.F. (1922): Weather Prediction By Numerical Process. London. 1922.

State of the Art of Multi-Processors[1]

OLIVER A. MCBRYAN[2,3,4,5]

Courant Institute of Mathematical Sciences, New York University, New York, NY 10012, USA

ABSTRACT

Scores of projects are underway world-wide to develop effective parallel processors. Some of these projects have already resulted in commercial machines, others will at most lead to interesting prototypes. We survey these activities, with special emphasis on those architectural features that are important for scientific computing.

As an illustration of the use of parallel processing in fluid mechanics, we present a number of parallel algorithms for the numerical solution of pressure equations and describe implementation on the Denelcor HEP Parallel Processor and on the Caltech Hypercube. We also describe the use of monitors as an effective tool for exhibiting parallelism in a machine-dependant way.

---

1 Invited address to the Workshop on using Multi-processors in Meteorological Models, European Centre for Medium Range Weather Forecasts, Reading, Dec. 3-5 1984.

2 Supported in part by DOE grant DEAC0278ER0303077
3 Supported in part by NSF grant DMS-83-12229
4 Supported in part by ARO grant DAAG29-84-K-0130
5 A.P. Sloan Foundation Fellow.

1. INTRODUCTION

Serial computers have dominated computer architecture for over 30 years. There has been a steady rate of increase in the processing power of these machines, culminating in supercomputers such as the CRAY and the CYBER machines. Because of limitations imposed by the speed of light, very substantial advances in computer powers are unlikely to be made in the near future using strictly serial architectures. This is already evident in the current range of supercomputers, which in additional to serial scalar computations provide their greatest computing performance by employing limited parallelism in the form of vector computational units.

Parallel computers will become essential to many computational programs over the coming decade, and will require the development of new algorithms, operating systems and programming methodologies. These machines will consist of many serial machines, joined together by some form of connection network and possibly sharing a global memory. Much of the development work on these machines is still at the research level. Significant problems remain in constructing efficient processor and memory inter-connection networks for these machines, and at the software level, in providing a manageable interface between the user and the complex architecture. Because there does not appear to be a unique optimal parallel computer design, a variety of distinct architectures are likely to become accepted, leading to substantial challenges in Designing portable applications software.

Parallel computers may be broadly categorized in two types - SIMD or MIMD. SIMD and MIMD are acronyms for Single Instruction stream - Multiple Data stream and Multiple Instruction stream - Multiple Data stream respectively. In SIMD computers, every processor executes the same instruction at every cycle, whereas in an MIMD machine, each processor executes Instructions independently of the others. The vector units of a CRAY-1 computer are an Example of SIMD parallelism - the same operation must be performed on all components of a vector. Most of the interesting new parallel computers are of MIMD type which greatly increases the range of computations in which parallelism may be effectively exploited using these machines. Any other categorization is between machines with global or local memories. In local memory machines, communication between processors is entirely handled by a communication network, whereas in global memory machines, communication between processors is entirely handled by a single high-speed memory is accessible to all processors. Beyond this, it becomes difficult to categorize parallel machines. There is an enormous variety in the current designs, particularly in the inter-connection networks. For a taxonomy of current designs, see the paper of Schwarz.[1]

In section II we begin by reviewing some of the parallel architectures that are currently available or are under development. In section III we look more closely at several of these architectures, emphasizing in particular issues most closely related to programming. As an example of the use of such parallel computers we describe in section IV several algorithms for fluid pressure solution which we have recently implemented on parallel computers. Finally, in section V we discuss the use of monitors as a step towards providing a portable programming environment for parallel computers.

2. OVERVIEW OF PARALLEL COMPUTING EFFORTS

2.1. A Partial List of Multi-processors

There are at least 50 to 100 parallel computer projects underway at this time worldwide. While many of these projects are unlikely to lead to practical machines, a substantial number will probably lead to useful prototypes. In addition, several commercial parallel computers are already in production (eg ICL DAP, Denelcor HEP, Intel iPSC), and more are under development. One should also remember that the latest CRAY computers, (eg CRAY-XMP and CRAY-2) involve multiple processors, and other vector computer manufacturers are working on similar strategies.

The following is a list of some of the parallel computers under development. This is just a sample of the projects mentioned above, but covers a wide range of different architectures chosen more or less at random.

Some Parallel Computer Projects

ICL DAP
Caltech Hyper-Cube
Intel iPSC Parallel Processor
NCUBE hypercube
Denelcor HEP-1 and HEP-2
NYU/IBM Ultra-computer
Connection Machine (Thinking Machines Inc.)
CRAY-XMP and CRAY-2, IBM 3084 etc
Cornell 16 FPS-164 System
Goodyear MPP
MIT Data-flow Machines
Wisconsin Database Machine
BBN Butterfly
ELXSI 16-processor
GMD SUPRENUM
IBM GF-11
Myrias 4000 (Myrias Research Corp.)
Cedar Project
Flex (Flexible Inc.)
Sequent

2.2. Machine Characteristics

Most of the machines currently under development have in the range of 2 to 1000 processors. In this section we will look briefly at the characteristics of a number of these machines, while in the next section we will look at a few of them in substantially more detail. As mentioned previously, parallel computers can be categorized either as global memory machines or as assemblies of independent loosely coupled processors. The Denelcor HEP is a prime example of the shared memory systems, while the Caltech and Intel Hypercubes are typical of the latter. Some of the machines are intermediate between these extremes - for example, the IBM/NYU Ultracomputer has a memory which is reconfigurable between global and local.

The machines listed above vary greatly in local processing power. The Intel iPSC has 128 processors, each essentially an IBM PC. The 512 processor GF-11 will have 512 20 Mflop processors for a combined peak rating of 11 Gflops.

The Connection Machine designed by Thinking Machines Inc. of Cambridge, Mass., has 65,536 1-bit processors, though this is simply an initial prototype for a machine with 1,000,000 processors. While designed primarily for artificial intelligence work, this machine also has potential applications to scientific computing applications.

The Myrias computer, built by Myrias Research Corp. of Edmonton, Alberta, will also have 64K processors with 8 Gbytes of memory and a peak rate of 16000 Mflops. This machine is definitely designed for scientific computing.

The NYU/IBM Ultracomputer ("Ultra" below) might have 4096 processors sharing an equal number of memory banks through a complex switch. In fact the design scales up well to at least 64000 processors. The initial configuration, the RP3, is a 512 processor machine which is being built at IBM, with a peak processing power of about 1 GIP and 500 Mflops. A smaller prototype with 8 to 16 processors is nearing completion at NYU, and will be used for software development for the RP3.

The ELXSI multi-processor is an assembly of up to 16 4 MIP processors connected by a very high-speed bus. While initially designed primarily to deliver throughput in a multi-user time-sharing environment, there is both software and hardware support for parallel processing.

The German SUPRENUM project involves coupling 256 processors with a network of busses, with the busses arranged as a rectangular grid - i.e. 16 horizontal and 16 vertical busses.

The GF-11 is another IBM parallel computer, designed to perform very specific scientific computations at Gflop rates. The GF-11 will have 576 processors, (includes 64 backup processors), coupled through a three stage Benes network which can be reconfigured at every cycle in 1024 different ways by an IBM 3084 control processor. Peak processing power of 11 Gflops will allow previously uncharted computational regimes to be explored. The machine has been designed primarily for solving quantum field theory programs and is not a general purpose computer - in particular very little software will be available. It is an SIMD architecture but with some flexibility in that the settings of local registers may be used to control the behaviour of individual processors.

Cornell University, with funding from NSF and IBM, is assembling a set of up to 16 FPS-264 array processors. These array processors are themselves powerful and cost-effective parallel computers. The 16 processors will be connected to a high-speed bus along with a very large quantity of memory. A top-end IBM machine, probably a 4-processor IBM 3084 or Sierra, will be used as control processor and will also provide the user interface to the assembly. Peak processing rates of several Gflops should be attainable for suitably tailored algorithms. A similar arrangement, based on the FPS-164 array processor is already running at IBM, though without a high-speed bus.

A variety of other important architectures are also under development. These include various dataflow machines (with bus, tree and grid structures), examples include the MIT Tagged Token machine, the NTT Dataflow grid machine and the Manchester Dataflow Machine; tree-structured machines (binary trees, trees with sibling or perfect shuffle connections), examples are the Columbia University DADO machine and the CMU Tree Machine; nearest neighbor machines such as the MPP and ring architectures, for example the University of Maryland's ZMOB (256 processors on a ring).

Most of the systems described above are designed to support substantial amounts of parallelism. There are also a host of new commercial machines under development that support limited parallelism, and are primarily designed to take advantage of economies of multi-processing without incurring the overheads required by massive parallelism - e.g. the need for complex data networks. The multi-processor ELXSI, CRAY and ETA computers are in this category, as are many smaller machines such as the 12 processor NSC-32032 based Sequent computer and the Flexible Computers System's FLEX computer which supports up to 20 processors with reconfigurable bus interconnects. Finally one should mention a range of new low-price vector computers such as the CONVEX C-1, the Scientific Computer System's SCS-40 and American Supercomputer's ASC, the latter two being more or less CRAY compatible although all of these machines are currently only single processor vector computers. However, with price tags around half a million dollars and processing power of 1/6 of a CRAY-1 up to 1 CRAY-1, multi-processor versions of these machines will likely appear before long.

3. DESIGN OF SOME REPRESENTATIVE SYSTEMS

Several parallel computers have already been marketed commercially while others have been developed as research projects at universities or other research centers. We review some of these systems here in more detail.

3.1. The Denelcor HEP

The Denelcor HEP-1 computer is an example of a parallel computer with a global shared memory, and three of these have been installed at various national laboratories for some years (BRL, ANL, LANL). These machines have provided researchers throughout the U.S. with a wealth of useful experience in the use of such an architecture. Because of the global memory, the HEP is relatively straight-forward to program, and most parallelizable algorithms are therefore easily implemented.

We will describe the HEP-1 computer briefly, referring the interested reader to the paper of Jordan [2] for details. The basic computational unit, called a PEM, is a pipelined MIMD processor rated at 10 MIPS. For the purpose of this paper, a PEM may be thought of as an 8-processor MIMD machine with a global shared memory. There is software/hardware support for creating virtual parallel processes in a PEM and up to 50 processes may execute concurrently. The physical parallelism involved is actually only about 8, the extent of pipelining in the instruction stream, although the memory system does have a higher parallel bandwith. An unusual feature of the HEP is that memory words have an extra **full/empty access bit** which is used for synchronization purposes between processes. Multiple PEM configurations utilize a complex data-switch to connect the processors and memory. However a disadvantage of this particular global memory implementation is that the machine is not easily extended to large numbers of processors. The largest configurations constructed so far have been 4-PEM machines.

Software for the HEP consists of UNIX System V along with FORTRAN or C primitives for implementing parallelism. These include a parallel subroutine call and primitives for reading and writing the **full/empty** access bits of the memory words. We now discuss these facilities in more detail.

3.1.1 HEP Programming

The high-level language user (FORTRAN, C) can exploit the available parallelism by executing parallel subroutine calls. For this purpose a new primitive called CREATE is added to FORTRAN, similar to the FORTRAN CALL statement, but implying that a subroutine is to be executed asynchronously. Processes spawned by CREATED routines perform concurrently and die on reaching a RETURN statement. Scheduling of tasks is the responsibility of the caller - for example in the common case, where N concurrent processes are spawned to execute the interior of a DO loop, it is essential that the calling program wait until all of the sub-tasks have completed. Such barriers are implemented using the natural locking mechanisms associated with memory variables through the extra access bit mentioned previously. For this purpose, special asynchronous access calls are provided that manipulate these bits. Thus a routine ASET is used to initialize a variable and mark it as FULL, while a function SETE is used to mark a variable as EMPTY. An asynchronous write operation to a variable (AWRITE) can only be performed if the access state of the variable is EMPTY, otherwise the write is delayed until the state becomes EMPTY. An asynchronous read, AREAD, can be performed on a variable only if the access state is FULL. In addition, logical functions (FULL(a) and EMPTY(a) can be used to test the access state of variables. All of these functions may be treated as indivisible operations by the user, the hardware supplying the underlying support necessary to deal with multiple simultaneous calls to the same memory location. Communication between processes other than for synchronization is accomplished through subroutine parameters and shared common blocks. Local variables of a subroutine are not shared with concurrently executing instances of the same routine.

3.2. The Caltech Cosmic Cube

The Cosmic Cube, developed by Geoffrey Fox and Charles Seitz[3,4] at Caltech is an example of a totally different architecture. Processors are located at the vertices of an n-dimensional hypercube and adjacent vertices of the cube are connected by a wire along the corresponding edge. There is no globally shared memory -all communication is through the inter-connection channels along edges. This architecture has the advantage of being easily fabricated from standard components, and may be scaled up to much larger sizes (in powers of 2) with almost no change in design. Because of these features, machines of this type are likely to become widely available in the immediate future, whereas development of global memory machines will take substantially longer.

The original Caltech design consists of a 64-node 6 dimensional hypercube utilizing Intel 8086/8087 processors with 128KB of memory at each node. All data exchange between processors occurs in 64-bit packets along the cube edges which are asynchronous full duplex channels. In addition there is an intermediate host machine which acts as a control processor for the entire cube and also provides the interface between the cube and a user.

The software for the cube consists of an operating system kernel, a copy of which resides in each processor, as well as a run-time library providing user access to the communication facilities. Typically, identical copies of a user program are down-loaded to all processors where they execute concurrently. All scheduling is accomplished through communication calls, so that some care is required to prevent locking situations from occurring.

The Caltech group have developed a simulator for the hypercube. This runs on a VAX or similar computer, using operating system facilities such as (under UNIX) pipes and process creation to simulate the cube. Availability of an effective simulator of this type is a tremendous advantage during code development since programs are much more easily debugged on the simulator than on the cube.

A more advanced Caltech cube called the Mark III is now under development. This will have much faster processors at the nodes (Motorola 68020) and local memory per node will reach several megabytes. Other enhancements will be incorporated based on the experience with the prototype.

3.2.1 <u>Caltech Hyper-Cube Programming</u>:

As discussed previously, the n-cube has 2^n vertices with n edges radiating from each. A processor and associated memory are located at each vertex (node) and the cube edges are data channels allowing data transfer between adjacent processors. It is extremely useful to employ a numbering of the cube nodes in which each processor is represented by a binary number such that adjacent processors differ only in 1 bit. Each processor sees n channels connecting it to its neighbours. These channels may then be numbered o, 1, .., n-1 according to which bit differs in the binary node numbers at either end of the channel. There is also an extra channel from the intermediate host (referred to as the IH below) to node 0 through which all communications to and from the cube pass. Data to be communicated between processors is sent in 8-byte packets, which are sufficient to encode all scalar data types. A set of system calls are available to node-resident programs which implement the required communication primitives for these packets. In order, to give the flavour of these system calls, we list some of the most important ones, along with their functions, in the following table:

Cube System Calls:

wtih(data,NODE_0)
  Write a data packet from the IH to node 0.

rdsig(data)
  Data sent from the IH is read by each node.

wtres(data)
  Data is sent from a node to the IH.

rdbufih(datas,NODE_0,CUBESIZE)
  Read the union of data sent by all nodes to the IH.

wtelt(data,chan)
  Send data to the cube neighbour on channel chan.

rdelt(data, chan)
  Read data from the cube neighbour on channel chan.

One additional system call is extremely useful in the simulation of many physically interesting problems - such as those derived from discretizations of partial differential equations on regular grids. The point here is that in such simulations there is typically only nearest-neighbour connectivity among the variables of interest. For efficient use of the hypercube, it is then very desirable to map the grid onto the cube in such a way that neighbouring grid points (in two or three dimensional space) are mapped onto adjacent nodes of the cube. Communications overhead will be minimized by such a mapping. Accomplishing such a mapping is non-trivial and in general impossible - for example there is no such mapping of a 3 dimensional grid onto a 5-cube since the grid requires a local connectivity of 6 at each node. However a general purpose routine called whoami has been developed by John Salmon at Caltech[5] which generates a suitable mapping of the above type in most cases where one is possible. The whoami call is usually executed at the start of any grid-oriented program, and in addition to creating a suitable mapping of the grid to the cube nodes it returns communication channel information for each of the grid neighbours of each processor. This allows the programmer to think entirely in grid space rather than in the less intuitive edge space of the cube.

3.3. The Intel iPSC Computer:

The Intel Corporation, a major manufacturer of microprocessors and computer chips, has just announced the first commercial realisation of the hypercube design, based largely on the Caltech Cosmic Cube. The machine, known as the iPSC, comes in three models -the d5, d6 and d7. These have respectively 32, 64 and 128 processors. The individual processors are the Intel 80286/80287 with up to 512Kb of memory, and the interconnections are provided by a high-speed Ethernet, again using an Intel Ethernet chip. The intermediate host machine, which is both the control processor and the user interface, is an Intel 310 micro-computer running a UNIX system (Xenix). In addition to the Ethernets along cube edges, a global communication channel is provided from the intermediate host machine to the individual processors. This feature is extremely useful for debugging and to a limited extent for control purposes. Besides the UNIX system on the host, software for the system consists of a node-resident kernel providing for process creation and debugging along with appropriate communications software for inter-processor exchanges, and for host to processor direct communication. Combined computing power of a 128-node system with all processors 80% used is about 100 MIPS or 8 MFLOPS, which along with the 64 Mbytes of memory available, provides a relatively powerful computer.

3.4. The NYU/IBM Ultracomputer

The NYU/IBM Ultracomputer is a collaborative effort between IBM and New York University to build a prototype shared memory parallel computer. Preliminary designs and specifications for the computer were developed at NYU over a period of years.[6,7,8,9]
Further design and refinement has been performed recently at IBM, and IBM has now begun construction of a prototype, the 512 processor

RP3. A smaller prototype has been built at NYU. The Ultracomputer consists of N processors and N memory modules interconnected by a complex switch. The switch is actually an Omega Network constructed out of 2 by 2 elementary switches. Every processor is at the root of a binary tree (of depth logN) connecting it to all N memory modules, and correspondingly every memory module is at the root of a binary tree connecting it to all N processors. Thus the network can be viewed as the superposition of 2N binary trees. The address of each machine or memory module is represented by a unique binary number, and specifying a destination address from a source processor or memory implicitly specifies a unique path through the nodes of the network to that destination. The cost of a memory access will be of order O(logN) on average, which is essentially optimal if unacceptable fan-in/fan-out hardware difficulties are to be avoided (a crossbar switch in which every processor is connected to every memory by a wire would require N^2 wires, which is not feasible for large N with any current technologies). Since even a long memory access time is not acceptable for some purposes, Ultra employs both local caches and a local memory at each processor.

A primary feature of the network is that the switch nodes are actually intelligent processors with their own local memories. If two different processors address the same memory location simultaneously, the requests for access will filter through the network until they meet at some switch node. The nodes are programmed to recognise multiple memory requests for the same memory location. The switch hardware combines the requests at that point and submits a single request to the rest of the tree that leads to the desired memory location. It also stores the two addresses of the originators of the requests. When the merged request to the addressed memory has returned a value to the switch node, the node then dispatches the result to the two originators. In the case of a merged write request one or the other requested value is actually written. Thus the result of simultaneous writes to a memory location is in general indeterminate. While general simultaneous writes to the same location are not likely to occur frequently in practice, the special case of simultaneous incrementing of a memory location is extremely important (consider for example the operation of computing the scalar product of two N component vectors using N processors). We discuss this case in the following paragraph.

A unique feature of Ultra is the Fetch_and_add instruction provided by the hardware. As indicated by the name, this instruction fetches a value stored in memory, simultaneously incrementing the stored value by a specified amount. Suppose that a processor issues the call:

Fetch_and_add (Sum,e),

where Sum is a variable in global memory and e is a desired increment to that variable. The call returns the old value of Sum and replaces Sum by Sum+e. The interesting feature of Ultra is that even if several processors all execute Fetch_and_add simultaneously to the same memory location, the calls will all return in one memory cycle as if they had been issued in some unspecified serial order. For example if the two calls

$$\text{Fetch\_and\_add}(\text{Sum}, e_1),$$
$$\text{Fetch\_and\_add}(\text{Sum}, e_2),$$

are issued simultaneously, then either the first call is returned the value Sum and the second is returned $\text{Sum}+e_1$ or else the first call is returned Sum and the second is returned $\text{Sum}+e_2$. In either case, the final value of Sum is $\text{Sum}+e_1+e_2$. In fact, even if all of the processors in Ultra issue Fetch_and_add calls simultaneously to the same memory location, the corresponding results will still be returned within one memory cycle. The key to this is that the Fetch and_add operation is designed into the memory interconnection network - colliding Fetch and_add requests are merged at switch nodes into single memory requests in a manner similar to the treatment of memory reads and writes discussed earlier.

The Fetch_and_add instruction is the key to avoiding critical sections (code segments that are inherently serial) on Ulta. It has been demonstrated that almost all critical sections in a distributed operating system may be avoided by using Fetch_and_add. For example the scheduling of tasks on a queue, a classical critical section, may be avoided by extracting queue entries using Fetch_and_add. This is the basis for the Ultracomputer Operating Systems, a distributed UNIX system which is currently under development. The Fetch_and_add primitive will also be available to high-level language programmers on the Ultra, allowing them to implement process coordination with minimal critical sections in applications programs.

For further details on the Ultracomputer we refer to the publications of the NYU Ultracomputer group.[6,7,8,9]

4. PARALLEL ALGORITHMS FOR FLUID PRESSURE SOLUTION

4.1 Introduction

We have recently begun a program to exploit parallelism in the solution of fluid pressure equations. We describe our initial studies and present results of actual computations run on the Denelcor HEP-1 multiprocessor. We have also implemented many of these algorithms for the Caltech Cosmic Cube. Our goal is to develop algorithms that make close to optimal use of available parallelism on a wide range of realistic problems. All of the work described here is joint work with Eric Van DeVelde.[10,11,12,13,14,15]

Our studies to date have been in two space dimensions for elliptic or implicitly solved parabolic equations. In order to study physically realistic equations which frequently involve shocks or other singularities, we allow irregular grids and refinements, though wherever possible we utilise rectangular indexing (i.e.

logically rectangular grids). In general, our grids consist of unions of rectangles and triangles, with triangles used primarily to fit boundaries and interior interfaces. We have used finite element methods, though similar studies could be applied to finite difference methods. To provide sufficient accuracy, we allow high order elements up to cubics (on triangles) or bicubics (on rectangles). Increased boundary resolution can be obtained by using isoparametric (curved-edged) triangles. We discuss below the implementation of fast solution methods on the HEP parallel processor, though the same codes also run on the CRAY and the FPS-164 array processor, as well as on various serial machines.

Our goal is to exhibit optimally parallel algorithms, which in the case of the HEP means that the algorithms should run at about 8 times their serial speed on a 1-PEM HEP whenever at least 8 parallel processes are used. In almost all cases studied below we have observed close to optimal performance, with speedups typically in the range 7-11 (speedups greater than 8 are possible because the memory banks of the HEP are more heavily pipelined than the processors). We have observed that performance degrades only negligibly if more than 8 processes are used (up to 50 are allowed) - in fact the optimal number of processes is usually substantially higher than 8. In the performance graphs presented below we have generally used from 1 to 19 processes, but using more does not significantly change the plots. The HEP design appears very robust from this point of view. On a 1-PEM HEP, performance eventually begins to degrade slowly as one goes above about 25 processes, so we have stayed below this limit. However we have noted a number of situations where several subroutines which individually yield optimal speedups, do not do so when combined serially into a larger program (overall speedup drops to perhaps 6). We could not explain this slowdown on the basis of serial control code, or anything else we could pin-point. For simplicity we will restrict our attention to the 1-PEM machine.

We have run most of the programs described here using the 4-PEM HEP at BRL, with comparable results (i.e. maximal speedups of order 32 or more). In this paper we will focus on the algorithms used to implement parallelism, referring elsewhere for details of the numerical analysis, see e.g. McBryan.[16]

4.2 Strategies for Parallelization

Our goal is to exploit parallelism at all levels of solution. At the highest level, we use spatial decomposition to exhibit parallelism. In our current work we have accomplished this through the use of overlapping grids. Here, refined subgrids in the form of rotated rectangles are introduced in regions of interest such as near singularities. The use of overlapping grids in hyperbolic problems was pioneered by Berger and Oliver.[17] The set of grids has a general tree structure. Within each subgrid we use a separate finite element formulation, using possibly different order elements in different subgrids, and in general we use a logically rectangular grid indexing within each subgrid. Boundary data for a subgrid are obtained from the solution on the parent grid. For some elliptic problems, an outer iteration may be required to transfer fine grid information back to coarse grids. All subgrids at a given level of refinement (i.e. at the same depth in the tree) may be solved in parallel, leading to a very effective use of MIMD parallelism. We present an example grid, which includes an interior interface, in figures 1 and 2. We note that boundary fitting at the interior interface leads to grids which are locally irregular although logically rectangular in each patch. As a more serious example, in an oil reservoir simulation there are singularities at each of the injection and production wells. Each of these wells would be surrounded by its own refined patch, all of which could be computed in parallel.

At intermediate levels, we use carefully tailored parallel algorithms within each individual spatial component. We will focus on these algorithms for the rest of the paper.

At the lowest levels, we use point-wise parallelism, though with no restriction to executing identical code at individual points. The simplest example here is the use of vector constructs. On machines with limited parallelism such as the HEP, it is not always possible to exploit all levels of parallelism simultaneously. However it is generally easiest to exploit parallelism at the higher levels as this requires a minimal of special programming. As more highly parallel machines with hundreds and even thousands of processors become available, parallelism at all levels will be used more effectively.

4.3. Software Overview

The software for pressure solution consists of a number of fairly independent packages. An Interface Package allows description and manipulation of complex interfaces and fronts including multi-component materials and bubbles. A Grid Package includes grid generators for fitting fronts and interfaces as described earlier, and generates grids of rectangles and triangles. A Finite Element Package provides routines for discretizing equations on such grids with variable order finite elements. A Solver Package provides routines for solving the resulting algebraic equations as efficiently as possible. Finally, a tree-structured control flow package handles the decomposition into rectangular sub-domains, and then makes calls to the other packages to perform the work inside each subgrid.

We have already discussed the parallelism deriving from subgrid decomposition (the equation is concurrently discretized and solved within all subgrids at the same tree depth). On a machine like the HEP with global shared memory, achieving this parallelism requires nothing more than the appropriate concurrent subroutine calls. The code that structures the decomposition is itself easily parallelized since it is tree based. Thus if necessary, subtrees may be handed to new processes for further processing. However since the decomposition involves very little actual computation, it can be run serially on moderately parallel machines such as the HEP without negative impact.

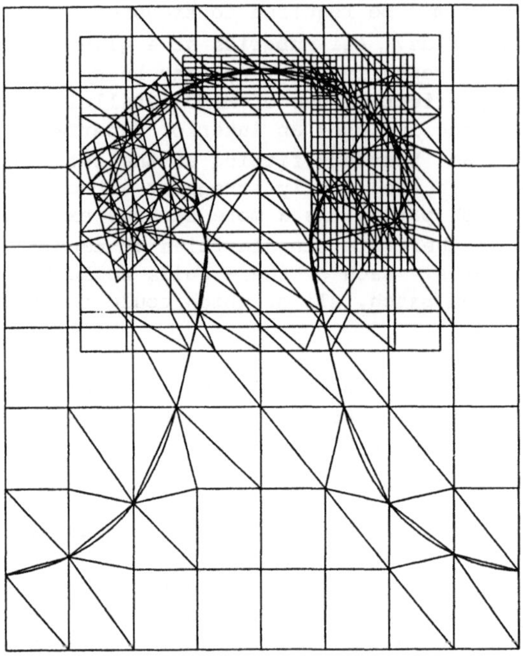

Figure 1 A sample overlapping grid with 3 levels

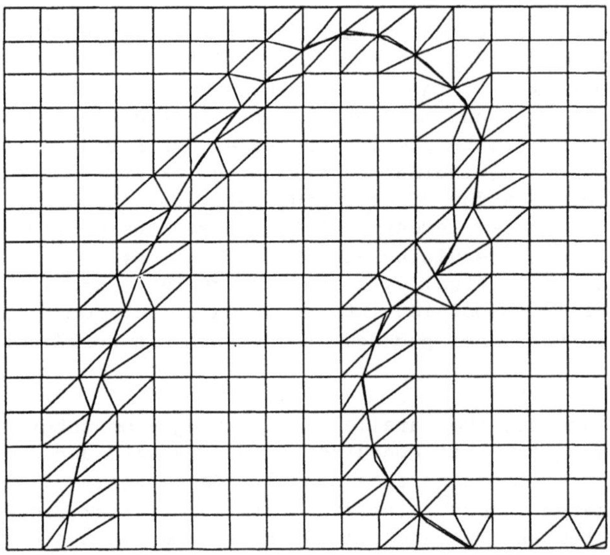

Figure 2 A subgrid from figure 1 shows interface fitting

Execution time is strongly dominated by the finite element and solver packages executed within each subgrid. The finite element construction is trivially parallelized, the construction of the element matrices proceeding on each element independently, fully utilizing available parallelism. Vectorizing such code is very difficult, because the operations to be performed at individual finite element nodes are complicated, and because the elements are neither regular nor homogeneous (rectangles and triangles inter-mixed).. This exemplifies the great advantages of true MIMD machines over SIMD machines such as the CRAY. Thus we will focus now on parallel solution of the resulting equations and refer to our papers for details of the above packages.[16,18,19,20,21]

4.4. Equation and Solution Methods

The basic equation we have studied is of the form:

$$\nabla.(-\vec{K}.\nabla P)\ (x,y) = F(x,y).$$

This is typical of a wide range of fluid pressure equations. Here K has discontinuities, possibly of order a thousand or more, across a given set of curves and in typical applications may represent a fluid density, permeability or dielectric constant. The quantity of most interest is not the pressure or potential P but the velocity or flux $-\vec{K}.\nabla P$. The right hand side F may contain arbitrary point and line sources. Boundary conditions may be Dirichlet, Neumann or mixed. We also allow a parabolic term on the left, provided an implicit time discretization is used. In the latter case, the following discussion will apply to the individual steps of the time discretization.

Discontinuities of coefficients imply discontinuities in the solution gradient. Discretization of the equation on a rectangular grid leads to bad pressure and velocity solutions at the front due to such discontinuities. As indicated earlier it is essential to locally adapt the grids to fit all discontinuity curves. In the resulting grids discontinuities lie only along edges of triangles. The cost of grid generation is negligible compared to equation solution. We allow the use of curved edged triangles (isoparametrics) to provide high order boundary fitting. For details of the grid construction methods used, we refer to our papers. [16,19,20]

We discretize using linear, quadratic or cubic functions on triangles and bilinear, biquadratic or bicubics on rectangles. We have used Pre-conditioned Conjugate Gradient Methods with FFT and (for high-order elements) ADI-based Fast Poisson Solvers which work well for logically rectangular grids. We have also used Multi-grid Methods for high order finite elements on irregular grids with isoparametric elements. Using either of these methods, the solution cost in total operations performed is essentially proportional to the number of unknowns, while at the same time allowing optimal use of parallelism. We now discuss these schemes in more details.

4.5. Accelerated Conjugate Gradient

The solution phase requires the solution of the discretization equations, which we represent as a matrix equation Ax = f, to a specified accuracy ε. The Pre-conditioned Conjugate Gradient Method [22,23] is the algorithm (starting with an initial guess for x):

```
r = f - Ax
p = Br
loop
    s = <r,Br> / <p,Ap>
    r = r - s*Ap
    x = x + s*p
    rbr = <r, Br>
    s = <r, Br>/old<r, Br>
    p = Br < s*p
until  converged.
```

Here B is an approximate inverse of A, which is assumed to be positive definite symmetric, and <x,y> denotes the inner product of vectors x and y.

This algorithm is parallelized by separately parallelizing each of the individual matrix or vector operations: y = Ax, y = Bx, x = y +a*z, with a a scalar, and <x,y>. The last two purely vector operations are trivial to handle. We will now turn our attention to the parallelization of Ax and Bx.

4.6. Parallelizing y = A x

We will represent A as set of r non-zero diagonals D at specified

4.7. The Asynchronous Algorithm

The idea here is to reduce the algorithm to a set of r potentially parallel vector operations based on the primitive:

```
add_vector_times_vector_to_vector(l,u,v,w),
```

which implements the vector operation $w_i = w_i + u_i v_i$ for vectors of length l. Thus the corresponding code is a loop of the form:

```
zero_vector(n,y)

for k in 1 ... r do
     add_vector_times vector_to vector(n_k,d_k,x(o_k),y)
```

where $x(o_k)$ denotes the vector x shifted by the appropriate offset. This code appears highly parallel _each add_vector () operation can be given to a different process - but in fact there is a problem in that two or more processes may try to update the value of the same component y_i simultaneously. We avoid this by providing a lock for every component of y. On the HEP this is accomplished using the extra full/empty bit available with all HEP variables. Prior to updating a variable we lock it, thereby excluding other processes, unlocking it after the value is stored. This algorithm relies explicitly on a locking mechanism for each element of the result y and might be quite inefficient on other parallel machines if they did not have hardware support for such locks. In addition, processes may wait for significant amounts of time for access to update individual components of y. As we will see, this leads to inferior execution speeds on the HEP. We note that this algorithm may also be parallelized by vectorizing each of the r vector operations while processing the diagonals sequentially. On vectormachines such as the CRAY-1, this form of the algorithm is quite attractive since the individual add_vector() calls vectorize perfectly. This approach may also of course be used on a parallel machine.

4.8. The Simple Algorithm

The Simple algorithm invoives simply evaluating (1) explicitly, for each component y_i in turn, giving each index i, or groups of such indices, to separate processes. This would be difficult to do on a vector machine since the code to be executed for each vector component is complex, (25 or maybe 49 diagonals), but is quite natural on a parallel machine like the HEP. However on some machines, especially those with data caches, there may be substantial memory overhead due to the fact that in order to compute y_i, r widely scattered components of x are needed because the offsets o_k are of order of the matrix bandwidth.

4.9. The Pieces Algorithm

The Pieces algorithm attempts to combine the best features of the previous algorithms. We decompose the result vector y into P pieces. The computation of Ax now consists of P*D units of work, each unit of the form y=y+dx though with shorter vector lengths, and consisting of application of the appropriate part of one diagonal to a corresponding part of x. This pool of tasks is handled in essentially random order by the available processes. As soon as a process completes the addition to a piece of y with some diagonal d, it then locates another undone pair and so on. There is an obvious synchronization constraint: two processes should not operate on the same piece of y with some diagonal d, it then locates another undone pair and so on. There is an obvious synchronization constraint: two processes should not operate on the same piece of y simultaneously. As a result it is clearly desirable to have P at least as large as the number of processes. It is also essential to ensure that two processes are not given the same piece of work to do.

Conceptually we may represent the work to be done with a P*D rectangular array Done of bits, initially initialised to all 0, and a P-dimensional bit-vector "Active" representing which pieces of y are in use, also initialised to 0. Processes search "Active" for an inactive piece, set the corresponding bit in "Active", and then look for an undone diagonal (0 bit) on the row of "Done" corresponding to that piece. After processing the piece with the diagonal, the "Done" entry is set to 1, and finally the Active entry for that piece is reset to 0. A naive implementation of this algorithm on a parallel machine will fail - the computation time, even for very large vectors, will be dominated by the critical sections described above due to the fact that searching the arrays for a new piece of work requires time of order O(P + D), and this search will be repeated P*D times. To handle these issues, we maintain a linked list of free pieces (of y) and for each piece, a linked list of undone diagonals for that piece - i. e. those diagonals not yet applied to that piece. These data structures reduce the cost of the critical sections to an O(1) operation, namely moving to the head of the appropriate list.

The corresponding code then looks like:

```
while(p=next_free_piece()) {
      d=next_undone_diag(p)
      process(y,p,d)
      set_done(p,d)
      if(not all_done(p)) free_piece(p)
}
```

4.10 A Parallel FFT-based Fast Solver

We will now describe a parallel implementation of an FFT-based Fast Poison Solver for the equation

$$-\Delta\Phi=F(x,y)$$

on a rectangular n*m grid. The algorithm involves applying the Discrete Fourier Transform to the columns of the above equation (i.e. Fourier transform in the variable y). The resulting equations will then be found to consist of m tridiagonal systems, one for each row. After solving these independent tridiagonal systems, a final inverse Fourier transform of each column yields the solution. Schematically the code looks like:

```
for col in 1,..,m  column_fft(n,col,rhs(col))
for row in 1,..,n  row_tridiag(m,t(row),rhs(row))
for col in 1,..,m  column_fft(n,col,rhs(col))
```

and gives the exact solution of the discrete Poisson equation in O (n m log(n)) operations. Here t(.,.) are the coefficients of the tridiagonal systems, obtained by Fourier Transforming the five-point matrix. We have found that such a solver is a useful pre-conditioner even for higher-order finite elements on logically rectangular grids - one can use the Poisson Solver for a rectangular grid corresponding in dimensionality to the nodal points of the finite elements, see McBryan.[16, 21]

This algorithm can be vectorised with considerable difficulty (it requires vectorising the FFT routines), but is much easier to parallelise. In fact each of the independent FFT calls may be processed in parallel, and similarly for the tridiagonal solutions. Barriers are required between the loops. All of the FFT calls must complete before any tridiagnonal call is started, and similarly all of the tridiagonal solutions must complete before the inverse FFT calls are started. On the HEP no more than a few extra lines of code are required to implement these barriers and generate the parallel subroutine calls. If there is residual parallelism beyond the numbers of rows or columns, which is not so on the HEP, then one could additionally program the individual column FFT and row tridiagonal routines to take advantage of such parallelism, though this is non-trivial to do.

4.11. ADI Fast Solver for Cubic Parametrics:

We have constructed a fast Poisson solver for higher order tensor product finite elements (e.g. bicubics), based on the decomposition into x and y directions suggested by the Laplacian:

$$\Delta u = \frac{d^2u}{dx^2} + \frac{d^2u}{dy^2}.$$

An explicit computation, using such elements, of the finite element energy matrix K yields the equation:

$$K_{i,j} = kx_{c(i),c(j)} my_{r(i),r(j)} + mx_{c(i),c(j)} ky_{r(i),r(j)}$$

where kx, mx are the corresponding one dimensional energy and mass matrices for elements of the same order, and c(i), r(i) are respectively the grid column and row containing point i. Thus we have expressed K as a direct sum of two tensor products.

The discrete Poisson equation Ku = f may now be solved using the basic iteration step:

$$u_{n+1}=(kx + a \cdot mx)^{-1}(ky + a \cdot my)^{-1}\{(kx-a.mx)(ky-a \cdot my)u_n + 2a \cdot f\}$$

This is an identity if u_n, U_{n+1} are replaced by the exact solution of K u=f, and it yields a convergent iteration scheme because the coefficient of u_n above is easily seen to be a contraction (at least when the parenthesised operators commute). To obtain an efficient convergence, we utilise a range of parameters a spanning the spectrum of the operator $mx^{-\frac{1}{2}}.kx.mx^{-\frac{1}{2}}$. The operators Kx + a·mx, ky + a·my, are just simple one-dimensional operators and are easily inverted (they are tridiagonal, pentadiagonal or septadiagonal symmetric matrices for the case of linear quadratic or cubic elements). As we have discussed elsewhere, McBryan[21], this iteration gives an effective fast preconditioner for higher-order tensor-product Lagrange elements, even with isoparametric elements or slightly irregular grids. In fact it is not necessary to allow the iteration above to converge - one or two sweeps through the parameter range suffices to give a good pre-conditioner*.

Because it involves only one-dimensional operations, the above iteration step is trivial to parallelise, as in the FFT example. Separate rows and columns in the individual operators involved in the iteration may be processed simultaneously by parallel processes.

4.12. The Complete Solver and Results

We now combine the three ingredients above: (a) conjugate gradient routine with parallel vector operations (b) a parallel algorithm for y = Ax and (c) a parallel algorithm for the pre-conditioner y = Bx to obtain a fully parallel solver for higher-order finite elements. Each of the components was tested and debugged separately on the HEP, and on combining them we found that the optimal speedups of each component translated directly into an optional speedup for the full solver. Figures 3-10 present results of these tests. The graphs shown are for a bi-quadratic finite

---

* We thank Jim Bramble for useful conversations on ADI methods.

element discretisation of a discontinuous coefficient pressure equation with 1089 unkowns where the conjugate gradient procedure converged in 12 iterations when pre-conditioned, or 100 iterations without pre-conditioning. (A diagonal scaling of the finite element matrix was used in each case). As a typical example, the speedup was 8.05 over serial for the complete solution using the Simple algorithm and an FFT fast solver as pre-conditioner.

In figures 3-7 we present speedups as a function of number of processes for the Asynchronous, Simple, and Pieces finite element operator (i.e. for the operator y = Ax discussed previously) where in the Pieces case we use 8, 16 and 32 pieces in successive figures. In figure 8 we compare the actual cpu times for the unaccelerated solver using the Asynchronous, the Simple, and the Pieces algorithm as in the previous 5 figures. In figure 9 we present the speedup for the FFT Fast Solver on a 32*32 grid. Figure 10 shows the corresponding speedup for a complete pre-conditioned finite element solution.

4.13 A Parallel Adaptive Multi-Grid Algorithm

The Multigrid Method introduced by Brandt[24], has become a very popular tool in recent years for the solution of elliptic equations. In particular, well-designed multigrid methods can yield solutions correct to the truncation error in O(n) operations, where n is the number of unknowns. We have implemented a multigrid solver for the Poisson equation on the HEP and successfully parallelised it.

Consider a typical algebraic system obtained by discretisation of an elliptic equation:

$$\sum_{i=1}^{n} A_{i,j}\, x_j = f_i$$

The Gauss-Seidel iteration:

$$x_i = (f_i - \sum_{j \neq i} A_{i,j}\, x_j)/A_{i,i}$$

is well known to converge slowly for a large class of matrices A. Detailed examination indicates that the slow convergence is due to poor behaviour on the low-frequency part of the solution. (By low-frequencies, we mean those which are resolvable on a grid approximately twice as coarse as the one in question, i.e. those with wavelengths of at least two mesh spacings). On the other hand, Gauss-Seidel iteration reduces the high-frequency error dramatically, typically by an order of magnitude in 2 or 3 iterations, independent of grid size. An operator with this property will be called a Smoothing Operator. The basic idea of multigrid methods is to use a smoothing operator to approximately compute the high-frequency part of the solution, then project the residual equation for the low-frequency error to a coarse grid, where some of the fine-grid low frequencies will actually be high frequencies relative to the coarse grid and consequently may be

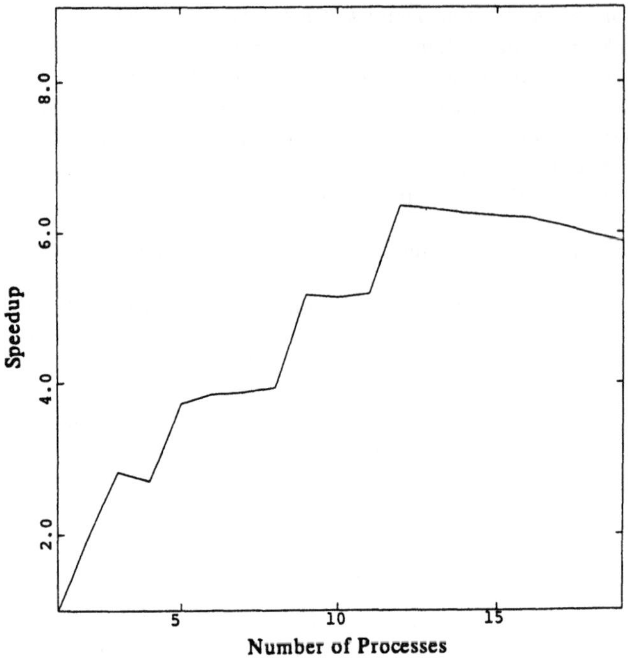

Figure 3 Speedup vs Number of processes for the asynchronous algorithm

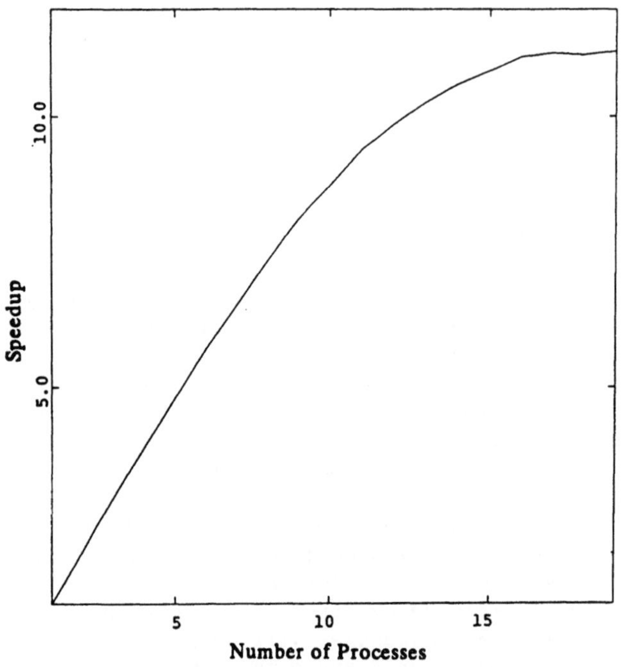

Figure 4 Speedup vs Number of processes for the simple algorithm

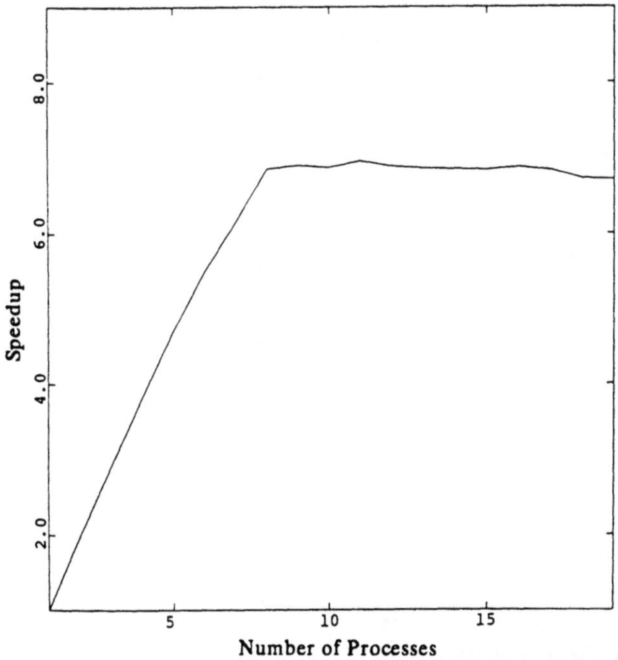

Figure 5 Speedup vs Number of processes for the pieces algorithm (8 pieces)

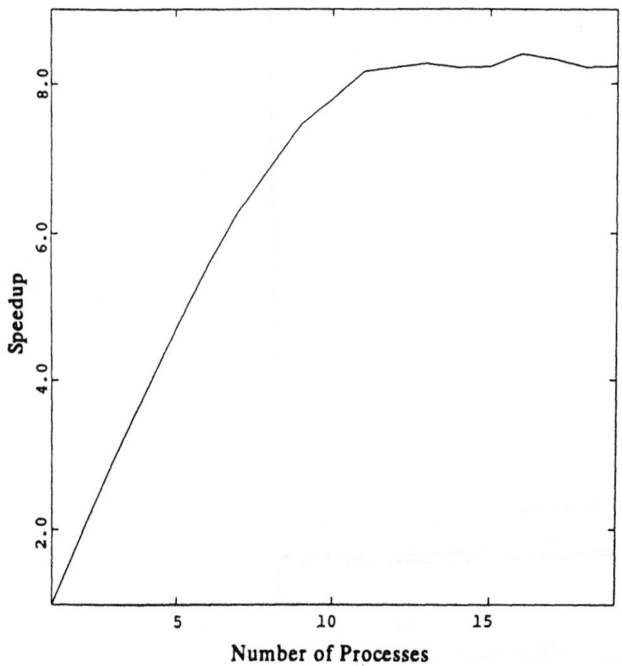

Figure 6 Speedup vs Number of processes for the pieces algorithm (16 pieces)

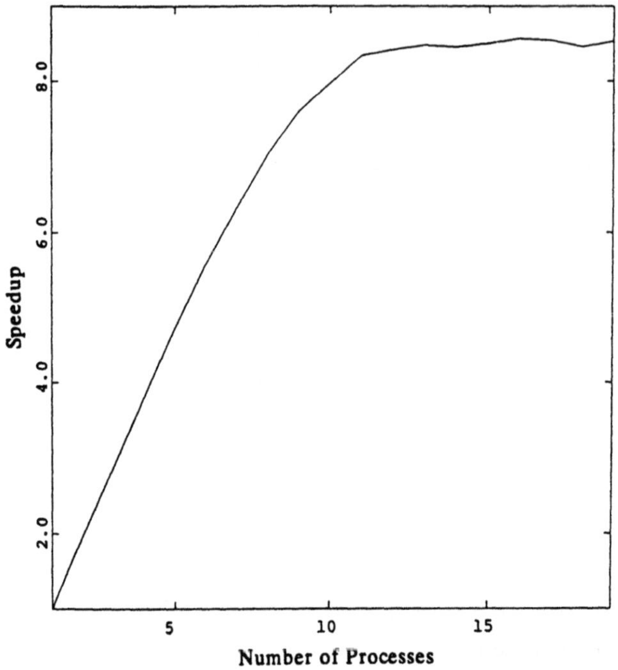

Figure 7 Speedp vs Number of processes for the pieces algorithm (32 pieces)

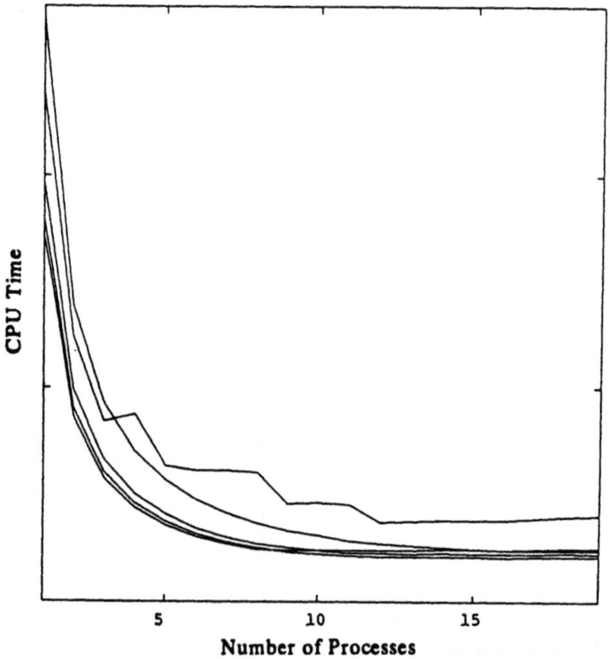

Figure 8 CPU time vs Number of processes for the cases in Figures 3-7

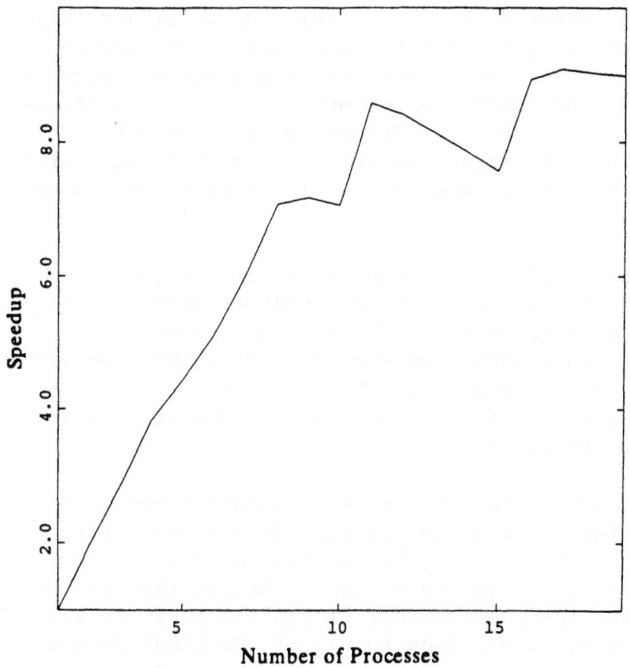

Figure 9 Speedup vs Number of processes for the FFT fast solver on 32*32 grid

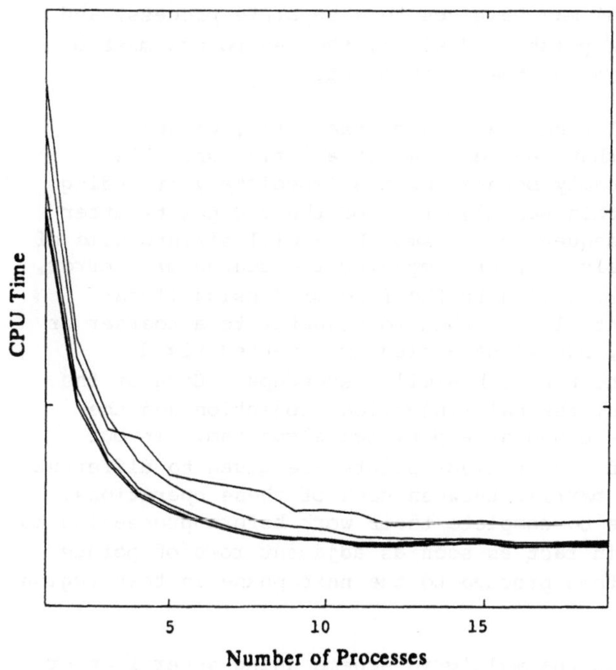

Figure 10 CPU time vs Number of processes for the accelerated conjugate method (FFT fast solver and each of the algorithms)

reduced by a smoothing iteration on that grid. One continues this procedure until all frequency components have their error reduced by an order of magnitude. These components are then projected back up to the finest grid and added into the current approximate solution. At this point, the solution has been improved by an order of magnitude (except for errors introduced by the intergrid transfers), and the whole procedure is then repeated until the required number of digits of precision are attained.

The three main steps in a multigrid computation are (a) the smoothing operation (b) the restriction to a coarser grid (c) the interpolation back to a finer grid. Each of these operations is local, affecting only grid points that are coupled by the non-zero matrix elements $A_{i,j}$, and consequently each operation is a candidate for parallelising. Exactly how this is done will depend on specific details of the problem.

In our HEP studies we have taken A to be the standard five-point discretisation matrix for the Poisson equation on a rectangular grid. We apply a red-black colouring to the grid points, alternating between the colors as one proceeds through the grid by rows. We smooth by applying Gauss-Seidel relaxation first to all of the red points, in some order, then to all of the black points. With this procedure all of the red points are uncoupled from each other for the 5-point operator - i.e. only black points contribute to the new value of x_i at a red point i. Similarly the black points are uncoupled from each other. Thus the Gauss-Seidel sweep over the red points is completely parallelizable, and each red point (or a set of such) may be given to a separate process, and similarly for the black points. However, the red points must be completed before starting on the black points.

To project the residual equation to a coarser grid, we use half-injection. Note that the residual at all fine-grid black points vanishes identically because of the immediately preceding smoothing operation - this was also true of the red points after smoothing, but the subsequent black smoothing will disturb some of these red point residuals. After computing the coarse grid error, we return the error corrections to the fine grid using linear interpolation. The control over when to transfer to a coarser or finer grid is adaptive, but we have also implemented fixed multigrid schemes on the HEP with similar speedups. Each of the red and black smoothers, the half-injection projection and the linear interpolation is coded as a parallel algorithm. In our implementation, rows of red or black points are given to different processes. There is a barrier between each of these operations, requiring all processes to complete their work before proceeding to the next step, though in fact as soon as adjacent rows of points are completed one can then proceed to the next phase in that region if desired.

We have also implemented the multigrid scheme using several other smoothing operations, including Weighted Point Jacobi with weight parameter w of .8 and also a Line Gauss Seidel smoother. For the

Point Jacobi operator, the solution at iteration step n+1 is defined in terms of that at iteration step n by:

$$x_i^{(n+1)} = \omega(f_i - \sum_{j \neq i} A_{i,j} x_j^{(n)}) / A_{i,i} + (1-w)x_i^{(n)}$$

Since only the values $x_i^{(n)}$ from the previous iteration are used, this operation is fully parallelizable. In our implementation we distribute all the points of each row to the same process, giving distinct rows to different processes whenever possible. In the Red-Black Line Gauss Seidel all points of each row are simultaneously updated. We alternately label rows as red or black. Distinct rows are given to distinct processes - all the red rows being treated first, then the black in order to achieve parallelism. Within each row we use a sequential tridiagonal solver to implement the simultaneous solution - another example of the use of high-level rather than low level parallelism. For each of the smoothing methods we have obtained more or less similar results in terms of successful use of available parallelism.

On very coarse grids, there is typically not enough work for all processes. For example, if the number of rows of grid points drops below the number of processes, then some processes will remain idle. At this point one can subdivide the rows into smaller pieces and finally assign individual points to processes. However sooner or later one encounters difficulty in using all of the processes effectively, which seems to be an inherent characteristic of multigrid methods. One solution is to terminate the grid coarsening at a higher level, computing the solution on that level by allowing the Gauss-Seidel iterations to proceed to convergence. This will provide a slower algorithm on a serial machine but may be faster on parallel machines.

We present our results for the Red-Black smoother, in the form of speedup versus number of processes, used in figures 11 and 12. Figure 11 displays results for a 128*128 grid computation on the HEP, where the iteration proceeds down to 2*2 grids. Speedups for 1 to 19 processes on a HEP-1 are presented, and as can be seen, are optimal. In figure 12, we compare the actual times for the same problem when the iterations are cut off at the 2*2, 4*4, 8*8, 16*16, 32*32 and 64*64 grid levels. Each curve is a plot of cpu time versus number of processes. As less levels are used the speedup over one process increases (the curve is steeper), but the actual timing becomes worse because relaxation is not an effective way to solve the equations on the coarsest grids. Using a direct solver on the coarsest grid would likely improve these results, provided it took advantage of parallelism.

Figures 13 and 14 repeat the same results, but this time using the Weighted Jacobi relaxation operator. As can be seen by comparing figures 12 and 14, the Red-Black smoother gives a much faster solution algorithm than Jacobi.

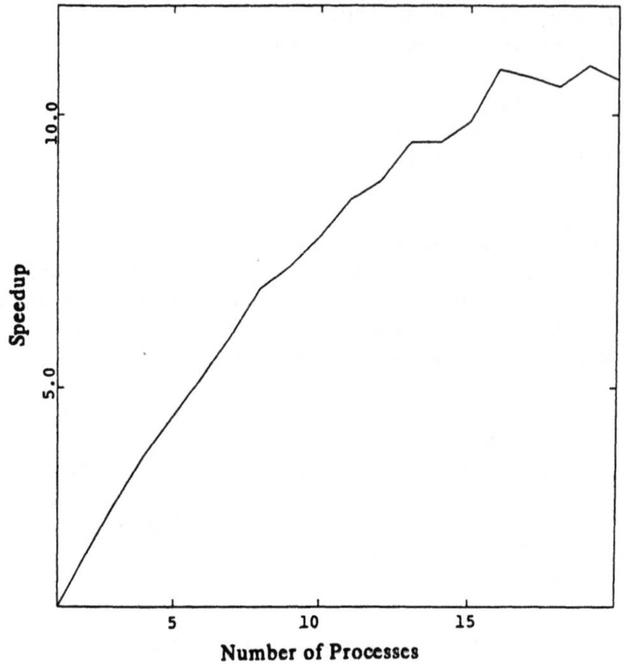

Figure 11 Speedup vs Number of processes for multigrid with Jacobi relaxation; 128 by 128 grid with 7 levels

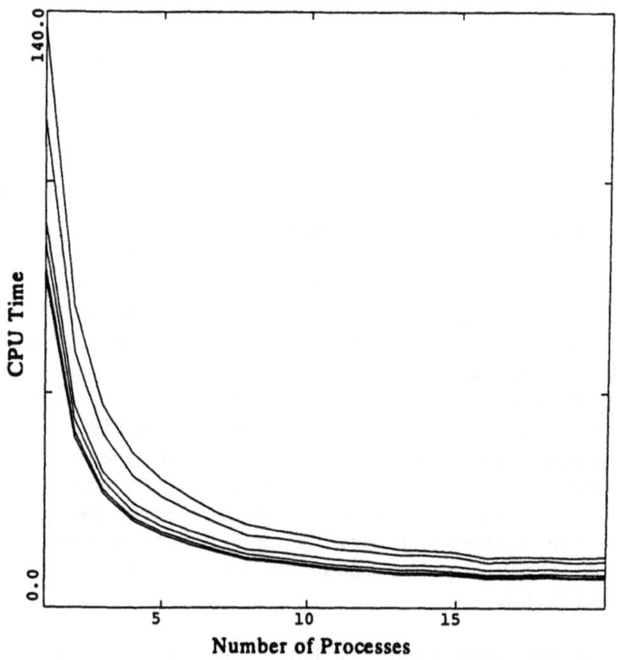

Figure 12 CPU time vs Number of processes for multigrid, with point Jacobi relaxation. We compare times for multigrid using 2 up to 7 levels on a 128 by 128 grid.

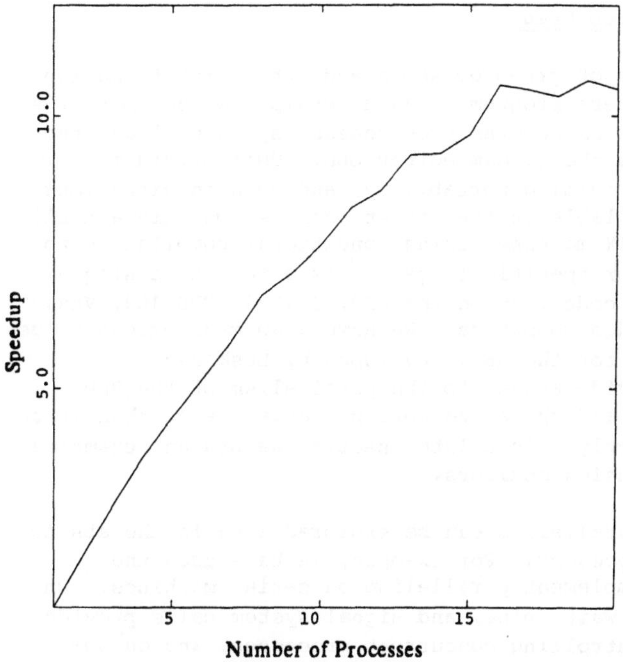

Figure 13 Speedup vs Number of processes for multigrid with Red Black Gauss Seidel relaxation; 128 by 128 grid with 7 levels.

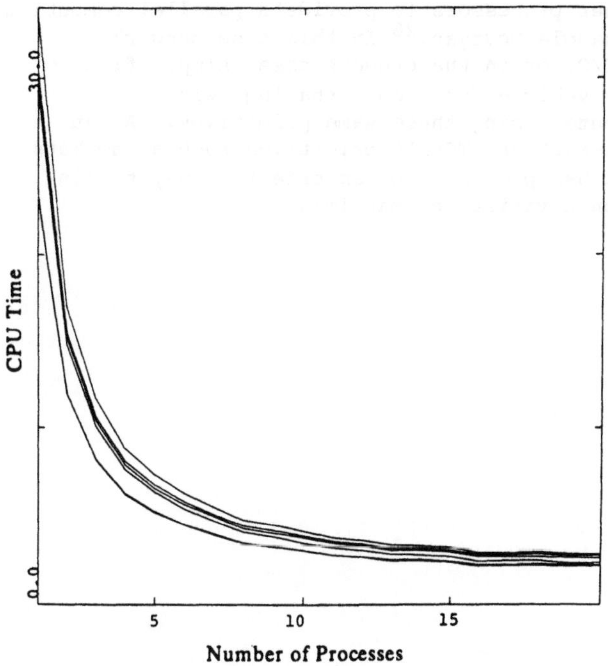

Figure 14 CPU time vs Number of processes for multigrid, with Red Black Gauss Seidel relaxation. We compare times for multigrid using 2 up to 7 levels on a 128 by 128 grid.

4.14 General Procedures Used

We have made heavy use of pre-processors and other transformational tools in developing these programs. As an example we use both the M4 and C preprocessors as FORTRAN pre-processors, as well as other UNIX utilities such as the stream editor SED. UNIX provides essential tools for providing portability, and even in situations where UNIX is not available on the target machine, we maintain all of the source on a UNIX machine, using conditional compilation to generate the source for specific targets. As a result, a single version of our source code runs on the HEP, CRAY-1, FPS-164, Vax, ELXSI, Pyramid and Ridge computers. We have also made extensive use of the monitor macros for the HEP, developed by Lusk and Overbeek.[25] These provide access to the parallelism of the HEP, at a somewhat higher (and therefore more portable) level than if one uses HEP FORTRAN directly. In a later section we present examples of programming style using monitors.

We also remark that parallelism can be explored even in the absence of a true parallel processor. For example, we have used the operating system to implement parallelism on serial machines. On UNIX, the fork, exec, wait, pipe, and signal system calls provide the possibility of controlling concurrent processes, and on some UNIX systems such processes can in fact share memory as well. This can be a useful tool for debugging parallel codes destined for other machines - though of course the programs now slow down instead of speeding up. More generally, we have used multiply networked VAXs, CRAYs, FPS-164s and other processors to provide a parallel computing environment, see for example McBryan.[26] In this case network primitives for stream I/O, or in the crudest cases simple file transfers, are the only vehicle for memory sharing, with synchronisation implemented using these same primitives. As an example, the use of Ethernet and TCP/IP primitives such as sockets in the UNIX environment has proven to be an effective way to link cooperating processes on a variety of machines.

5. MONITORS AND PORTABLE SYNCHRONIZATION

5.1 Introduction to Monitors

We now give some simple examples of the style of programming we have been using on the HEP, utilizing monitors. For a full exposition we refer to the paper of Lusk and Overbeek.[25] A monitor is a combination of data structures, initializers and code used to prevent interference between processes. Only one process at a time may be in a monitor. Typically monitors are used to implement critical sections. By implementing the monitors as macros, they can be designed to be as portable as possible and, yet capable of generating efficient machine-specific FORTRAN after pre-processing.

Some typical monitors we have used are:

```
create(func,arg1,arg2,...,)
getsub(name,sub,max_sub,nprocs)
barrier(name,nprocs)
lock(j)
unlock(j)
```

The create () monitor is used to execute a concurrent function or subroutine call. It behaves similarly to an ordinary subroutine call except that a separate process is generated to run the subroutine. On completion of the subroutine, the process dies.

The other monitors are used to control concurrently executing processes. For example, the getsub () monitor is used by "nprocs" cooperating processes to acquire a unique subscript "sub" in the range $1 \leq sub \leq max\_sub$. When all subscripts in this range have been returned, further calls to getsub () will return a "sub" of 0. Because more than one subscript may be required, instantiations of getsub () are allowed with different values for "name". The getsub () monitor ensures that two processes are never given the same subscript, and also arranges that processes that have been returned a 0 subscript are held until all the processes have reached that state. This approximates as closely as possible the semantics of a DO loop on a serial machine. Using getsub (), a self-scheduling parallel DO loop may be written, with arbitrary code inside the loop, and with no static connection between the number of processes to be used and the size of the loop upper bound.

The barrier () monitor holds processes that encounter it until "nprocs" processes are waiting, at which point all the processes are released. Again barriers are named since multiple instances may be required simultaneously. As mentioned above, the getsub () monitor includes an implicit barrier simulating the exit from a DO loop, and consequently explicit barrier () calls are not required between loops (e.g. as in the FFT Fast Solver) when programming with getsub ().

On the HEP these monitors are written in terms of shared global variables. The extra read/write access bit available with all HEP variables is easily used to ensure the indivisibility of the basic operations underlying these monitors - for example the incrementing of the sub variable. On other machines, these monitors may be implemented quite differently. The monitors can even be sensibly defined on serial machines so that parallel codes may first be debugged serially, in the same form that they will be used on the parallel processor. We give such an example below. On message based machines such as hypercubes, send() and receive() primitives will also be required, and these may simply reduce to trivial operations (noops) when implemented on shared memory computers.

5.2. Sample Example of Monitor Use

To illustrate the use of monitors, we present here a routine to add two vectors of length "n" using "nprocs" processes. After generating the processes using a create() call, each process uses getsub() to obtain a unique index "i" and then adds the corresponding components of the vectors. The actual getsub() calls and vector addition are performed by the routine called work(), and clearly the same mechanism allows an arbitrary segment of code to be executed for each component of the vector. In order to provide the synchronization required by getsub(), the macros gsdec() and gsinit() are called to define and initialize underlying common blocks used by getsub(). The number of processes to use, nprocs, is supplied by the user in a common block.

```
subroutine add_vectors(n,u,v,w)
real u(n),v(n),w(n)
common /parall/ nprocs
gsdec(gs)

gsinit(gs)

do i=1,nprocs-1
     create(work(n,u,v,w))
continue

call work(n,u,v,w)
return
end
```

The work() routine must include the same declaration gsdec() for the internal getsub() variables, and then proceeds to an indefinite loop, calling getsub() until it returns a 0 subscript.

```
subroutine work(n,u,v,w)
real u(n),v(n),w(n)
gsdec(gs)
common/parall/nprocs

start:

getsub(gs,i,n,nprocs)
if(i.eq. 0) return
w(i) = u(i) + v(i)
goto start

end
```

Note that even if "nprocs" is 1, the work routine is still called once. This single instance of work() will then obtain all of the subscripts from 1 to n. Consequently this code can behave correctly even on serial machines.

5.3 Remarks on Message Based Programming

As an illustration of how architecture and algorithm design interact, consider again the FFT-based fast Poisson Solver described in section 4.10 and represented by the three steps:

```
for col in 1,..,m  column_fft(n,col,rhs(col))
for row in 1,..,n  row_tridiag(m,t(row),rhs(row))
for col in 1,..,m  column_fft(n,col,rhs(col))
```

The implementation described in section 4.10 for the HEP involved simply distributing separate rows or columns to different processes in each step. If we wish to implement the same algorithm on a message-based machine like the Hypercube, we can again parallelize each of the three steps above in the same way. However in the absence of shared memory a new issue arises. The first step of the algorithm is column oriented, whereas the second step is row oriented. Thus there is an implicit matrix transpose of the 2 dimensional matrix rhs. On a hypercube, this implicit transpose becomes very explicit - it is necessary to reshuffle data among all of the processors between the first and second step of the algorithm in order to arrange that individual rows are continuously stored in processors at the start of the second step. A similar transpose is if course required between the second and third steps.

The point to be learned from this is that in writing a portable version of the algorithm, two extra steps should be included - the transpose operations. On a shared memory machine these operations may reduce to no code but should still be included in the software to provide portability.

Another example of a similar issue is the problem of computing a scalar product of length N efficiently using N processors. Because summing the contributions from different processors involves a critical section, the work is best organized over a binary tree of processors, which reduces the critical section cost from O(N) to O(logN). Implementing this tree in terms of communication primitives on a hypercube or similar machine involves a decidedly non-trivial amount of work considering the goal.

The need to program communication primitives is a substantial disadvantage of the hypercube and similar designs. In time it is probable that these difficulties will be overcome by providing a sufficiently powerful run-time system and associated compiler support. In current designs the extra work involved includes the need to embed grids into cubes and the need to program the intermediate host.

ACKNOWLEDGEMENTS

We would like to thank both ANL and BRL for allowing us access to their HEP computers. The operations staffs at both sites gave us unstintingly of their time, even when called on after hours. This research, on an at times quite temperamental machine, would never have been completed without their help. We would also like to thank ANL for the excellent 2-day HEP introduction provided us prior to the start of this research. We wish to thank M. Lusk and R. Overbeek of ANL for providing us access to their HEP macros. Finally we thank G. Fox of Caltech for access to the Cosmic Cube and to its software simulator.

References

1. Schwartz WO (1983) "A Taxonomic Table of Parallel Computers, Based on 55 Designs", Ultracomputer Note No. 69. Courant Institute, New York.
2. Jordan HF (1984) "HEP Architecture, Programming and Performance". University of Colorado Preprint.
3. Seitz, CL (1985) "The Cosmic Cube", Communications of the ACM, 28:22-33.
4. Fox GC, Otto SW (1984) "Algorithms for concurrent processors". Phys. Today, 37,5:50-59.
5. Salmon J. (1983) "Binary Gray Codes and the Mapping of a Physical Lattice into a Hypercube". Caltech Concurrent Processor Report (CCP)Hm-51.
6. Schwartz JT (1980) "Ultracomputers". ACM Trans. Program. Lang. Syst., 2:484-521.
7. Gottlieb A, Grishman R, Kruskal CP, McAuliffe KP, Rudolph L, Snir M (1983) "The NYU Ultracomputer - Designing an MIMD Shared Memory Parallel Computer". IEEE Transactions on Computers, C-32: 175-189.
8. Kalos MH (1983) "The NYU Ultracomputer". Ultracomputer Note No. 48 Courant Institute, New York.
9. Gottlieb A. (1980) "Washcloth - The logical successor to soapsuds". Ultracomputer Note No. 29, Courant Institute, New York.
10. McBryan O, Van de Velde E. (1984)" Parallel Algorithms for Elliptic Equations". Proceedings of the 1984 Army Novel Computing Environments Conference, Stanford University.
11. McBryan O, Van de Velde E, Vianna P (1984) "Parallel Algorithms for Elliptic and Parabolic Equations". Proceedings of the Conference on Parallel Computations in Heat Transfer and Fluid Flows, University of Maryland.
12. McBryan O, Van de Velde E (1985) "Parallel Algorithms for Elliptic Equation Solution on the HEP", Proceedings of the First HEP Conference, University of Oklahoma.
13. McBryan O, Van de Velde E (1985) "Parallel Algorithms for Elliptic Equations". Submitted to Commun. Pure and Appl. Math.
14. McBryan O, Van de Velde E (1985) "Elliptic Equation Algorithms on Parallel Computers". Proceedings of the Conference on Parallel Computers and Partial Differential Equations, University of Texas, Austin. To appear.

15. McBryan O, Van de Velde E (1985) "Multiple Grid Algorithms on Parallel Processors". Proceedings of the Second International Multigrid Conference, Copper Mountain, Colorado. To appear.
16. McBryan O (1985) "Computational Methods for Discontinuities in Fluids". Lectures in Applied Mathematics, vol. 22, AMS, Providence.
17. Berger M, Oliger J (1984) "Adaptive mesh refinement for hyperbolic partial differential equations". J. Comp. Phys., 53: 484-512.
18. Glimm J, McBryan O (1985) "Planar Interfaces", Courant Institute, Preprint.
19. McBryan O (1980) "Elliptic and Hyperbolic Interface Refinement in Two Phase flow". In Boundary and Interior Layers, ed. J.J.H. Miller, Boole Press, Dublin.
20. McBryan O (1982) "Shock Tracking for 2d Flows". In Computational and Assymptotic Methods for Boundary Layer Problems, ed. J.J. Miller, Boole Press.
21. McBryan O (1984) "Fluids, Discontinuities and Renormalization Group methods". In Mathematical Physics VII, ed. Brittin, Gustafson and Wyss, pp. 481-494, North-Holland Publishing Company, Amsterdam.
22. Hestenes MR, Stiefel E (1952) "Methods of conjugate gradients for solving linear systems". J. Res. Nat. Bur. Standards, 49: 409-436.
23. Golub GH, Van Loan CF (1984) Matrix Computations. John Hopkins Press, Baltimore.
24. Brandt A (1977) "Multi-level adaptive solutions to boundary-value problems". Math. Comp. 31:333-390.
25. Lusk EL, Overbeek RA (1983) "Implementation of Monitors with Macros". Technical Memorandum ANL-83-97, Argonne National Lab. Mathematics and Computer Science Div.
26. McBryan O., "Using CRAY super-computers as Attached Processors". Courant Institute, Preprint.

Programming Languages for Multiprocessor Systems

R.H. PERROT

Department of Computer Science, Queens's University, 23 Belfast, N. Ireland

ABSTRACT

High level languages for multiprocessor systems are required to provide features which control the concurrent activities (called processes) which can occur in an application program. Specifically the activities which need to be regulated arise from mutual exclusion situations where it is required that only one process can access a critical resource at any time without interruption and synchronisation situations where processes wish to co-operate to their mutual benefit.

To illustrate these problems this paper considers a solution to the bounded buffer problem in two different concurrent programming languages, viz, Pascal Plus and Ada. These languages represent the two different approaches which have been used to regulate concurrent activities in a programming language. Pascal Plus, like Concurrent Pascal and Modula, uses a monitor structure while Ada uses message passing primitives.

The solutions indicate the different design techniques which must be used depending on which synchronisation technique is employed.

1. INTRODUCTION

During the 1970's it was the design and implementation of operating systems which led the research into techniques to regulate concurrent activities. At the same time the price of hardware components was decreasing enabling the construction of truly parallel configurations, such as multiprocessor and distributed systems. As these systems became more widely available it was necessary to introduce concurrent programming languages to enable applications other than operating systems to be programmed.

The term process is used to describe a sequence of program instructions that can be performed in parallel with other program instructions. The point at which a processor is withdrawn from one process and given to another is dependent on the progress of the processes and the algorithm used to assign the processor(s). The simple and well-defined processor allocation strategy of a sequential system is replaced in order to achieve greater processor utilisation. The nett effect is that processes are independent and thus capable of interacting in a time-dependent manner. The series of states which the system passes through is not necessarily identical when the same batch of programs is presented with the same data for execution.

Thus, in the concurrent programming environment a programmer requires not only program and data structures similar to those required in a sequential programming environment but also tools to control the interaction of the processes - processes which are proceeding at fixed but unknown rates.

The situations in which the processes interact can be divided into two categories. The first occurs whenever processes wish to update a shared variable or a resource at the same time (or in an interleaved fashion).

For example, when one process wishes to use a resource which is currently reserved by another process. A process must be able to carry out such actions without interference from the other processes; this is described as **mutual exclusion**. The second category occurs when processes are co-operating in some task, they must be correctly interleaved in time. For example, when one process requires a result not yet produced by another process. The processes are communicating or scheduling one another and are now aware of each other's existence and purpose; this is described as **process synchronisation**.

In this paper two concurrent programming languages, namely Ada (Ada, 1983) and Pascal Plus (Welsh and Bustard, 1979) are considered. These languages are chosen as being representative of the two main synchronisation techniques that have evolved; message passing primitives and monitors with condition variables. Only the concurrent features of each language are described in detail and these are then illustrated by means of the bounded buffer problem. The solutions illustrate the main considerations which must be addressed in a concurrent programming environment as well as enabling a comparison of the two languages to be made.

2. PASCAL PLUS

Pascal Plus, as its name implies, is an extension of Pascal. The extensions are as follows:

i) the envelope structure which is an aid to program modularisation and data abstraction;

ii) the process, monitor and condition structures which provide a means of representing parallel processes and controlling any subsequent interaction.

Processes are used to identify independent actions which may take place in parallel. After a process has been defined, instances of it can be declared. Once activated the processes proceed conceptually in parallel.

A process is defined as a block with a suitable heading which may include parameters. For example:

```
process  PRODUCER;
    (*local data definitions and declarations*)
begin
    (*statements of PRODUCER*)
end ; (*PRODUCER*)
instance
    BEE1,BEE2 : PRODUCER;
```

The last statement activates two processes which may subsequently run in parallel.

One solution to the mutual exclusion problem is to gather together any variables which are shared and the operations which can be performed on them; such a construct is called a monitor or secretary (Brinch Hansen, 1973; Dijkstra, 1972; Hoare, 1974). The monitor therefore collects all the critical pieces of code into a single structure and only one process can have access to this structure at any time. Such a language construct takes the form:

```
monitor  MONITORNAME;
    (*declaration of local data*)
  procedure  PROCNAME (parameter list);
    begin  (*procedure body*)  end ;
    (*declaration of other local procedures*)
begin
    (*initialisation of local data*)
end ;
```

On declaration the monitor initialises its local data and all subsequent calls use the values of the local variables obtained on completion of the previous call. To invoke a monitor procedure the monitor name and the required procedure and its parameters are specified as

```
MONITORNAME.PROCNAME (actual parameters);
```

If a process enters a monitor it may have to be suspended pending the action of another process; this is achieved by means of CONDITION queues. The user can declare these queues in the form

```
instance  FULL:CONDITION;
```

To suspend itself on a CONDITION queue a process performs a WAIT operation expressed as

```
FULL.WAIT
```

Such an action deactivates the process and appends it to the queue identified by FULL. It also releases the exclusion on the monitor, otherwise other processes would be prevented from entering the monitor.

To release a process from a queue another process performs a SIGNAL operation, indicating to the signalled process that the reason for its delay no longer holds. This is expressed as

FULL.SIGNAL

Control is immediately passed to the process at the head of the FULL queue. The signalling process is delayed until the signalled process releases the exclusion of the monitor; a SIGNAL operation has no effect if there are no processes waiting. This form of enforced politeness is necessary in case the condition for resuming a delayed process is subsequently changed by the signalling process or another process intervening.

To illustrate the use of a monitor with condition variables consider the situation where several processes (producers) wish to communicate a series of items to other processes (consumers). This can be achieved by a buffer of a finite capacity into which the producers deposit items and from which the consumers remove items when ready. The processes must be synchronised in such a way that the producers will not produce when the buffer is full and the consumers will not consume when the buffer is empty. This is known as the bounded buffer problem.

The necessary process synchronisation can be achieved by associating a queue with each possible waiting condition, namely a full or an empty buffer. A producer should perform a SIGNAL operation on the appropriate queue after it has placed an item in the buffer. A consumer should join this same queue if the buffer is empty. Another queue is required for the situation when the buffer is full. In this case the producer should wait and the consumer should signal.

By this means the processes can co-operate to their mutual benefit as shown in the following Pascal Plus program fragment.

```
type   MESSAGE = definition;
monitor   BOUNDEDBUFFER;
var  BUFFER :  array[ O..(N-1)  ] of   MESSAGE;
  POINTER : O..N-1;
  COUNT : O..N; (*number of items in buffer*)
 instance   EMPTY, FULL : CONDITION;
procedure   *DEPOSIT (ITEM : MESSAGE);
  begin
     if   COUNT = N  then  FULL.WAIT;
  BUFFER   [  (POINTER + COUNT)   mod   N  ]   := ITEM;
  COUNT := COUNT + 1;
  EMPTY.SIGNAL
  end ;  (*deposit*)
    procedure   *REMOVE (   var   ITEM : MESSAGE);
  begin
    if   COUNT = O   then   EMPTY.WAIT;
  ITEM:= BUFFER   [  POINTER  ];
  COUNT := COUNT - 1;
  POINTER := (POINTER + 1)   mod   N;
  FULL.SIGNAL
  end ;  (*remove*)
begin
    COUNT := O; POINTER := O
end  . (*bounded buffer*)
```

The procedures DEPOSIT and REMOVE are starred indicating that they may be called from outside the monitor. The value of the variable COUNT determines whether a process should or should not be delayed.

An instance of this monitor can then be declared and the starred procedures called as shown:

```
instance   BUFFER1 : BOUNDEDBUFFER;

process   PRODUCER;
var  ITEM : MESSAGE;
begin
    repeat
      (*produce item*)
      BUFFER1.DEPOSIT (ITEM);
    until   FINISHED
end   ;(*producer*)

process   CONSUMER;
var ITEM : MESSAGE;
beginn
    repeat
      (BUFFER1.REMOVE (ITEM);
      (*consume item*)
    until   FINISHED
end   ; (*consumer*)
instance
  P :   array [  1..X  ]of   PRODUCER;
  C :   array [  1..Y  ]of   CONSUMER;
  (* X producers and Y consumers are activated*)
begin
end  .
```

3. ADA

Ada was designed for the U.S. Department of Defense with three main objectives:

i) a recognition of the importance of program reliability and maintenance;

ii) a concern for programming as a human activity;

iii) efficiency.

To realise these objectives Ada has insisted on the following:

a) readability,

b) strong typing,

c) programming in the large features,

d) exception handling,

e) data abstraction,

f) generic units,

g) tasking.

In Ada the word task rather than process is used to indicate a sequence of actions. Every task is written in the declarative part of some enclosing program unit, which is called its parent. A program unit is either a subprogram, package or task.

The definition of a task consists of two parts, its specification and its body. The specification part is the interface presented to other tasks and may contain entry specifications which are a list of the services provided by the task. The task body contains the sequence of statements to be executed when each of the services is requested; it represents the dynamic behaviour of the task. For example, -- indicates a comment in Ada,

```
        task  PRODUCER  is
         -- specification
        end;
        task body  PRODUCER  is
         -- body
        end  PRODUCER:
    begin  --active here
      -- parent
    end  ; --terminates here
```

The task automatically becomes active, its body is executed, when the **begin** of the parent unit is reached. The task then executes in parallel with the statements of its parent, i.e. those statements between the **begin** and **end** . The task terminates normally by reaching the final **end** .

The program unit in which a task (or tasks) is declared cannot be left until all tasks in that unit have terminated. If several tasks are required with similar properties it is possible to define a task type and then to declare as many instances as are necessary. For example:

```
task type   PRODUCER_TASK   is
   -- specification
end ;
task body   PRODUCER_TASK   is
--body
end   PRODUCER_TASK;
```

defines a type called PRODUCER_TASK which is a task and

```
BEE2, BEE1: PRODUCER_TASK;
```

declares two instances of the type PRODUCER_TASK. These tasks will be activated at the **begin** of the parent unit in which they are declared.

In Ada tasks can be regarded as either active or passive. Passive tasks have the function of providing some service (or services) which are used by other active tasks. An example of a passive and several active tasks occurs in the bounded buffer problem. In Ada the buffer itself must be described as a task - a passive task which will accept calls from the other active tasks, the producers and consumers.

The communication method used is that one task calls a procedure defined in another task and the parameter list is used to provide for the transfer of data.

The task specification is used to declare the procedures which other tasks can call, that is, the services available from this task. For this reason they are referred to as entry procedures. For example, the specification of a task BOUNDED_BUFFER which allows items to be deposited and removed from a buffer could have its specification defined as

```
task   BOUNDED_BUFFER   is
  entry   DEPOSIT (ITEM :   in    MESSAGE);
  entry   REMOVE (ITEM :   out   MESSAGE);
end   ;
```

indicating that it will accept calls to the procedures DEPOSIT and REMOVE from other tasks. Thus the specification contains a list of entry procedures which are the parts of a task which are available to other tasks to call and thus use. If the task provides no services then it requires no entry procedures in its specification.

In the body of the task BOUNDED_BUFFER more details of the entry procedures must be given in the following form:

```
        accept   DEPOSIT (ITEM :   in   MESSAGE)   do
         -- statements
        end   ;
      -- statements

        accept   REMOVE (ITEM :   out   MESSAGE)   do
         -- statements
        end   ;
```

The **accept** statement must use the same identifier name and a list of parameters of the same number and type as appeared in the specification part.

The statements between the **do** and **end** perform the service associated with the entry procedure. To use any of the entry procedures another task must make a call of the form

```
BOUNDED_BUFFER.DEPOSIT (ITEM);
```

The task name qualified with an entry procedure name and a list of actual parameters. Thus the structure of a solution for the bounded buffer problem takes the following form:

```
task   BOUNDED_BUFFER   is
  entry   DEPOSIT (ITEM :   in   MESSAGE);
  entry   REMOVE (ITEM :   out   MESSAGE);
end   BOUNDED_BUFFER;
task type   BOUNDED_BUFFER   is
-- depends on the representation of the buffer
begin
--statements
  accept   DEPOSIT (ITEM :   in   MESSAGE)   do
--statements
  end   ;
end   BOUNDED_BUFFER;
task type   PRODUCER is
-- no entries required
end   ;
  task body   PRODUCER   is
  ITEM : MESSAGE;
```

```
    begin
      loop
       --produce item
       BOUNDED_BUFFER.DEPOSIT (ITEM);
       exit when  FINISHED;
      end loop   ;
  end  PRODUCER;
  task type  CONSUMER  is
  -- no entry procedures required
  end  ;
  task body  CONSUMER  is
   ITEM : MESSAGE;
  begin
    loop
    BOUNDED_BUFFER.REMOVE (ITEM);
    -- consume item
    exit when  FINISHED;
  end loop  ;
  end  CONSUMER;
```

Several instances of these task types can be declared, for example:

```
P :  array  (1..X)  of  PRODUCER;
C :  array  (1..Y)  of  CONSUMER;
```

so that the X producers, the Y consumers and the BOUNDED_BUFFER task are all operating independently and in parallel.

The question now is what happens when they wish to communicate. The information accepted by the BOUNDED_BUFFER task is that of a deposit or removal of an item as indicated in the specification part. Thus the part of the task body which will receive the information is the statements following the **accept**. Another task requests the information by calling one of the relevant entry procedures of the task.

Whichever task reaches its communication statement first, that is, the **accept** or **call** statement, waits for the other task, when both tasks are ready to communicate we have a rendezvous. At the rendezvous the parameters are transferred and the body of the **accept** statement executed. The calling task is held up and can only proceed when the execution of the **accept** statement has been completed. The two tasks then proceed independently. Thus the synchronisation of the tasks is implicit.

The BOUNDED_BUFFER task does not know which of the consumer tasks has called it.

Associated with each entry procedure there is a single queue of waiting tasks which are processed on a first come first served basis.

Non-determinism has been introduced by means of the **select** statement. The **select** statement enables a choice among several entry calls to be specified. For example, the BOUNDER_BUFFER task is not programmed to accept producer calls before consumer calls, the order of selection is non-deterministic and programmed as

```
select
    accept DEPOSIT (ITEM : in MESSAGE) do
        -- statements
    end;
or
    accept REMOVE (ITEM : out MESSAGE) do
        -- statements
    end;
end select;
```

If neither of the entry procedures has been called the BOUNDED_ BUFFER task waits; if one of them has been called it is executed; if both have been called then either one is selected and executed. It is not known which call will be selected, the choice is at random and the programmer cannot rely on the selection algorithm used.

A conditional execution of a **select** alternative can be -introduced by means of a **when** statement which specifies what is called a guard. For example:

```
when COUNT<N=>
  accept DEPOSIT (ITEM : in MESSAGE) do

    -- statements
  end ;
```

will ensure that the value of COUNT is less than N before the **accept** statement is activated. The guards are evaluated at the beginning of the **select** statement, if an alternative of the **select** statement does not have a guard then it is regarded as being true. One of the true guards is selected at random for execution. If all the guards evaluate to false an error is reported. It is possible for a **select** statement to have an else part which will be executed if other alternatives are false. A solution for the bounded buffer problem is as follows:

```
task BOUNDED_BUFFER is
  entry DEPOSIT (ITEM : in MESSAGE);
  entry REMOVE (item : out MESSAGE);
end BOUNDED_BUFFER;

task body BOUNDED_BUFFER is
  BUFFER : array (INTEGER range O..N-1) of MESSAGE;
  POINTER : INTEGER range O..N-1 :=0; -- initially O
  COUNT : INTEGER range O..N :=O;
begin
  loop
    select
      when COUNT<N=>
         accept DEPOSIT (ITEM : in MESSAGE) do
           BUFFER ((POINTER + COUNT) mod N) :=ITEM;
         end;
      COUNT := COUNT + 1;
      or
      whenCOUNT>O=>
         accept REMOVE (ITEM : out MESSAGE) do
           ITEM := BUFFER (POINTER);
         end;
      POINTER := (POINTER + 1) mod N;
      COUNT := COUNT - 1;
    end select;
  end loop ;
end BOUNDED_BUFFER;
```

The accept statements are the critical sections and the calling task is delayed while an **accept** statement is executed. However, the updating of the variables POINTER and COUNT is not part of the critical sections which was the case in the monitor solution.

4. CONCLUSION

The Pascal Plus and Ada solutions to the bounded buffer problem illustrate the essential differences between these two synchronisation methods. In Ada the transfer of data is direct and synchronised while in Pascal Plus it is through a passive abstract data structure. These two different communication techniques require different program design methods.

In Ada the buffer is represented as a task with the synchronisation specified in terms of **entry, select** and **accept** statements while in Pascal Plus the CONDITION, WAIT and SIGNAL primitives are used to control the buffer. In both cases the buffer is accessed in a mutually exclusive manner. In Ada the tidying up operations, for example the adjustment of the value of the variable COUNT can be performed without delaying the calling process.

5. REFERENCES

Ada (1983), Reference Manual for the Ada Programming Language (ANSI/MIL-STD-17815A). United States Department of Defense, Wahsington D.C.

Brinch Hansen P (1973) Operating Systems Principles. Prentice-Hall, New Jersey.

Dijkstra E W (1972) "Hierarchical Ordering of Sequential Processes" in Operating Systems Techniques, pp 72-79. (Eds.) C.A.R. Hoare and R.H. Perrott, Academic Press, London.

Hoare C A R (1974) "Monitors: An Operating System Structuring Concept" pp 549-557. Comm. ACM 10.

Welsh J, Bustard D W (1979) "Paascal Plus - Another Language for Modular Multiprogramming". Software Practice and experience 9, 947-957.

Practical Concerns in Multitasking on the Cray X-MP

JOHN L. LARSON

Cray Research Inc., Chippewa Falls, WI, USA

SUMMARY

Multitasking and vectorization are both optimizations which exploit program and machine parallelism. Many of the goals for obtaining increased performance through multitasking are identical to those for vectorization, only the terminology and underlying structures are changed. High performance on the CRAY X-MP is achieved by addressing familiar concerns in a new setting. Practical objectives in programming as well as limitations in performance are described for a multitasking environment.

1. INTRODUCTION

The exploitation of parallelism for increased performance is commonplace. Architectural features such as multibanked memory, instruction execution overlap, multiple functional units, and asynchronous I/O are but a few examples of techniques which use parallelism to decrease program execution time. The CRAY-1, with its segmented functional units, exploits pipeline parallelism to increase performance through vectorization (Johnson, 1978). The factors which influence vector performance have been studied in detail. Relationships between performance and quantities such as percentage of time vectorized, vector length, and balanced resource utilization seem well understood. See, for example, Hockney and Jesshope (1981).

With the introduction of the vector multiprocessor CRAY X-MP (2 processors in 1982, 4 processors in 1984), simultaneous exploitation of vectorization and multitasking became possible (Hwang and Briggs, 1984). At first glance, many of the concerns for increased performance through multitasking seem new. However, closer inspection reveals that the goals in multitasking are identical to those in vectorization, only the terminology and underlying structures are different. This paper investigates the issues in obtaining optimal performance in a multitasking environment and relates them to familiar concerns in vectorization optimization.

2. CRAY X-MP/48 HARDWARE OVERVIEW

Figures 1 and 2 illustrate, respectively, the overall system organization and typical data flow of the CRAY X-MP/48. The mainframe consists of 4 identical processors which share 8 million 64-bit words of ECL bipolar memory. The clock cycle time is 9.5 nanoseconds, and the memory bank cycle time is 38 nanoseconds. Processors may be configured to operate independently on separate jobs, or assigned in any combination to operate jointly on a single job. For a multitasked job, the processors may use a shared set (cluster) of additional registers for synchronization and communication.

The computational capabilities of the mainframe are complimented by the I/O capabilities of the DD-49 disk, the I/O Subsystem (IOS) and the Solid-state Storage Device (SSD). The DD-49 disk has a capacity of 1200 megabytes, a 10 megabytes/second transfer rate, and an access time of 20 milliseconds. The IOS contains up to 8 million words of buffer memory, and is connected to the mainframe and SSD by several 40 megabytes/second channels. The SSD provides up to 128 megawords of secondary storage, and is connected to the mainframe by 2 very high speed channels with an aggregate transfer rate of 2000 megabytes/seconds. The rates given here represent actual, rather than peak performance.

Data flow for a typical job moves input data files from disk to central memory or SSD. Disk transfers may use the parallel disk striping capabilities of the IOS to enhance the disk transfer rate. Temporary Files may be created and repetitively manipulated on the SSD during program execution. Output data files are migrated from central memory or SSD to disk at the end of the job.

3. CRAY X-MP/48 SOFTWARE OVERVIEW

Software support for multitasking programming on the CRAY X-MP/48 is available at the FORTRAN level, where basic multitasking operations are implemented as user callable subroutines. Table 1 contains the key multitasking subroutines in the multitasking library. The capability of starting an asynchronous task to be executed on another processor is provided by the TSKSTART routine. Synchronization, communication and mutual exclusion functions are provided by the other routines using various programming styles. See Larson (1984), and CRAY (1985) for more details.

Category	Subroutine	Function
TASK CONTROL	CALL TSKSTART(TASKID, SUBNAME,ARGS)	Creates a task with identification TASKID and entry point SUBNAME.
	CALL TSKWAIT(TASKID)	Suspends the calling task until the task with identification TASKID has completed.
EVENT CONTROL	CALL EVPOST(EVENT)	Changes EVENT to 'posted'.
	CALL EVWAIT(EVENT)	Suspends the calling task until the status of EVENT is 'posted'.
	CALL EVCLEAR(EVENT)	Changes EVENT to 'cleared'.
LOCK CONTROL	CALL LOCKON(LOCK)	Suspends calling task until LOCK is 'unlocked', then changes LOCK to 'locked'.
	CALL LOCKOFF(LOCK)	Changes LOCK to 'unlocked'.

TABLE 1: Key multitasking subroutines

The FORTRAN multitasking environment is also supported by enhancements in other system software. Code re-entrancy is provided by CFT, the FORTRAN compiler, through the generation of stack-based object code and a compatible calling sequence. The operating system, COS, allows a job to create new tasks belonging to the job, and handles tasks as the fundamental unit of schedulable work. The multitasking library also performs task scheduling at the user level.

4. MULTITASKING PERFORMANCE ISSUES

In this section we investigate several concerns pertaining to multitasking performance. Upper limits of achievable speedup and lower limits of applicability are studied. Programming techniques are given to maximise load balance and minimise memory contention. Finally, the distinction between time and work is clarified.

4.1 Theoretical Speedup and Amdahl's Law

Multitasking is an optimization which changes the apparent execution time of code segments to which it is applied. The result of having two different execution speeds is that the overall execution time of a multitasked program behaves according to Amdahl's Law based on the percentage of time which is multitasked (Amdahl, 1967).

Let T1 be the execution time of a non-multitasked program. If multitasking is applied to a fraction, f, of the original execution time, then the theoretical execution time (assuming no overhead or delays) is the time to do the sequential portion, Ts, plus the time to perform the multitasked part, Tm. These quantities are a function of the original execution time, the number of processors (p), and f.

T1 = original execution time
f = fraction of T1 multitasked
p = number of processors
Tp = wall clock time of multitasked execution
Ts = (1-f)*T1 = wall clock time of sequential part
Tm = (f/p)*T1 = wall clock time of multitasked part

The theoretical speedup attainable with p processors, S(p,f), is a ratio of the original execution time to the total execution time of the multitasked program.

$$S(p,f) = \frac{T1}{Tp} = \frac{T1}{Ts + Tm} = \frac{T1}{T1 * (\ (1-f) + (f/p)\ )}$$

For a given, fixed number of processors, a plot of speedup versus f produces the familiar Amdahl's Law curve in figure 3.

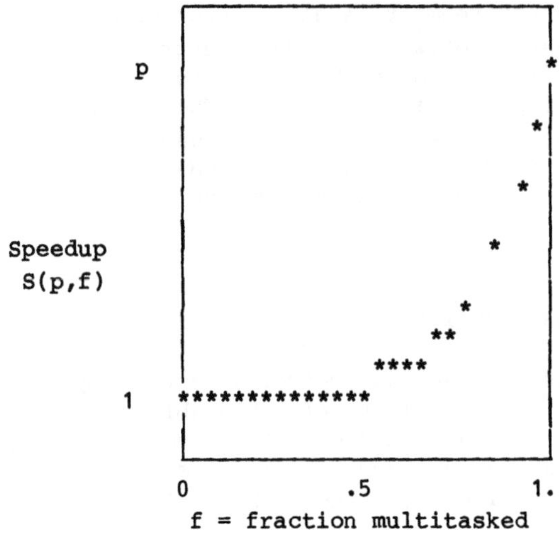

FIGURE 3: Amdahl's Law

It is interesting to look at the theoretical speedup for different values of p and f. See Table 2. Several important observations can be made from this table. The small entries in the lower part of the table show that significant speedups are not possible unless significant portions of the program are multitasked. For example, if 50 percent of the time in a code is multitasked on 4 processors, it should be no surprise that the actual speedup attained is, say, "only 1.5". The best possible speedup which could have been achieved is 1.6. The tendency to think that with p processors, speedups should always be p, is a myth.

Fraction of time multitasked	Number of Processors						
f	p=1	p=2	p=4	p=8	p=16	p=32	p=64	p=infinity
1.00	1.00	2.00	4.00	8.00	16.00	32.00	64.00	infinity
.99	1.00	1.98	3.88	7.48	13.91	24.43	39.26	100.00
.98	1.00	1.96	3.77	7.02	12.31	19.75	28.32	50.00
.97	1.00	1.94	3.67	6.61	11.03	16.58	22.14	33.33
.96	1.00	1.92	3.57	6.25	10.00	14.29	18.18	25.00
.95	1.00	1.90	3.48	5.93	9.14	12.55	15.42	20.00
.94	1.00	1.89	3.39	5.63	8.42	11.19	13.39	16.67
.93	1.00	1.87	3.31	5.37	7.80	10.09	11.83	14.28
.92	1.00	1.85	3.23	5.13	7.27	9.19	10.60	12.50
.91	1.00	1.83	3.15	4.91	6.81	8.44	9.59	11.11
.90	1.00	1.82	3.08	4.71	6.40	7.80	8.77	10.00
.75	1.00	1.60	2.28	2.91	3.37	3.66	3.82	4.00
.50	1.00	1.33	1.60	1.78	1.88	1.94	1.97	2.00
.25	1.00	1.14	1.23	1.28	1.31	1.32	1.33	1.33
.10	1.00	1.05	1.08	1.09	1.10	1.11	1.11	1.11
.00	1.00	1.00	1.00	1.00	1.00	1.00	1.00	1.00

TABLE 2: Theoretical speedup

Another message contained in the table comes from the upper rows. For a fixed percentage of multitasked execution time, the speedup does not increase as fast as the number of processors. Indeed, the speedup converges to a finite limit for p=infinity. The cause of this phenomenon is the percentage of time not multitasked. For a large number of processors, the execution wall clock time is dominated by the non-multitasked code. Speedups approaching the number of processors are possible, but for larger number of processors, much more of the code must be multitasked. This encourages future application development to strive, using a top-down approach, for high level parallelism.

The use of Amdahl's law in explaining general speedup limitations for vectorisation is commonplace. In this environment f represents the fraction of the original sequential execution time which is vectorizable, and p denotes the ratio of vector to scalar speed. An alternative formulation would relate (1-f) to the ratio of vector startup time to total vector execution time, and p would represent the number of pipeline stages.

Multitasking speedup performance is often compared to the "obvious" speedup maximum (the number of processors), although the correct comparison should be against the theoretical speedup given by Amdahl's Law. Ironically, actual vectorization speedup performance is seldom compared in the same way against any theoretical maximum.

4.2 Granularity and Overhead

At one end, Amdahl's Law limits the maximum performance possible for a multitasked program. At another extreme, the overhead incurred in multitasking limits the smallest granularity which can be profitably exploited. By granularity we mean the length of time to execute a multitaskable segment of the original program on a single CPU. Granularity is measured in time units, rather than floating point operations, since the amount of work that can be done in a given time depends on whether the computation is in scalar or vector mode.

On the CRAY X-MP, a granularity of one millisecond is considered large. For this granularity size, the overhead of the FORTRAN multitasking calls is often negligible. Very small granularity (on the order of microseconds) can also be profitably exploited through assembly language coding for synchronisation which directly accesses the hardware supporting multitasking. See Chen, Dongarra, and Hsiung (1984).

The effect of overhead on performance can be modelled in a simple way. Let T1 = X be the time to execute a multitaskable program segment on one processor. Then for a given overhead, OH, the time to execute the code on p CPUs will be Tp = OH + X/p. The ratio T1/Tp gives the speedup, Sp, of the multitasked code over the original program. Figure 4 shows the relationship of speedup to granularity for a fixed overhead.

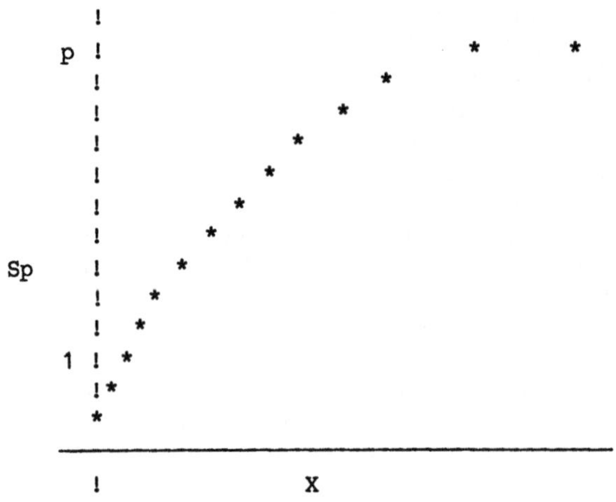

Figure 4: Speedup vs. granularity for fixed overhead

By simple manipulation of the formula for speedup, one obtains a relationship for the minimum granularity required to obtain a given speedup in the presence of overhead.

$$X = (Sp * p * OH) / (p-Sp)$$

For a given OH on p processors, a speedup Sp can be obtained only if the granularity is at least X.

The curve in figure 4 is identical to one produced in vectorisation analysis. For this environment, granularity is measured in terms of vector length (VL), and the overhead is denoted as vector startup time. As Kuck (1978) describes, a scalar computation may take time T1 = k(VL) where k is the clock time for the operation. For a pipelined computation using a functional unit with k stages, the vector time is Tk = (k-1) + VL, where the overhead of filling the pipe is (k-1). The speedup curve of Sk = T1/Tk vs. VL shows the minimum vector length required for profitable vectorisation speedups.

4.3 Load Balancing

Multitasking is most often applied to parallel work found in the independent iterations of DO loops. If the loop has N iterations, we call N the extent of the parallelism of the loop. Load balancing is the technique of mapping N onto p processors or tasks so that each task has the same amount of work to do. See figure 5.

There are two major cases to consider: static and dynamic partitioning. Static partitioning is used when the time for each iteration of the loop is approximately the same. This technique assigns a fixed subset of iterations to each task. The iterations which as task is to perform can be computed from the task, or processor number. One strategy assigns a subset of continuous iterations to each processor. The jth processor computes iterations (j*N)/p+1 to ((j+1)*N)/p. Another static strategy assigns the iterations in an interleaved fashion. The jth processor computes every pth iteration from j+1 to N. See figure 6.

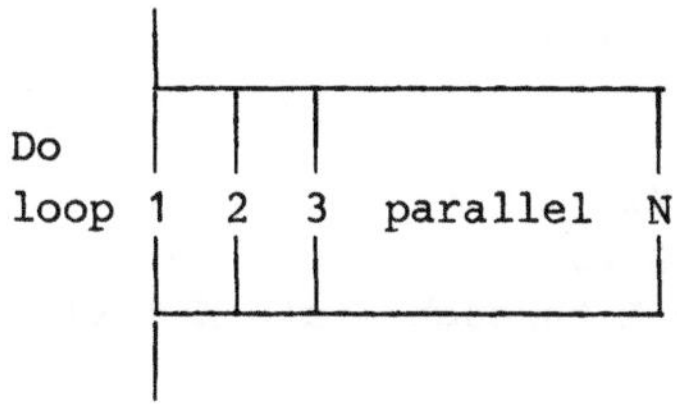

N = extent of parallelism

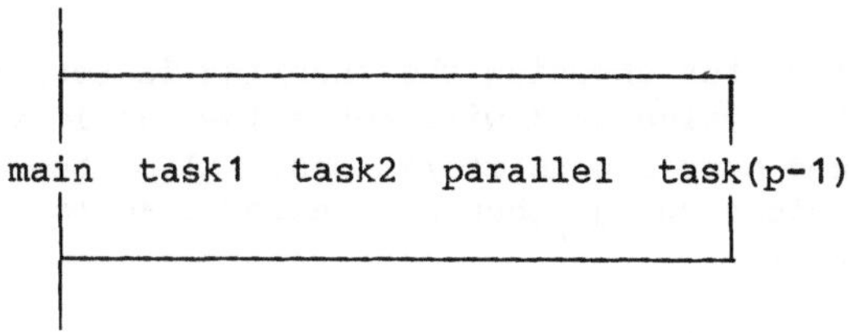

multitasked code on p processors

FIGURE 5: Mapping parallelism to processors

Processor			
p0	I = 1, N/4		I = 1, N, P
p1	I = N/4+1, 2*N/4	or	I = 2, N, P
p2	I = 2*N/4+1, 3*N/4		I = 3, N, P
p3	I = 3*N/4+1, 4*N/4		I = 4, N, P

FIGURE 6: Static partitioning with p = 4

If the time for an iteration of the loop varies significantly then a dynamic partitioning of work will tend to balance the load on each processor. This technique maintains a shared counter which indicates the next unprocessed iteration. Each processor accesses and updates the counter to commit itself to one or more iterations. Those processors which commit to short iterations will return more often to find more work, while processors committed to long iterations will look for new work less frequently. This technique incurs an overhead to protect the counter during the update process. If the average granularity of the iterations is large compared to this overhead, then the iterations may be distributed one at a time. If the average granularity of the iterations is small compared to the overhead of protecting the iteration counter, then the iterations can be distributed to processors in chunks. Each chunk contains K iterations. See figure 7, where I, K, and N are shared, and L is private to each task.

```
      COMMON /TASKS/ I, K, N
      I = 1
```

Each processor executes

```
10    CONTINUE
      CALL LOCKON ( LOCKI )
      L = I
      I = L + K
      CALL LOCKOFF ( LOCKI )
      IF ( L.GT.N ) GO TO 20
      compute iterations L through min(L+K-1,N)
      GO TO 10
20    CONTINUE
```

FIGURE 7: Dynamic partitioning for large (K=1) or small (K>1) granularity iterations

Heuristics can be devised for choosing the chunking factor, K. On the one hand, too small a value of K will not allow the lock overhead to be amortized over enough iterations. Also, too large a value of K will not produce enough chunks to allow load balancing to take place. See figure 8.

```
      |*               *
      | *             *
      |  *           *
wall  |    *       *
clock |      *   *
time  |       ***
      |
      |too            no
      |much         load
      |overhead balancing
      |___________________    K

              K
                opt
```

FIGURE 8: Choosing the optimal chunking factor

To reduce the relative overhead, K should be at least Kmin. To facilitate load balancing, K should be less than Kmax. The value of K should be chosen in the range (Kmin, Kmax), where

$$Kmin = \text{smallest } K \text{ such that } \frac{(\text{lock overhead})}{K*(\text{ave.iter.time})} < .01$$

$$Kmax = \text{largest } K \text{ such that } \frac{N}{K} > 10*p$$

The balancing of computational work on several processors in multitasking has the same goal as overlapping asynchronous I/O with computation in a uniprocessing environment, namely, to maximise the simultaneous utilisation of available resources.

4.4 Minimising Memory Contention

One of the factors influencing performance in a shared memory architecture is memory contention. Memory references made with the three computational ports of each processor may result in resource conflicts. These conflicts are resolved by having some references wait until the required resource becomes available. This waiting can cause a computation on a single CPU to take longer in a system with all processors busy, than in an otherwise quiet environment.

The user has control over the intensity of memory references as reflected in the programming style. The intensity of vector memory references can be measured in units of memory references per floating point operation (memrefs/flop). Also important is the layout of the memory references across the interleaved memory banks.

Programming styles which reduce the memory contention produce an increased performance both in vector uniprocessing and multitasking environments.

The following is a list of program optimisations which reduce the memory contention for the current compiler. A complete description can be found in CRAY (1985).

1. Vertical Inner Loop Unrolling
2. Horizontal Inner Loop Unrolling
3. Vertical Outer Loop Unrolling
4. Horizontal Outer Loop Unrolling

   For example, (Dongarra, 1984),

```
              DO 40 J = 1, 4*M+1           3 MEMREFS / 2 FLOPS
                DO 40 I = 1, N
       40         Y(I) = Y(I) + X(J) * M(I,J)
              DO 41 J = 1, 4*M+1, 4        6 MEMREFS / 8 FLOPS
                DO 41 I = 1, N
       41         Y(I) = (((Y(I) + X(J  )  * M(I,J  ))
        1                        + X(J+1) * M(I,J+1))
        1                        + X(J+2) * M(I,J+2))
        1                        + X(J+3) * M(I,J+3)
```

5. Dimensioning of Arrays

   For example,

```
           DIMENSION X(64,100), XX(65,100)
           DO 50 I = 1, 64              X(I,*) IN 1 BANK
             DO 50 J = 1, 100
       50      X(I,J) = X(I,J) + 1.0 .
           DO 51 I = 1, 64              XX(I,*) IN 64 BANKS
             DO 51 J = 1, 100
       51      XX(I,J) = XX(I,J) + 1.0
```

6. Loop Interchange

7. Padding Between Common Block Arrays

These techniques are commonly considered vectorisation optimisations since they enhance vector performance in a uniprocessing environment. However, they are also multitasking optimisations which minimise -inter-CPU memory contention by minimising the intra-CPU conflicts.

4.5 Measuring Time and Work

For a single CPU system used in a dedicated mode, measuring work is often done with CPU charges in time units. For computational jobs, the CPU time and the wall clock time are frequently quite close. No distinction between time and work is necessary. The ability of several CPUs to perform work at the same time creates a distinction between work and the time to complete the work. In a multitasked job, several CPUs may perform more work in less real time than an equivalent program executed on a single processor. The presence of multitasking overhead raises the question of what work is to be credited and measured.

Kuck (1978) defines some basic quantities which clarify the situation (see figure 9). From an external viewpoint, we observe the work (CPU charges), 01, and the wall clock time, T1, for a single processor execution of the program. Often these quantities are defined to be equal. We also record the wall clock time, Tp, for the execution on a p processor system. The ratio of T1 to Tp gives the speedup, Sp. A comparison of Sp to p quantifies the efficiency, Ep. This is an equitable comparison only when the theoretical speedup is p (the program is 100% multitaskable).

From an internal viewpoint, we observe the work (CPU charges), Op, and the wall clock time, Tp, of the multitasked job. Naturally, Tp has the same value as was recorded externally. However, additional work may be done in the parallel execution. Such work may include charges for multitasking calls, synchronisation delays, and memory contention. The redundancy, Rp, measures the extra work performed by the multitasked program. The internal speedup, "Sp", gives credit for all the work performed. Finally, the utilisation, Up, shows how busy the system really is.

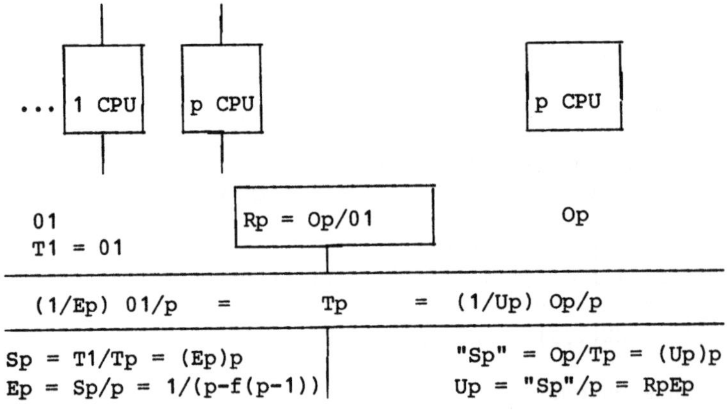

FIGURE 9: Measuring time and work

Some algebraic manipulation shows that the two viewpoints define similar quantities within a factor of Rp. The redundancy factor Rp is the link which relates and separates the two perspectives.

The situation is illustrated by an example code. A lattice gauge program, which is 100% multitaskable, requires 3.28 hours on a single X-MP-4 processor. When executed on four CPUs, the CPU charges are 3.45 hours, and the run completes in 0.87 hours. We have O1 = T1 = 3.28, Op = 3.45, and Tp = 0.87. From these quantities all others may be computed. The realised speedup is Sp = 3.77 with an efficiency Ep = 0.94. Internally, the system speedup is "Sp" = 3.96 and utilization is 0.99, where more work (Rp = 1.05) was performed than was in the original program.

One way to interpret these results is in the spirit of a backward error analysis for assessing roundoff error. Here we relate the performance of the multitasked job with overhead to a hypothetical multitasked job with no overhead, but with a smaller fraction of time multitasked. By doing this, all of the overhead can be accounted for in a single, simple way. From the formula, $Ep = 1/(p-f(p-1))$, we can see that our 100% multitaskable job with overhead gave the same performance as an equivalent program with no overhead, but which was 98% multitaskable. Amdahl's Law shows that for p=4 and f=0.98 the theoretical speedup with no overhead is Sp=3.77, which was the speedup recorded for our actual program.

The quantities presented here can be defined in a vectorization environment. For example, the value of p may be taken as the vector speed to scalar speed ratio. The vectorized work, Op, may account for vector set up or for redundant operations such as those which occur in masked operations. Efficiency, Ep, may take into account the fraction of time which was vectorized in computing the theoretical speedup possible.

5. APPLICATION PERFORMANCE

Multitasking is utilised as a performance optimisation for a variety of application codes on the CRAY X-MP. The results are summarised in table 3. Comparisons are made against runs executed on the same system, but employing only one processor. The percent multitaskable column shows the fraction of single processor execution time which was multitasked. This percentage limited the theoretical speedup, based on Amdahl's Law, which could be obtained. These multitasking percentage values are greater than typical vectorisation percentages, illustrating the common presence of extensive high level parallelism in many application areas. The nearness of the actual speedups to the theoretical speedups shows the high efficiency delivered by the CRAY X-MP.

	(4 CPUs)		
	Percent multitaskable	Theoretical speedup	Actual speedup
1. Particle-in Cell	97 %	3.67	3.48
2. Weather Forecast	98 %	3.77	3.55
3. Seismic Migration	98.7%	3.85	3.45
4. Monte Carlo	99 %	3.88	3.75
5. Lattice Gauge	100 %	4.00	3.77
6. Seismic Forward Modelling	98.2%	3.80	3.50
7. Aerodynamics	98.8%	3.86	3.22
8. Chess	variable	n/a	3.15*

* CPU time divided by wall clock time

TABLE 3: Application performance

6. CONCLUSION

In a sense, multitasking is not new. It is only the exploitation of parallelism along a different dimension than vectorisation. Multitasking urges us to open our eyes wider, beyond the innermost loop, to see parallelism at a higher level, where often a natural independence exists along geometric or other problem attributes. Even at this higher level, however, the concerns in the exploitation of multitasking parallelism are basically the same as for vectorisation, only the details are different. Additionally, multitasking offers an optimisation alternative, often being applicable when vectorisation is not possible. When applied on top of vectorisation, multitasking gives the performance benefits of both worlds.

7. REFERENCES

Amdahl G (1967) The Validity of the Single Processor Approach to Achieving Large Scale Computing Capabilities. AFIPS Conference Proceedings, SJCC 3:483-485.

Chen S S, Dongarra J, Hsiung C C (1984) Multitasking Linear Algebra Algorithms on the CRAY X-MP-2: Experiences with Small Granularity. Journal of Parallel and Distributed Computing 1:22-31.

Cray Research Inc (1985) Multitasking User Guide. CRI internal technical note, SN-0222, 2nd printing.

Dongarra J J, Eisenstat S C (1984) Squeezing the Most Out of an Algorithm in CRAY FORTRAN. ACM TOMS 10:221-230.

Hockney R W, Jesshope C R (1981) Parallel Computers. Adam Hilger Ltd., Bristol, pp 47-95.

Hwang K, Briggs F A (1984) Computer Architecture and Parallel Processing. MacGraw-Hill, New York, pp. 714-728.

Johnson P M (1978) An Introduction to Vector Processing. Computer Design 17:89-97.

Kuck D J (1978) The Structure of Computers and Computations. John Wiley and Sons, New York, pp. 100 and 255.

Larson J L (1984) Multitasking on the CRAY X-MP-2 Multiprocessor. IEEE Computer 17:62-69.

(r_∞, $n_{\frac{1}{2}}$, $s_{\frac{1}{2}}$) Measurements on the 2-CPU Cray X-MP

ROGER W. HOCKNEY

Computer Science Department, Reading University, Whiteknights, Reading, Berkshire, United Kingdom

SUMMARY

We report performance measurements made on the 2-CPU Cray X-MP at ECMWF, Reading. Vector (SIMD) performance on one CPU is interpreted by the two parameters (r_∞, $n_{\frac{1}{2}}$), and we find for dyadic operations using FORTRAN r_∞ = 70 Mflop/s, $n_{\frac{1}{2}}$ = 53 flop. All vector triadic operations produce r_∞ = 107 Mflop/s, $n_{\frac{1}{2}}$ = 45 flop; and a triadic operation with two vectors and one scalar gives r_∞ = 148 Mflop/s and $n_{\frac{1}{2}}$ = 60 flop. MIMD performance using both CPUs on one job is interpreted with the two parameters (r_∞, $s_{\frac{1}{2}}$), where $s_{\frac{1}{2}}$ is the amount of arithmetic that could have been done during the time taken to synchronize the two CPUs. We find, for dyadic operations using the TSKSTART and TSKWAIT synchronization primitives, that r_∞ = 130 Mflop/s and $s_{\frac{1}{2}}$ = 5700 flop. This means that a job must contain more than ~6000 floating-point operations if it is to run more than 50% of the maximum performance when split between both CPUs by this method. Less expensive synchronization methods using LOCKS and EVENTS reduce $s_{\frac{1}{2}}$ to 4000 flop and 2000 flop respectively. A simplified form of LOCK synchronization written in CAL code further reduces $s_{\frac{1}{2}}$ to 220 flop. This is probably the minimum possible value for synchronization overhead on the Cray X-MP.

1. INTRODUCTION

It has been recognised for years (Calahan, 1977; Hockney, 1977; Calahan and Ames, 1979) that the single parameter Mflop/s (megaflops) is inadequate to measure the performance of a vector computer, because it takes no account of the vector startup overhead. Nevertheless there has been a reluctance to use a second parameter to overcome this deficiency, and manufacturers still do not publish a two-parameter description of the performance of their vector pipelines, even though some (notably Cray, Inc) could do so to their competitive advantage. The two-paramter (r_∞, $n_{\frac{1}{2}}$) description measuring the importance of the startup overhead in terms of how much useful arithmetic (in fact $n_{\frac{1}{2}}$ floating-point operations) could have been done during the time of the overhead. In a similar spirit this description has been extended (Hockney, 1985) to MIMD computing by measuring the overhead of synchronizing multiple instruction streams in terms of $s_{\frac{1}{2}}$: the amount of useful arithmetic that could have been performed during the time taken for synchronization. The three parameters r_∞, $n_{\frac{1}{2}}$ and $s_{\frac{1}{2}}$ are defined mathematically in Section 2, and measurements of their values on the 2-CPU Cray X-MP at ECMWF are reported in Section 3.

2. THE PARAMETERS (r_∞, $n_{1/2}$, $s_{1/2}$)

An SIMD computer is one in which a single instruction initiates many identical operations on multiple data items, in short one in which vector instructions are included. The most successful SIMD computers execute a vector instruction by pipelining the successive elements of the vectors (usually one per clock period) through a high-performance multi-stage pipeline. The Cray X-MP is an example of such a computer with a six-stage floating-point adder and a seven-stage floating-point multiplier. The performance of a vector pipeline can be characterized by measuring the time to execute a single vector instruction, t, as a function of the length of the vector, n, and fitting the results to the linear relationship

$$t = r_\infty^{-1}(n + n_{1/2}) \qquad (1)$$

where:

r_∞ - is the asymptotic (i.e. maximum) performance in millions of floating-point operations per second (Mflop/s),

$n_{1/2}$ - the half-performance length, is the vector length necessary to achieve half the asymptotic performance.

Expressing Eqn. (1) as a startup time in microseconds, t_0, and a time per result, τ we have

$$t = t_0 + n\tau \qquad (2)$$

where $t_0 = n_{1/2}/r_\infty$ and $\tau = r_\infty^{-1}$.

Comparing Eqns. (1) and (2) one sees that $n_{1/2}$ is the number of floating-point operations which could have been done in the time of a vector startup. The parameter $n_{1/2}$ therefore measures the importance, in terms of lost floating-point operations, or vector startup to the user. Not surprisingly one finds that, within the approximations of this timing model, $n_{1/2}$ is the parameter that determines the best choice of vector algorithm on a particular computer (see Hockney, 1983, 1984). When the vector length equals $n_{1/2}$ then the two terms in Eqn. (1) are equal, and half the time is being used to perform useful arithmetic (first term) and half the time is being lost in vector startups (second term). In this case the average performance is obviously only 50% of the maximum.

Using Eqn. (1) we find the average performance, r, as a function of vector length to be

$$r = n/t = r_\infty/(1 + n_{1/2}/n) \qquad (3)$$

which gives the functional form of the approach of the average Mflop/s to the maximum Mflop/s, r_∞, quoted by the manufacturer, as the vector length increases. One should note that Eqn. (3) demonstrates a slow approach to the asymptotic performance. By definition a vector length of $n_{\frac{1}{2}}$ is required to produce an average performance of half the maximum, but also a vector length of ten times $n_{\frac{1}{2}}$ is required to reach 91% of the maximum performance. Equation (3) also shows how the long tables of average performance against vector length often published by computer manufacturers, can be represented much more compactly by quoting equivalently the values of the two-parameters $(r_\infty, n_{\frac{1}{2}})$.

The values of r_∞ and $n_{\frac{1}{2}}$ are likely to depend to some extent on the software being used (e.g. assembler code or FORTRAN) and the types of vector operations (e.g. dyadic, triadic, register-to-register or memory-to-memory). For an accurate description separate pairs of values need to be measured for these different cases. The timing relation (1) can also be used to estimate the time, T_V, for the vector part of an algorithm; that is to say the part that is executed using vector instructions. If the parameters r_∞ and $n_{\frac{1}{2}}$ are approximately constant for the code in question then

$$T_V = r_\infty^{-1}(s_V + n_{\frac{1}{2}}q) \qquad (4)$$

where:

s_V - is the total amount of useful arithmetic (i.e. floating-point operations between pairs of numbers) in the vector part of the algorithm

q - is the number of vector instructions comprising the vector part of the algorithm.

If the parameters r_∞ and $n_{\frac{1}{2}}$ vary too much to be considered constant over the whole vector part of the algorithm, then the vector operations must be grouped so that the parameters are constant within each group. Equation (4) may then be applied to each group using a different pair of values $(r_\infty, n_{\frac{1}{2}})$ for each group.

The performance of scalar instructions, which initiate only a singe floating-point operations, can similarly be characterized by the two-parameters $(r_\infty, n_{\frac{1}{2}})$, and one finds that usually the value of $n_{\frac{1}{2}}$ is sufficiently small that it can be ignored. Hence the time for the scalar part of an algorithm is given by

$$T_S = r_{\infty S}^{-1} s_S \qquad (5)$$

where:

s_S - is the number of floating-point operations between pairs of numbers in the scalar part of the algorithm

$r_{\infty S}$ - is the performance in Mflop/s of scalar instructions.

Combining Eqns. (4) and (5), we have the time, T, for the complete algorithm

$$T = T_v + T_s \tag{6}$$

Usually one finds that the scalar performance is less than the vector performance by a factor of about ten, showing the importance of executing as much of the arithmetic as possible in vector instructions - that is to say achieving a high level of vectorization.

The critical path through a multi-instruction-stream (i.e. MIMD) program can be considered as a sequence of work segments, between which program synchronization must occur. That is to say, all work must be completed on a segment before the next can begin. Within a segment, however, several independent tasks exist that may be assigned to different instructions streams (i.e. executed by different CPUs in the Cray X-MP). For synchronization to take place correctly it is necessary for the control program to initiate each instruction stream, and to recognize when all the instruction streams have finished. The time taken to perform these operations is called the synchronization overhead. It is important in multi-instruction stream programming to be aware of this overhead, and to be able to assess its magnitude quantitatively. The parameter $s_{\frac{1}{2}}$ has been introduced for this purpose (Hockney and Snelling 1984, Hockney 1985), and measures the synchronization overhead in terms of how many floating-point operations could have been performed in the time of the overhead. Accordingly the time, t, to execute a work segment on multiple CPUs becomes

$$t = r_\infty^{-1}(s_i + s_{\frac{1}{2}}) \tag{7}$$

where:

s_i - is the total number of floating-point operations between pairs of numbers in the ith work segment

If the critical path through an MIMD algorithm comprises a sequence of q work segments, each obeying Eqn (7), then the total time, T, for an MIMD algorithm becomes

$$T = r_\infty^{-1}(s + s_{\frac{1}{2}}q) \;, \; s = \sum_{i=1}^{q} s_1 \tag{8}$$

In Eqn. (8) we have implicitly assumed that it is possible to schedule the work, s, between the multiple instruction streams in such a way that each stream has the same amount of work to do. This would be perfect scheduling, or load balancing. If this is not possible then the effect of imperfect scheduling can be included into Eqn. (8) by introducing the average efficiency, $\bar{E}_q$, as defined by Kuck (1978).

$$T = r_\infty^{-1}(s/\bar{E}_p + s_{\frac{1}{2}}q) \tag{9}$$

As with the case of the vector parameter $n_{\frac{1}{2}}$, Eqn. (6) shows that when $s = s_{\frac{1}{2}}$, half the time is spent on useful arithmetic and half on synchronization overhead, resulting in an average performance of half the maximum, r_{∞}. Thus $s_{\frac{1}{2}}$ is a yardstick that can be used to judge whether a work segment is large enough to be divided between multiple CPUs. As with the vector case, the work in a segment must exceed ten times $s_{\frac{1}{2}}$ in order to achieve more than 91% of the maximum performance. In the next section we describe some simple benchmarks that have been used to measure values for the three parameters r_{∞}, $n_{\frac{1}{2}}$ and $s_{\frac{1}{2}}$ on the ECMWF 2-CPU Cray X-MP.

3. MEASUREMENTS

3.1 Values of (r_{∞}, $n_{\frac{1}{2}}$) for one CPU

Figure 1 gives the program used to measure r_{∞} and $n_{\frac{1}{2}}$ on one CPU of the Cray X-MP. Initially the overhead of making the timing measurement is obtained by measuring the time TØ for two successive calls to the timing routine SECOND. This overhead is subsequently subtracted from all measurements. The vector length, N, is varied in steps of two, up to a maximum of 400, and for each vector length NREPEAT = 100 trials are made. The minimum time, TMIN, maximum time, TMAX, and average time (TSUM/NREPEAT) of the 100 trials are recorded. The DO-1Ø loop is replaced by a vector instruction when compiled, and is the subject of the timing, statement 10 being changed for the different types of loops considered below. Values of (r_{∞}, $n_{\frac{1}{2}}$) are obtained by fitting the best straight line through the values of TMIN versus N as is shown in Fig. 2. The asymptotic performance, r_{∞}, is the inverse slope of this line, and the half-performance length, $n_{\frac{1}{2}}$, is the negative intercept of the line with the vector-length axis. In the case of triadic operations TMIN is divided by two, so that the time axis correctly records the time <u>per</u> vector operation. In a multi-user operating system it is necessary to make multiple trials at each vector length and take the least time, in order to minimize interference from other jobs. One hundred trials appears sufficient for this purpose, as the resulting values in Fig. 2, except for a few rogue points, are quite well clustered along the timing line. The results obtained are given in Table 1.

```
        PROGRAM MULT
        PARAMETER (NMAX = 4ØØ)
        PARAMETER (NMAXA = NMAX, NMAXB = NMAX, NMAXC = NMAX)
        PARAMETER (NREPEAT = 1ØØ)
        DIMENSION B(NMAXB), C(NMAXC), A(NMAXA)
        DATA B/NMAX*1.Ø, C/NMAX*1.Ø/
        DATA S/1.Ø/

        CALL SECOND(T1)
        CALL SECOND(T2)
        TØ = T2-T1

        DO 2Ø N = 2, NMAX,2
        TMIN = 1ØØØØØØ.
        TSUM = Ø.
        TMAX = Ø.

        DO 111 JR = 1, NREPEAT
        CALL SECOND(T1)
CDIRØ   IVDEP
        DO 1Ø I = 1, N
1Ø         A(I) = B(I)*C(I)
        CALL SECOND (T2)

        T = T2-T1-TØ
        TMIN = MIN(TMIN,T)
        TMAX = MAX(TMAX,T)
        TSUM = TSUM + 1
111     CONTINUE

        WRITE (6,1ØØ) N, TMIN, TMAX, TSUM/NREPEAT
 2Ø     CONTINUE
1ØØ     FORMAT (' N: ', I4, 3FI5.8)
        STOP
        END
```

Fig. 1 The program used to measure (r_∞, $n_{\frac{1}{2}}$) on a single CPU of the Cray X-MP. The statement 1Ø is changed for the different cases.

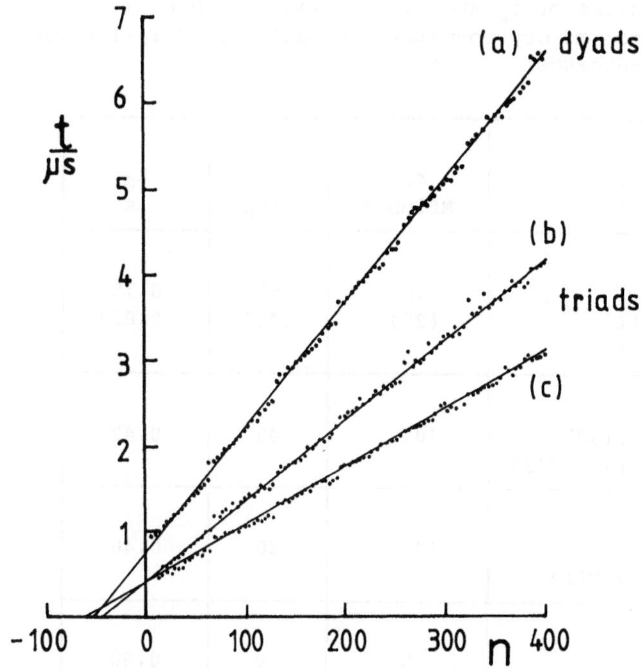

Fig. 2 Measurement of r_∞ and $n_{\frac{1}{2}}$ on one CPU of the Cray X-MP. Time, t, as a function of vector length, n, for a single vector operation. The asymptotic performance, r_∞, is the inverse slope of the best fit line, and the half-performance length, $n_{\frac{1}{2}}$, is the negative intercept of the line with the n-axis. (a) dyadic operations $\underset{\sim}{A}=\underset{\sim}{B}*\underset{\sim}{C}$, (b) all vector triadic operation $\underset{\sim}{A} = \underset{\sim}{D}*\underset{\sim}{B} + \underset{\sim}{C}$, (c) Cyber 205 triad $\underset{\sim}{A} = s*\underset{\sim}{B} + \underset{\sim}{C}$.

Table I - Measured values of r_∞ and $n_{\frac{1}{2}}$ on a single CPU of a Cray X-MP for memory-to-memory operations. Cray-1 values are in parentheses.

Operation: Statement 10	r_∞ Mflop/s	$n_{\frac{1}{2}}$ flop	t_0 μs
Dyadic A(I) = B(I)*C(I) (Cray-1 values)	70 (22)	53 (18)	0.75 (0.82)
All vector Triad A(I) = D(I)*B(I) + C(I)	107	45	0.42
Cyber 205 Triad A(I) = s*B(I) + C(I)	148	60	0.40
Scalar code A(I) = B(I)*C(I)	5	4	0.80

The first three cases in Table I are measurements of vector instructions. The dyadic case uses only a single vector pipeline with all vectors stored in main memory, and is to be compared with values of $r_\infty = 22$ Mflop/s and $n_{\frac{1}{2}} = 19$ flop previously obtained on the Cray-1 (Hockney and Jesshope, 1981). We find a three fold increase in r_∞ due, primarily to the provision of three memory ports on the Cray X-MP compared with one on the Cray-1. The startup time in microseconds, $t_0 = n_{\frac{1}{2}}/r_\infty$, has not changed significantly between the two machines: the extra complexity of memory access on the X-MP being compensated by a reduction in clock period. However the importance of this overhead, which is what $n_{\frac{1}{2}}$ measures, is three times greater on the X-MP, because three times as much arithmetic could have been done during this time, as compared to the Cray-1. The maximum rate at which one floating-point pipeline can deliver results is one result per clock period of 9.5 ns, that is to say 105 Mflop/s. The measured value of $r_\infty = 70$ Mflop/s is less than this because of the time taken to refill the vector registers from main memory. Because the vector registers hold 64 elements, this is an overhead that is incurred every 64 elements, and is just visible in Fig. 2.

The second two cases are measurements for triadic vector operations, which involve the chaining of two vector instructions, and the simultaneous use of both the floating-point multiply and add pipelines. At best we can expect a doubling of r_∞ which is achieved if one of the arguments is a scalar, but is not achieved in the all vector case. The value of $n_{\frac{1}{2}}$ is not materially altered in the triad cases, however the startup time in microseconds is halved because a single startup of 0.8μs is shared between two instructions.

For comparison purposes, we have run the dyad benchmark with instructions to the compiler to use only scalar instructions, and obtain r_∞ = 5 Mflop/s and $n_{\frac{1}{2}}$ = 4 flop. The startup time t_0 remains at 0.8 μs but this is now of negligible importance because the arithmetic performance is about twenty times slower than when a vector instruction is used. This is made clear by the fact that $n_{\frac{1}{2}}$ is reduced to a negligible value compared to any vector lengths that are likely to be used, thus substantiating the statement in Section 2 that $n_{\frac{1}{2}} \cong 0$ for scalar code.

3.2 Values of $(r_\infty, s_{\frac{1}{2}})$ for two CPUs

Four methods of synchronizing the operation of the two CPUs of a Cray X-MP on a single job have been considered, and the programs used are given in Figs. 3 to 6. They are the use of the TSKSTART and TSKWAIT primitives which we refer to as the TASKS method; the use of LOCKON and LOCKOFF primitives which we refer to as the LOCKS method; the use of the EVPOST and EVWAIT primitives which we refer to as the EVENTS method; and finally the use of a simplified LOCKS method written in CAL code. In all cases the programs were run on the computer in stand-alone mode, and timing was performed using the real-time clock function RTC(DUM). In this way we ensure that the second physical CPU is assigned to a second logical CPU in the programs, and that we are measuring the wall-clock time for the complete job.

In the TASKS method (Fig. 3), after calling the timer (RTC) at the start of the measurement (T1), the second CPU is given a copy of the subroutine DOALL by the TSKSTART statement, and begins to execute it. The first CPU, which is performing the control program MULTI, then executes another copy of the subroutine DOALL in the CALL DOALL statement. The TSKWAIT statement ensures that both CPUs have finished their share of the work before the timer is called again to record the end of the measurement (T2). The parameters to DOALL are used to ensure that the two CPUs do different elemental operations (N/2 each) from the total of N operations. In this method the overhead of starting a new task occurs at every FORK into a work segment that is divided between the two CPUs, and is therefore included in the measured time. By choosing an element-by-element vector multiply for even vector lengths, as the work to be done in a work segment, we ensure that the workload of N/2 floating-point operations can be equally balanced between the two CPUs, so that the efficiency of scheduling E_p = 1. Hence the measured time, t, can be compared with the standard form of Eqn. (7)

```
      PROGRAM MULTI
      COMMON/GLOBAL/A(4ØØ), B(4ØØ), C(4ØØ)
      DIMENSION IDT(2)
      EXTERNAL DOALL
      DATA B/4ØØ*1.Ø/, C/4ØØ*1.Ø/

      NMAX = 4ØØ
      IDT(1) = 2

      T1 = 9.5E-9*RTC(DUM)
      T2 = 9.5E-9*RTC(DUM)
      TØ = T2-T1

      DO 2Ø N = 2, NMAX, 2
      T1 = 9.5E-9*RTC(DUM)
      NHALF = N/2,
      NH1 = NHALF + 1

      CALL TSKSTART (IDT,DOALL, NH1, N)
      CALL DOALL (1, NHALF)
      CALL TSKWAIT (IDT)

      T2 = 9.5E-9*RTC(DUM)

      T = T2-T1-TØ
      WRITE (6, 1ØØ) N, T
 2Ø   CONTINUE

1ØØ   FORMAT (' N: ', I4, 4X, 'TIME IN SECONDS:' F16.12)
      STOP
      END

      SUBROUTINE DOALL (N1, N2)
      COMMON/GLOBAL/A(4ØØ), B(4ØØ), C(4ØØ)

      DO 1Ø I = N1, N2
 1Ø      A(I) = B(I)*C(I)

      RETURN
      END
```

Fig. 3 Program for measuring r_∞ and $s_{\frac{1}{2}}$ when a job is split between the two CPUs of the Cray X-MP, using the TASK method of synchronization.

Table II - Measured values of r_∞ and $s_{\frac{1}{2}}$ when dyadic memory-to-memory operations are split between two CPUs on the Cray X-MP/22. The overhead is separately measured using TASKS, LOCKS, EVENTS and CAL code for synchronization. $t_0 = s_{\frac{1}{2}}/r_\infty$ is the synchronization overhead in microseconds. Equivalent values for a single PEM Denelcor HEP1 computer are given in parentheses (Hockney and Snelling, 1984; Hockney 1985)

METHOD	r_∞ Mflop/s	$s_{\frac{1}{2}}$ flop	t_0 μs	$\pi_0 = t_0^{-1}$ k/s
TASKS (HEP1)	130 (1.7)	5700 (820)	45 (490)	22 (2)
LOCKS	140	4000	28	36
EVENTS (HEP1)	140 (1.7)	2000 (230)	14 (140)	71 (7)
simplified LOCKS CAL code	110	220	2	500

Note: Results are deliberately rounded to two significant figures only. Greater precision would suggest spurious accuracy.

$$t = r_\infty^{-1}(s + s_{\frac{1}{2}}) \quad (10a)$$

or $$t = t_0 + s/r_\infty \quad (10b)$$

where $s = N$ of the program in Fig. 3, is the total amount of work split between the two CPUs,

and $t_0 = s_{\frac{1}{2}}/r_\infty$ is the synchronization overhead time in seconds.

The measured time fits the formula

$$t = 45 + 3.2\ s/400\ \mu s \quad (11)$$

which leads to the values of r_∞ and $s_{\frac{1}{2}}$ given in Table II.

We also show in Table II the result of exactly the same measurements performed on a single PEM, Denelcor HEP1 (Hockney and Snelling 1984, Hockney 1985). The comparison raises some important issues concerning overheads and their interpretation. One sees that, although the overhead time in seconds, t_0, on the Cray X-MP is only about one tenth that of the Denelcor HEP1, the value of $s_{\frac{1}{2}}$ of the Cray X-MP is almost ten times more than that of the HEP1. So which is the "best" machine from the overhead point of view? If we interpret "best" to mean least execution time, then no single parameter - neither t_0 alone nor $s_{\frac{1}{2}}$ alone - can answer this question. It is necessary to know two parameters, either the pair $(r_\infty, s_{\frac{1}{2}})$ or the pair (r_∞, t_0). Furthermore Eqns. (10) show that the answer may depend on the amount of work, s, that is split between the CPUs, which is often referred to as the grain of the MIMD calculation. In analogy with Eqn. (3) for vector calculations, one has from Eqn. (10a)

$$r = s/t = r_\infty/(1 + s_{\frac{1}{2}}/s) \quad (12)$$

for the average performance, r, as a function of grain size, s. Taking the limit grain size ($s/s_{\frac{1}{2}} >> 1$), we have $r \to r_\infty$. Thus r_∞ is the parameter that characterizes the performance of large-grain MIMD programs. However, in the limit of small grain-size ($s/s_{\frac{1}{2}} << 1$), we have the performance

$$r = r_\infty s/s_{\frac{1}{2}} = \pi_0 s \quad (13)$$

where $\pi_0 = t_0^{-1} = r_\infty/s_{\frac{1}{2}}$ is called the specific performance.. Thus π_0, rather than r_∞, characterizes the performance of small-grain MIMD programs. Using the numbers in Table II, one finds that both the small and large-grain performance of the Cray X-MP is greater than that of the HEP1. In this case, what then is the point of the parameter $s_{\frac{1}{2}}$, which suggests that in some sense the HEP1 is better, not worse, than the Cray X-MP at synchronization? The point is that, for any given grain size, the HEP1 makes more efficient use of the available computing capability: that is to say, given s, the fraction of the maximum Mflop/s that is actually achieved is always higher on the HEP1 than on the Cray X-MP. This follows from Eqn. (12), because $s_{\frac{1}{2}}$ is smaller on the HEP1, and therefore $s/s_{\frac{1}{2}}$ is always larger. Put another way, if the HEP1 were implemented in the same technology as the Cray X-MP, that is to say if it had the same clock period, its r_∞ would become equal to that of the Cray X-MP, but its value of $s_{\frac{1}{2}}$ would remain unchanged at about one tenth of that of the Cray X-MP. Thus one may say that $s_{\frac{1}{2}}$ is a characterization of the computer architecture (in the general sense of including the efficiency of the synchronization software), whereas r_∞ is a characteristic of the technology in which the architecture is implemented. In this sense the $(r_\infty, s_{\frac{1}{2}})$ characterization of a computer nicely separates architecture from technology. This separation is also evident from the fact $s_{\frac{1}{2}}$ does not contain any physical units, whereas r_∞ contains inverse seconds in its units.

The conclusion from the comparison is that the HEP1 is architecturally better at synchronization than the Cray X-MP, however the latter is better in absolute terms because of its faster circuitry. The Denelcor HEP2 is said to be architecturally similar to the HEP1, but implemented in high-speed technology. This machine could therefore prove also to be better than the Cray X-MP at synchronization in absolute terms.

In other cases of comparison one may find that one computer has the better small-grain performance and the other the better large-grain performance. It is worth noting that since the vector length equation (1) for $n_{\frac{1}{2}}$ is identical in form to the grain size equation (7) for $s_{\frac{1}{2}}$, vector length n in SIMD programs and grain size s in MIMD programs are analogous quantities. Hence the previous statement about MIMD performance is analogous to saying of two SIMD computers that one has the better short-vector performance whilst the other has the better long-vector performance.

The second virtue of the $s_{\frac{1}{2}}$ parameter for measuring the synchronization overhead, is that it is the yardstick by which a programmer may judge the grain at which to multi-task a program. If we regard 50% of the asymptotic performance rate, r_∞, as the minimum acceptable efficiency, then we know that the grain size s must exceed the $s_{\frac{1}{2}}$ of the computer before multi-tasking is worthwhile. Thus the figures in Table II tell us that the HEP1 is suitable for multi-tasking problems with about ten times finer grain than is the Cray X-MP. Furthermore, $s_{\frac{1}{2}}$ is measured in units of floating-point operations, to which a numerical analyst devising an algorithm can directly relate, because he knows the number of such operations in different parts of his algorithm. If however the overhead is measured in physical units, e.g. by t_0 in microseconds, then these are not the units in which the complexity of an algorithm is most naturally expressed or known.

The next least expensive method of synchronization proves to be the LOCKS method, the code for which is shown in Fig. 4. In this case we observe (Table II) an $s_{\frac{1}{2}}$ = 4000 flop about 2/3 of the value

```
      PROGRAM MULTI
      COMMON/GLOBAL/A(4ØØ), B(4ØØ), C(4ØØ)
      COMMON IVT1, IVT2
      DIMENSION IDT(2)
      EXTERNAL DOALL
      DATA B/4ØØ*1.Ø, C/4ØØ*1.Ø/

      NMAX = 4ØØ
      IDT(1) = 2

      CALL LOCKASGN(IVT1)
      CALL LOCKASGN(IVT2)
      CALL LOCKON(IVT1)
      CALL LOCKON(IVT2)

      T1 = 9.5E-9*RTC(DUM)
      T2 = 9.5E-9*RTC(DUM)
      TØ = T2-T1
      CALL TSKSTART (IDT, DOALL)

      DO 2Ø N = 2, NMAX, 2
      T1 = 9.5E-9*RTC(DUM)
      NHALF = N/2

      CALL LOCKOFF(IVT1)
 1Ø   A(I) = B(I)*C(I)

      CALL LOCKON(IVT2)

      T2 = 9.5E*RTC(DUM)
      T = T2-T1-TØ
      WRITE (6,1Ø,Ø) N, T
 2Ø   CONTINUE

1ØØ   FORMAT(N' N: ', I4, 4X, 'TIME IN SECONDS:', F16.12)
      STOP
      END
```

Fig. 4(a) Program for measuring r_∞ and $s_{\frac{1}{2}}$ when a job is split between the two CPUs of the Cray X-MP, using the LOCKS method of synchronization. Above code executed in CPU1.

```
        SUBROUTINE DOALL
        COMMON/GLOBAL/A(4ØØ), B(4ØØ), C(4ØØ)
        COMMON IVT1, IVT2

        NMAX = 4ØØ

        DO 2Ø N = 2, NMAX, 2
        NH1 = N/2 + 1

        CALL LOCKON(IVT1)

        DO 1Ø I = NH1, N
  1Ø    A(I) = B(I)*C(I)

        CALL LOCKOFF(IVT2)

  2Ø    CONTINUE

        RETURN
        END
```

Fig. 4(b) The subroutine DOALL used with Fig. 4(a), and executed in CPU2.

```
      PROGRAM MULTI
      COMMON/GLOBAL/A(4ØØ), B(4ØØ), C(4ØØ)
      COMMON IVT1, IVT2
      DIMENSION IDT(2)
      EXTERNAL DOALL
      DATA B/4ØØ*1.Ø/, C/4ØØ*1.0/

      NMAX = 4ØØ
      IDT(1) = 2

      CALL EVASGN(IVT1)
      CALL EVASGN(IVT2)

      T1 = 9.5E-9*RTC(DUM)
      T2 = 9.5E-9*RTC(DUM)
      TØ = T2-T1
      CALL TSKSTART (IDT, DOALL)

      DO 2Ø, N = 2, NMAX, 2
      T1  9.5E-9*RTC(DUM)
      NHALF = N/2

      DO 1Ø I = 1, NHALF
1Ø    A(I) = B(I)*C(I)

      CALL EVPOST(IVT1)
      CALL EVWAIT(IVT2)
      CALL EVCLEAR(IVT2)

      T2 = 9.5E-9*RTC(DUM)
      T= T2-T1-TØ
      WRITE (6, 1ØØ) N, T
2Ø    CONTINUE

1ØØ   FORMAT (' N; ', I4, 4X, 'TIME IN SECONDS:', F16.12)
      STOP
      END
```

Fig. 5(a) Program for measuring r_∞ and $s_{\frac{1}{2}}$ when a job is split between the two CPUs of the Cray X-MP, using the EVENTS method of synchronization. Above code is executed in CPU1.

```
      SUBROUTINE DOALL
      COMMON/GLOBAL/A(4ØØ), B(4ØØ), C(4ØØ)
      COMMON IVT1, IVT2

      NMAX = 4ØØ

      DO 2Ø N = 2, NMAX, 2
      NH1 = N/2 + 1

      DO 1Ø I = NH1,N
 1Ø   A(I) = B(I)*C9I)

      CALL EVPOST(IVT2)
      CALL EVWAIT(IVT1)
      CALL EVCLEAR(IVT1)

 2Ø   CONTINUE

      RETURN
      END
```

Fig. 5(b) The subroutine DOALL used with Fig. 5(a), and executed in CPU2.

```
      PROGRAM MULTI
      COMMON/GLOBAL/A(4ØØ), B(4ØØ), C(4ØØ)
      COMMON IVT1, IVT2
      DIMENSION IDT(2)
      EXTERNAL DOALL
      DATA B/4ØØ*1.Ø/, C/4ØØ*1.Ø/

      NMAX = 4ØØ
      IDT(1) = 2

      CALL INIT

      T1 = 9.5E-9*RTC(DUM)
      T2 = 9.5E-9*RTC(DUM)
      TØ = T2-T1
      CALL TSKSTART(IDT, DOALL)

      DO 2Ø, N = 2, NMAX, 2
      T1  9.5E-9*RTC(DUM)
      NHALF = N/2

      CALL POST1

      DO 1Ø I = 1, NHALF
 1Ø   A(I) = B(I)*C(I)

      CALL WAIT2

      T2 = 9.5E-9*RTC(DUM)
      T = T2-T1-TØ
      WRITE (6, 1ØØ) N, T
 2Ø   CONTINUE

1ØØ   FORMAT (' N; ', I4, 4X, 'TIME IN SECONDS:', F16.12)
      STOP
      END
```

Fig. 6(a) Program for measuring r_∞ and $s_{\frac{1}{2}}$ when a job is split between the two CPUs of the Cray X-MP, using the CAL code for synchronization.

```
      SUBROUTINE DOALL
      COMMON/GLOBAL/A(4ØØ), B(4ØØ0, C(4ØØ)
      COMMON IVCT1, IVT2
      NMAX = 4ØØ

      DO 2Ø N = 2, NMAX, 2
      NH1 = N/2 + 2

      CALL WAIT1

      DO 1Ø I = NH1, N
 1Ø   A(I) = B(I)*C(I)

      CALL POST 2

 2Ø   CONTINUE

      RETURN
      END

        IDENT   CALIB
        ENTRY   INIT, POST1, POOST2, WAIT1, WAIT2
INIT    =       *
        A1      Ø
        SBØ     A1
        J       BØØ
POST1   =       *
        A1      1
        SBØ     A1
        J       BØØ
WAIT2   =       *
        AØ      SBØ
        JAN     *-1
        J       BØØ
WAIT1   =       *
        AØ      SBØ
        JAZ     *-1
        J       BØØ
POST2   =       *
        A1      Ø
        SBØ     A1
        J       BØØ
        END
```

Fig. 6(b) The subroutine DOALL and the CAL simplified LOCKS synchronization routines used with Fig. 6(a). Code kindly supplied by John Larson (Cray Research Inc).

found for the TASKS method. The code for synchronization using the EVENTS methods is given in Fig. 5, and we find this to be half as expensive as the LOCKS method with $s_{\frac{1}{2}} = 2000$ flop. In order to determine the least overhead possible, a simplified form of the LOCKS method has been programmed in CAL by John Larson of Cray Research Inc. and his code is given in Fig. 6. The overhead is thereby reduced by a factor of ten to $s_{\frac{1}{2}} = 220$ flop. An examination of the CAL code in Fig. 6(b) shows that there is no wasted time, and it is unlikely that synchronization can be achieved on the Cray X-MP with less overhead. However, it must be said that in the CAL code, one CPU waits for the other to finish by continually testing one of the synchronization registers. This prevents the waiting CPU from doing any other work during this time, and hence this code would hardly be acceptable as a general method of synchronization.

4. CONCLUSIONS

Measurements have been reported of the vector startup overhead on a single CPU of the Cray X-MP, and the overhead of splitting a job between the two CPUs of a Cray X-MP. Interpreted in terms of the parameters $(r_\infty, n_{\frac{1}{2}})$ we conclude that the Cray X-MP is about three times faster than the Cray-1 in asymptotic performance r_∞, however this has the effect of similarly increasing the half-performance length, $n_{\frac{1}{2}}$. One of the principle differences between the Cray-1 and the Cyber 205 is the fact that their values of $n_{\frac{1}{2}}$ differed by about a factor of ten (being ≅10 to 20 and ≅100 to 200 respectively). Now that we find the Cray X-MP with $n_{\frac{1}{2}}$ ≅50 it has become more like the Cyber 205 from the vector length point of view. Other differences, like the need for contiguous vectors on the Cyber 205, of course, remain.

Measurements of job splitting between two CPUs interpreted in terms of the parameters $(r_\infty, s_{\frac{1}{2}})$ show that the cost of job synchronization in terms of lost floating-point operations is about ten times more expensive on the Cray X-MP than on the Denelcor HEP1. This means that the grain size of jobs split between two CPUs needs to be about ten times larger on the Cray X-MP than on the HEP1 in order to achieve the same fraction of the maximum performance.

5. ACKNOWLEDGEMENTS

The author wishes to acknowledge his debt to David Dent of ECMWF who ran the programs described in this paper in a very short time, and made many valuable suggestions. The author is also endebted to John Larson of Cray Research, Chippewa Falls, for his helpful criticism of the first draft of the paper, and for supplying the CAL code for simplified LOCKS synchronization. The opportunity to discuss synchronization and its characterization with Paul Swarztrauber, Tor BLock and others at the workshop is much appreciated. Dr Lennart Bengtsson, Director of ECMWF, and Mr Geerd-R. Hoffmann, Head of the Computer Division, are also thanked for granting their permission to use the ECMWF computer centre for these measurements. The author's work is supported by the UK Science and Engineering Research Council grant GR/B/39336.

6. REFERENCES

Calahan, D.A., 1977: Algorithmic and architectural issues related to vector processors, Proc. Intl. Symp. Large Eng. Syst., New York, Pergamon.

Calahan, D.A. and W.G. Ames, 1979: Vector Processors: models and applications, IEEE Trans. Circuits and Systems, CAS-26, 715-726.

Hockney, R.W., 1977: Supercomputer architecture, in Infotech State of the Art Conference: Future Systems, F. Sumner (ed.), Maidenhead, Infotech Int., 277-305.

Hockney, R.W. and C.R. Jesshope, 1981: Parallel Computers - architecture, programming and algorithms. Bristol, Adam Hilger, pp 423. (Distributed in North and South America by Heyden and Son, Philadelphia).

Hockney, R.W., 1983: Characterizing computers and optimizing the FACR(ℓ) Poisson-solver on parallel unicomputers. IEEE Trans. Comput., C-32, 933-941.

Hockney, R.W., 1984: The $n_{\frac{1}{2}}$-method of algorithm analysis, in PDE Software: Modules, Interfaces and Systems, B. Engquist and T. Smedsaas (eds.), Amsterdam, Elsevier Science Publ. (North-Holland), 429-444.

Hockney, R.W. and D.F. Snelling, 1984: Characterizing MIMD computers: e.g. the Denelcor HEP, in Parallel Computing 83, M. Feilmeir, G.R. Joubert and U. Schendel (eds.), Amsterdam, Elsevier Science Publ., North-Holland, 521-526.

Hockeny, R.W., 1985: Performance characterization of the HEP, in MIMD Computation: HEP Supercomputer and its applications, J.S. Kowalik (ed.), Cambridge, MIT Press.

Kuck, D.J., 1978: The Structure of Computers and Computation, vol. 1, New York, Wiley, pp 611.

MIMD Processing and the Denelcor HEP

DAVID F. SNELLING and BURTON J. SMITH

Denelcor Inc., Avrora, CO, USA

INTRODUCTION

The weather community has long provided the forefront of computational science in that their need for high-speed computation tends to grow as order n^3 on the size of the problem.

As a result the weather community first entered into the use of vector processors and were shortly followed by the rest of the scientific community. This has also been the case with the multi-processor industry; the European Centre for Medium-Range Weather Forecasts has taken early membership in the community of multi-processor users with their CRAY X-MP dual processor system.

With the increase in the number of multi-processors available on the market and the fast-growing need for high-speed computation in the MIMD industry, it is important to consider the issues involved in multi-processing such as synchronisation, create processing, process termination, control of large numbers of parallel processors and questions as to the number of parallel streams versus size and computational power of the processors executing these streams. In view to addressing some of these issues this paper will discuss briefly the Heterogeneous Element Processor architecture from Denelcor as an example of a very efficient high-performance MIMD architecture. Based on the architecture of the HEP we will further discuss some aspects of parallel computation and how they are particularly related to the weather industry.

THE HEP ARCHITECTURE

The Denelcor HEP is a modular multi-processor mainframe with 3 types of modules: PEM (Process Execution Module) and Data Memory Module, MSS (Mass Storage System). These are connected to each other via a high-speed packet switching network called, appropriate enough, the Switch.

A HEP system may have 1 to 16 PEMs each capable of running at 10 million instructions per second. The PEM consists of a self-contained program memory of 1 to 8 megabytes. A 16 PEM HEP system therefore could have 128 megabytes of program space (this is independent of the data memory). Each PEM contains 2048 general purpose registers and 4096 system constants. Constant memories are used in the same way as registers except that they are read only to user programs. They contain frequently used values like the small integers, pi and other math library constants.

Up to 128 data memory modules may be included in the system allowing for a total memory of 1 Gigabyte. The memory is organised into 64 bit words with 8 error correction bits. The memory may be addressed as words (64 bits), half-words (32 bits), quarter-words (16 bits), and bytes. This memory is one continuous address space and may be interleaved or blocked by module.

The MSS is a solid state buffer that is addressable as part of the data memory and is used as a buffer for disk transfers, a swap space for rolled out programs, and/or a solid state disk. There may be up to four MSSs in a HEP system each with a maximum space of 128 megabytes.

All the above modules are interconnected by the Switch. The Switch is composed of nodes (the number of which increases with the complexity of the system). Each node has 3 bidirectional ports with a bandwidth of 80 megabytes per second per port. To illustrate, when a load data memory instruction is executed in the PEM, a request message is inserted into the Switch; is passed from node to node; has data added to it at the data memory; is passed back node to node; and is returned to the PEM. At each node along the way a preset routing map determines which exit port will lead, with the shortest path, to the data memory. The optimal routing from any module to another is set at system initialisation time, and hence a PEM may be partitioned out of the network for routine maintenance while other PEMs in the system continue to operate. The same holds for other modules in the system.

Each PEM is itself a pipelined parallel processor. Each operation is subdivided into 8 segments and separate parts of the hardware handle each segment. With the exception of divide and data memory accesses, all machine instructions require eight stages. Unlike most pipelined machines, the HEP need not be executing the same kind of instruction in all segments of the pipeline as in SIMD (Single Instruction Multiple Data stream) computer architectures.

When an instruction stream or process is executing, an instruction from it enters the pipeline, passes through all the stages, and 8 cycles later the next instruction from that process enters the pipeline. This leaves 7 pipeline stages unused. These 7 stages are available to other processes which run in parallel with the original process. Control of these parallel processes is achieved by hardware queues. For each process in the machine there is a PSW (Process Status Word). A PSW is comparable to the program counter in a conventional Von-Neumann machine. It contains the address of the next instruction to execute and some other information pertaining to the process to which it belongs. Each PEM in a HEP system has 128 PSWs (processes). These PSWs are queued by the hardware for execution.

The final topic in this basic architecture discussion is that of process synchronisation. If two processes are acting as producer and consumer, some form of synchronisation is necessary to avoid one outrunning the other. In the HEP every location in data memory has a access bit associated with it. This bit tells whether the location is empty of full. The access state of a memory location may be tested and changed indivisibly by use of the special access modes in the memory addressing scheme. In the example below subroutine ABC is sending a stream of values to subroutine XYZ. XYZ is consuming them and the transfer is protected from either process overrunning the other. The example is written in Fortran 77 with the parallel routines provided with the HEP system.

```
      program doitforever
      external abc, xyz
      call create (abc,x)
      call xyz (x)
       :
       :

      subroutine abc (x)
10    continue
      t=bigfunction()          generate the value t
      call awrite (x,t)        write x only if empty and set full
      goto 10
      end

      subroutine xyz (x)
10    continue
      t=aread (x)              read x only if full and set empty
      call uset (t)            use the value t
      goto 10
      end
```

x, when used in this way, is called an asynchronous variable.

PARALLEL PROGRAMMING TECHNIQUES

In parallel processing there are two basic approaches to handling parallel streams: The first is called the fork-join in which parallel streams are created, perform their assigned work, and then are joined back together to form a serial stream of execution. The other alternative is up-front creation of many parallel streams. Once the streams are created work is passed around among the various executing streams throughout the execution of the program.

Both the schemes have their advantages. Fork-join first of all provides very clear and precise definition of where the parallelism is and also provides a more simplified mode of programming. On the other hand, if the particular machine in question has relatively inefficient create processing, the overheads involved can limit the size of parallel streams allowed. For example, if a machine is inefficient at parallel process creation, then only very large segments of the program can be efficiently parallelized. The second alternative, that of up-front process creation, has advantages and disadvantages. When a process is created at the beginning it is not known what routines it may actually execute. As a result the parallelism is not apparent to the programmer or to someone debugging the program. A point of synchronization somewhere in the program, a barrier of some sort would indicate that at this point some form of synchronization is taking place and there would be no indication as to what other routines may be involved. The corresponding situation in a fork-join can only be a fork-join statement in an isolated place in the program. At this point all synchronization would be apparent to someone debugging or trying to understand the program. The key trade-off between using fork-join process creation and up-front process creation is the trade-off between efficiency and readability. As a result some form of measure of the overhead involved in synchronization is required in order for program designers to determine which form of process creation is most suited to the application.

One such measure has been put forward by Dr. Roger Hockney of Reading University with his S-½. S-½ is a measure of the amount of overhead required to perform a parallel process creation. To be more precise it is a measure of the amount of parallel computation required for a particular architecture to make use of half the number of instruction streams being considered.

For example, in a HEP1 program with two instruction streams, sixty floating point operations are required to guarantee half utilisation of the two instruction streams. As a result S-½ is the overhead for creating another instruction stream. The S-½ for the HEP1 processor running 14 instruction streams is 820. The S-½ for CRAY X-MP dual processor running two instruction streams is 5700. (Reference Dr. Hockney's paper in this publication).

Given the large difference in the efficiencies of the two machines mentioned above, it will be of interest to describe how the two generate their parallel instruction streams.

In the CRAY X-MP there is a system routine called Task-Start that actually generates another virtual job within the CRAY X-MP dual processor, and then the scheduling between the two processors is determined by the operating system. On the HEP, however, the approach is much simpler. The execution stream that will be doing the create performs a score-boarded store into a demon location. This uses the empty/full facility of the HEP architecture. Concurrently with this stream is another stream already set-up by the system at load time.

This stream then becomes the second instruction stream. As a result some of the create processing itself is running in parallel, and the number of processes that can be created is much greater, and the efficiency much higher than on a system that involves operating system control and operating system scheduling of parallel streams.

Once many instruction streams have been created it would be nice if all of the streams could execute independently throughout the execution of a job. This, however, is rarely the case and some form of synchronization between instruction streams is virtually always required. As a result, various different schemes for controlling the process synchronization have been developed. Two will be discussed here:

One is pre-scheduling in which what work a particular stream does is determined prior to the creation of that stream. The other scheme is called self-scheduling in which processes are created at the on-set and then work allocated dynamically as a program runs. In the pre-scheduling scheme it is ultimately the programmer who decides what instruction streams will execute what code. As a result issues such as load-balancing and efficiency of parallel processing are left entirely up to the user and not necessarily a function of the actual load on the machine at a given time. Whereas, in self-scheduling, when the process becomes available (i.e. it has just finished a previous assignment) it goes to a queue of work and picks up additional computation. In experience it has been found that self-scheduling tends to provide a more efficient load balancing of parallel process execution.

In general it is the feeling of the authors that self-scheduling with fork-join process creation is the most self-explanatory and efficient mode of parallel programming. This, however, does require that the architecture in consideration have an efficient way of first creating parallel instruction streams and second of allocating self-scheduled process modules.

OTHER MIMD ISSUES

There are two major MIMD issues that have become quite familiar to the industry as a result of experience in vector processing.

The first is Amdahl's Law which states that unless the entire program can be run in parallel, total performance cannot be achieved. To the extent that, if a program is 90% parallelizable or vectorizable, the program can run a maximum of 10 times faster, in which case the 90% parallelizable section would be running in zero time.

It is well known that Amdahl's Law cannot be changed or avoided. However, by investigating the contributing factors one can get an idea of how programming practice can avoid Amdahl's Law to some extent. The goal in avoiding Amdahl's Law is to have as much of the program parallelizable as possible. What typically happens in a program once it is parallelized is that different instruction streams wind up doing different amounts of work. The problem is typically called load-balancing, in which case a large segment of code has been allocated to one instruction stream whereas another instruction stream is dealing with a much smaller section. When a smaller section has completed, the longer section must continue running on only one processor. As a result, Amdahl's Law comes into play and the program does not run at full efficiency. The load-balancing problem is particularly pronounced when pre-scheduling is used, if a dynamic scheduling algorithm on a large number of parallel streams is not possible. The best solution to this problem is to have a large number of smaller parallel code segments executing in a self-scheduling manner. In this case busy processes continue executing while non-busy processes return to the queue and acquire more computation.

In a very large problem one would be able to achieve high degrees of parallelism by using a large number of very small pieces for parallel execution. The problem that arises here is that the synchronisation associated with selecting a parallel segment is significant to the total performance of the system. As a result, for any given system, a balance between the number and size of executable grains, and the scheduling algorithm (either self-scheduling or pre-scheduling) must be chosen.

Another issue that contributes to Amdahl's Law is memory latency. As the systems are built with larger and larger memories, the memory latency or access time to memory becomes significantly large. Even when facilities for masking the memory latency (i.e. overlapping of instructions and other techniques used by various different architectures) are employed, the memory reference time becomes a significant factor in the performance of a parallel system. It becomes even more important as the number of parallel streams increases, as memory conflicts will begin to occur between several different processes in the system.

The Heterogeneous Element Processor from Denelcor has a facility for handling this memory latency issue. The way it works is by ensuring that there is more work to do than there are actual processors in the system. The HEP, in a single processor version, is essentially a 12 instruction stream machine requiring typically 12 instruction streams to keep the processor busy.

By queueing up a larger number of instruction streams, say 20 to 30, the HEP is able to fill in the time it takes to go to memory, with more work. As a result, when more processors are added to the HEP system, the increased conflicts and latency associated with a multi-processor system are also masked by creating more instruction streams in the overall system.

Based on a preliminary study of the weather model at ECMWF it appears to the authors that there are some 900,000 potential parallel streams in the standard weather model currently being run at the Weather Centre. It should be noted, of course, that each one of these streams has associated with it enough instructions that it would land within the efficiency range of a fork-join synchronization approach on the Heterogeneous Element Processor. Hence the potential is high for parallelizing weather models to a very high degree on a large number of processors. As the weather community comes closer and closer to using parallel processing as a production tool, there are still many questions that have to be considered.

One issue that would require a little closer examination would be an extension to Amdahl's Law. Amdahl's Law has typically been stated for one or all processors executing. It would be interesting to investigate what happens to Amdahl's Law when one considers the performance achieved when a certain percentage of the code is running with full parallelism, a smaller percentage is running at 50% parallelism and then a smaller section running in serial or scalar mode.

Such a study is currently being carried out by Denelcor Limited in the United Kingdom along with a study into a parallel run time load-balancing package in which, at execution time, the load is balanced dynamically at a very fine grain level.

One issue being addressed by Denelcor Inc. in the United States with regard to their second generation processor, is the development of a parallel memory management system for programs executing many parallel streams simultaneously.

Studies underway at Los Alamos National Laboratory in conjunction with the University of California at San Diego are providing tools for global data dependency analysis and a simplified package of portable parallel programming tools.

These tools provide a collection of macro facilities and synchronization modes that allow them to generate code for HEP processors and multiple VAX processors, all using one standard syntax.

Parallel processing provides a rich and efficient mode of achieving high degrees of performance in various different applications, the weather industry in particular. However, many questions remain to be considered and one hopes that academic activity in the area continues and that manufacturers and industry continue to co-operate towards that end.

Using an IBM Multiprocessor System

A.L. LIM and D.B. SOLL

IBM Corporation, Data Systems Division, Kingston, New York, USA

ABSTRACT

This paper presents an overview of the evolution of IBM multiprocessor products and current product offerengs. Programming language solutions for parallel program execution are discussed and their suitability for the implementation of scientific and engineering applications is assessed.

A set of groundrules governing the provision of parallel programming capabilities is proposed. Results from applying these groundrules to real scientific application codes executing on a multiprocessor are provided.

Open questions for future investigation are posed.

SECTION 1: OVERVIEW OF IBM MULTIPROCESSING

The trend in large computer systems has been driven by desires for higher performance, increased reliability and lower cost per unit of computation. Much has been written about advances in computing technology from the mid 1940's to today. Figure 1 highlights some of the major accomplishments in large system developments.

In fact, parallelism, exploited in machine design, has been a significant contributor to these improvements. For example, even uniprocessor configurations may incorporate parallelism at various levels such as interleaving of storage, signal multiplexing, pipelining of functional units and high speed cache. Today, large, high end systems consist of multiple processor complexes.

Multiple processor configurations may be utilized in various ways, depending on the manner in which the processors are connected. Loosely coupled processors, that is, processors which are connected to each other by channel-to-channel attachments, for example, need not be identical, since there is no sharing of storage, channels, or controls. Loosely coupled processor systems have existed since the 1960's.

HISTORICAL REVIEW

FACILITY	TIME
* FIRST ELECTRONIC COMPUTER	1944-46
* FIRST STORED PROGRAM	1949
* CONTROL UNITS AND CHANNELS	1958
* INTERLEAVED STORAGE	1961
* PARALLEL/PIPELINE PROCESSORS	1961
* LOOSELY COUPLED SYSTEMS	1964
* HIGH SPEED BUFFERS (CACHE)	1969
* TIGHTLY COUPLED SYSTEMS	1969

FIGURE 1: HISTORICAL REVIEW

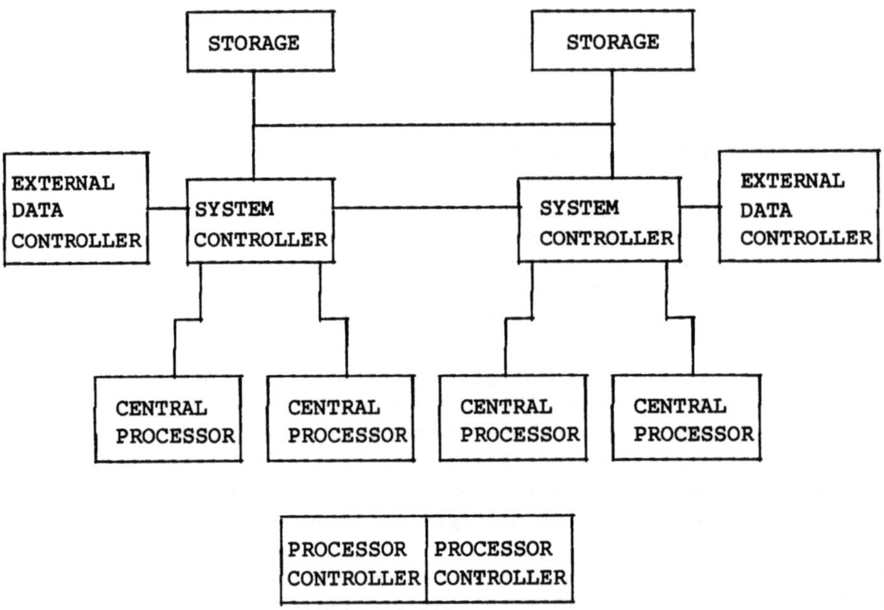

FIGURE 2: IBM 3084 ORGANIZATION

Tightly coupled processors, on the other hand, have a much greater connectivity and generally share memory, channels and other functional control. This arrangement also implies a high degree of interaction between the processors, which can be utilized to achieve either optimal throughput or optimal turn-around. Tightly coupled processors such as the IBM 370/168 MP were available in 1975. Today, the IBM 3084 is a 4-way, dyadic based multiprocessor complex, as shown in Fig. 2, featuring 48 to 128 Mbytes of real storage and 48 channels.

Dividing the work between processors at, say, the JOB level, means that the configuration is utilized as a multiprocessor, in which throughput is the primary consideration. Allocating the work within a single job to multiple processors, then, indicates that the configuration is used as a parallel processor, with job turn-around as the primary concern. In fact, the IBM 3084 can be used in both modes of operation.

Concurrently with the development of hardware, there has been a similar evolution of operating system architecture. Operating systems evolved from simple batch systems to sophisticated, multi-purpose systems supporting mixed interactive and batch environments with virtual storage and multiple address space management. The IBM 3084 is supported by MVS (Multiple Virtual Storage) and MVS/XA (Extended Architecture) operating systems. MVS/XA gives the user virtual and real addressing to 2 Gbytes via 31-bit addressing, whereas MVS provides 16 Mbytes of virtual addressing.

Both MVS and MVS/XA contain integrated support for loosely and tightly coupled multiprocessing. Their libraries include macro instructions for multi-task management such as ATTACH and DETACH of Tasks, POST and WAIT of Events, and ENQUEUE and DEQUEUE of Resources. Further information is contained in Arnold et al (1974). A higher level user interface that is callable from FORTRAN is available as a program offering. This facility is called the Application Program Performance Extender, see IBM (1984). It is the purpose of this paper to report on a series of experiments in using an IBM tightly coupled multiple processor configuration such as the IBM 3084 as a parallel processor. Prior to this discussion, a review of the history of programming language support for parallelism is appropriate.

SECTION 2: PROGRAMMING LANGUAGE SOLUTIONS FOR PARALLELISM

Programming language features for expressing parallel execution in programs were actively proposed twenty years ago and actual implementations were realized in widely used compilers. The concept of FORK and JOIN was introduced in the early 1960's. One of the earliest proponents of parallel processing was M. E. Conway. In Conway (1963), he said, "Parallel processing is not so mysterious a concept as the dearth of algorithms which explicitly use it might suggest". He proposed the following instructions to permit an adequate specification of parallelism:

FORK A	Fork to location A and the next statement.
FORK A,J	as above, in addition set counter J to 2, denoting 2 parallel tasks.
FORK A,J,N	as above, except set counter at location J to N, where N denotes N parallel tasks.
JOIN J	Decrement counter at J by 1. If result is zero jump to location J+1; else detach task.
JOIN J,B	If counter at J equals 1, set it to zero and jump to location J+1; else jump to location B.

A. Opler (1965) proposed extensions to high level languages such as FORTRAN to facilitate parallel processing. The two new statements were "DO TOGETHER" and "HOLD" and they are used as shown in the following example:

```
START          DO TOGETHER  TASK1, TASK2, TASK3 (MEET)

TASK1          statement

               statement         PATH 1

               statement

                . . .

TASK2          statement

               statement         PATH 2

               statement

                . . .

TASK3          statement         PATH 3

                . . .

MEET           HOLD
```

Fairly explicit rules for DO TOGETHER and HOLD were defined covering notions of sharing variables and nesting.

J. P. Anderson (1966), proposed extensions to ALGOL60 based on the work of Conway and Opler. He introduced three additional statements, namely, TERMINATE to explicitly deactivate tasks, OBTAIN to have exclusive use of selected variables, freeing them from interference by other tasks, and RELEASE to free variables previously locked onto by an OBTAIN.

Similar constructs were proposed by Dennis and Van Horn (1966) as shown in their example program which evaluates the product of two vectors, A and B:

```
Begin    Real Array A(/1:n/),B(/1:n/)
         Boolean w; Real s; Integer t;
         Private Integer i;
         t: = n ;
         For i: = 1 Step 1 Until n Do      Create n processes with
            FORK e;                        instructions at label e
         Quit;
      e: Begin Private Real x;
         x: = A(/i/) * B(/i/) ;
         LOCK w;                           For each e-process, add x
         s: = s + x;                       to s only when no other
         UNLOCK w;                         e-process is accessing s
         JOIN t,r;                         When t tasks complete,
         Quit;                             branch to r
         End;
      r:
End;
```

The LOCK and UNLOCK pair basically solves the 'critical section' problem by providing protection for common data from simultaneous access and change by two or more processes. Similar solutions were provided by Dijkstra (1965) namely, the semaphore concept; by IBM OS/360 with ENQUEUE and DEQUEUE macro-instructions for resource sharing and WAIT and POST macro-instructions for process synchronization.

Programming languages that have practical implementations of parallel programming constructs include Burroughs' extended ALGOL, PL/I, Concurrent PASCAL and ADA. PL/I provides task initiation via a CALL statement of the form:

```
CALL P(AL,a2,...,an) TASK(taskname) EVENT(eventname) PRIORITY(N);
```

The call initiates a new task, whose execution is represented by the procedure P. The status of the task may be interrogated via the event variable. This task has an associated priority of N which provides explicit scheduling of task execution when there are more tasks than processors.

The WAIT statement is the PL/I version of a JOIN. It has the form:

```
WAIT (E1,E2,...,En) K;
```

where Ei are event names and K is an integer number or null. This statement will hold program execution at this point until any K events signal completion or until all events signal completion, if K is null.

In summary, programming language features for parallel processing have been in existence for twenty years. Practical implementations have also been available, although they are mostly for ALGOL-based languages.

SECTION 3: GROUNDRULES FOR PARALLEL PROGRAMMING

Robert E. Kahn of the Defence Advanced Research Project Agency, in his paper "A New Generation in Computing", (see Kahn (1983)), said,

"...It is difficult to estimate how much of the existence of sequential machines has affected the way we think about problems. It clearly affects the way we implement them. A given algorithm implemented on a multiprocessor system will surely bear little resemblance to its counterpart on a sequential machine. And multiprocessor-based LISP systems need bear no relation to a single processor version. Conversely, we still know very little about multiprocessor architectures, concurrent programming, or parallelism. Our understanding of task decomposition strategies is limited and current languages, both natural and computer based, are inadequate to represent concurrency. .... Some problems do not lend themselves to much parallelism and cannot make good use of a multiprocessor system. An outstanding generic research question is how to determine the amount of performance speedup (or parallelism) possible in an application..."

Although Dr. Kahn was primarily addressing the world of artificial intelligence, his comments are appropriate for scientific computing in general. Parallel programming approaches and concepts that have existed for two decades have had little impact. Specifically, the scientific/engineering programmer or analyst is concerned with the problem at hand, and views the intricacies of detailed control and communication required to allow the desired parallel execution as a burden. Furthermore, while the trend is towards higher level descriptions of problems, these approaches are actually proceeding toward low level programming techniques involving functions that are part of operating system technology.

It would certainly be an immense task to re-solve and re-implement most of the scientific programs that are in use today. How do we make progress toward answering the question of how much parallelism is possible in applications? The approach proposed here is to experiment with converting existing serial applications to run parallel on multiprocessors. For this to be successful, some groundrules have to be established. They are:

- Control of parallelism should be accomplished through a minimal number of new commands or statements.

- The new commands or statements should conform to or be compatible with the current syntax and semantics of FORTRAN.

- The algorithm structure and execution flow of the application should be preserved as much as possible.

- System parameters and multitasking control complexity should be hidden from the end user.

The above groundrules were applied in constructing an experimental environment to permit parallel execution of FORTRAN application programs using currently available hardware and software, including operating system and compiler. Results for converting three VS FORTRAN programs for the IBM 3084 running MVS/XA are discussed below.

SECTION 4: APPLICATION CASE STUDIES

Three scientific application programs were analysed and modified to execute in parallel. These applications were adapted for parallelism following the groundrules described above. The modified versions of these programs were measured on an IBM 3084 to determine the elapsed time speed-up achieved when executed on a 4 processor system.

APPLICATIONS MODIFIED

The first application program, TBLADE, is a turbine blade analysis application. The application uses a finite difference algorithm to solve for the three dimensional air flow between the blades of a rotating turbine stage. The application requires a 38 megabyte private region to execute, and runs for approximately 24 hours when executed on an IBM 3084 using a single processor.

The second application program, VA3D, is a fluid dynamics application from a national laboratory using methods developed by Pulliam (1978) and Beam (1978). The program simulates the subsonic or supersonic flow field around a hemispherical-nosed body. It solves the Euler equations or the Navier-Stokes equations with a thin-layer approximation using a three dimensional, implicit, approximate-factored algorithm. This is a form of the Alternating Direction, Implicit (ADI) technique as reported in Douglas (1964).

The third application program, BOAST, is a three phase black oil reservoir simulation program developed by the U.S. Department of Energy and reported in Keplinger (1982). The program can simulate both the primary depletion and the secondary recovery operations in a two- or three-dimensional black oil reservoir. The solution uses the Implicit Pressure, Explicit Saturation (IMPES) method to solve for the fluid flow in a reservoir. The pressure and saturation equations for the oil/gas/water systems are approximated using a finite difference method. The resulting system of linear equations are solved using an iterative Line Successive Over-Relaxation (LSOR) technique.

REORGANISATION EFFORT

Each of the applications described above was modified for parallel execution. The major portion of the reorganisation time was spent understanding the original application, locating the data and functional independence, and introducing the appropriate forks and joins in the application. Once this was accomplished, the process of creating parallel subroutines was straightforward and mainly mechanical. Code sequences that could execute in parallel were packaged as separate subroutines to be scheduled for dispatching by the operating system.

The level of effort required to perform these modifications is indicated by Table 1. For each of the three application programs considered, the size, in lines of code, (LOC) of the original source program, the amount of code which actually performed the operations which could be executed in parallel, the percentage represented, and the resulting number of newly created parallel subroutines are listed. The modifications of these three application programs was performed without changing the original algorithm, rather, the inherent parallelism of each application was exploited.

Application	Size (LOC)	Parallel LOC	%LOC PARALLEL	≠ Parallel Routines
TBLADE	3500	677	19	13
VA3D	3800	1350	36	15
BOAST	4200	1021	24	5

Table 1: Application Code Modified for Parallelism

RESULTS

The results of executing the modified application programs on a 4 processor IBM 3084 are summarised in Table 2. All of the measurement runs were performed on a dedicated system with no other jobs executing concurrently with the measurement jobs.

For each application program, both the original version and the parallel version of the program (using 4 subtasks) were executed, and the elapsed time for the program was measured. The speed-up ratio is the ratio between the elapsed time for the original serial run and the elapsed time for the 4-way parallel run for each application.

Application	Serial Time (Min)	Parallel Time (Min)	Speedup Ratio	Percent Parallel	Scope of Run
TBLADE	80.39	31.84	2.52	81	200 Time Steps
VA3D	70.18	21.24	3.30	93	200 Time Steps
BOAST	50.23	15.50	3.24	92	3000 Simulated Days

Table 2: Application Measurement Results

The percentage of parallelism for each application was calculated from the serial and parallel elapsed times using the following simple model:

Time serial/Time parallel = 1/((1-P) + (P/N))

or

P = (N/N-1) * (1 - Time parallel/Time serial)

where P is the effective parallelism and N is the number of processors being used.

This simple model provides a rough estimate of the amount of work that was done with some degree of parallelism and takes into account the overhead that was introduced.

SECTION 5: CONCLUDING REMARKS

The results as shown in Table 2 are significant with respect to performance improvement for the applications concerned. More importantly, they were achieved with little investment in program modification. In fact, between 1 and 2 person months were spent in each case with program behaviour analysis time included.

Other experiments with parallelism for scientific applications have been conducted within IBM. An example is the work of Blaine and Wang (1976) on the IBM 370/168, at the IBM Palo Alto Scientific Center. Another effort is by Meck (1984) on the IBM 308X.

The experiences above provide considerable motivation to further investigate the issues in advancing the state of the art in the exploitation of parallelism. Some questions deserving exploration are:

- Should general purpose scientific processing be limited to low levels of parallelism or conversely, are highly parallel machines inherently application specific?

- How is the effective cost for running a program in parallel to be determined?

- Is automatic detection of parallelism possible and can efficient tools be constructed?

- What extensions should be made to FORTRAN to facilitate parallel algorithm construction?

- How do we complement the efforts to exploit parallelism and vectorisation?

Though it is premature to conclude that our groundrules are valid, the results do provide evidence of their effectiveness. Much remains to be done to advance the exploitation of parallel or multiprocessing systems by scientific applications.

ACKNOWLEDGEMENT

The authors wish to acknowledge the contributions of J. M. Gdaniec and R. J. Sahulka for making the above experimental work possible.

REFERENCES

Anderson J.P. (1966) Program Structures for Parallel Processing. Comm. A.C.M. 8:12

Arnold J.S. Casey D.P. McKinstry R.H. (1974) Design of Tightly Coupled Multiprocessing Programming. IBM System Journal 1

Beam R.M. Warming R.F. (1978) An Implicit Factored Scheme for the Compressible Navier-Stokes Equations. AIAA Journal 16:4

Blain R.A. Wang H.H. (1976) Private Communication. IBM Palo Alto Scientific Center, California 94304

Conway M.E. (1963) A Multiprocessor System Design. Proceedings Fall Joint Computer Conference 24, Spartan Books, Baltimore

Dennis J.B. Van Horn E.C. (1966) Programming Semantics for Multiprogrammed Computations. Comm. A.C.M. 9:3.

Dijkstra E.W. (1965) Solution of a Problem in Concurerent Programming Control. Comm. A.C.M. 8:9

Douglas J. Gunn J. (1964) A General Formulation of Alternating Direction Methods. Numerical Mathematics 6:5

IBM Documentation (1984) Application Program Performance Extender, 5798-DNL

Keplinger and Associates (1982) BOAST: A Three Dimensional Three Phase Black Oil Applied Simulation Tool (Vers. 1.1) DOE/BC/10033-3

Kahn R.E. (1983) A New Generation in Computing. IEEE Spectrum 20:11

Meck D.L. (1984) Parallelism in Executing Fortran Programs on the 308X: Systems Considerations and Application Examples. IBM Kingston NY, Laboratory Technical Report, TR-21.942

Opler, A. (1965) Procedure-Oriented Language Statements to Facilitate Parallel Processing. Comm. A.C.M. 8:5

Pulliam T.H. Steger J.L. (1978) On Implicit Finite-Difference Simulations of Three Dimensional Flow. AIAA paper.

Computer Requirements for Atmospheric Modelling

LENNART BENGTSSON

European Centre for Medium-Range Weather Forecasts, Shinfield Park, Reading, Berkshire RG2 9AX, United Kingdom

1. INTRODUCTION

It is a great pleasure for me to open this Workshop on 'The usage of Multiprocessors to integrate atmospheric models'. As most of you know, the meteorological forecasting problem was identified already by von Neumann as an ideal application for computers. Under his guidance and supervision a special group was set up at the Institute for Advanced Studies in Princeton in 1984 under J. Charney to undertake an integration of the vorticity equation using real atmospheric observations. The result was successful and was one of the very first attempts to use electronic computers to solve non-linear equations by numerical methods. Since then, the meteorologists have stayed in the forefront in applying the fastest available computers to solve the forecasting problem using increasingly more realistic models of the atmosphere. The meteorological community is an active user and a very substantial user for supercomputers and discussions between the meteorological modellers and the experts within the computing industry are important and I am sure mutually beneficial.

What makes the problem of weather prediction so intruiging is the fact that the atmospheric system is essentially non-linear and cannot be decomposed into independently acting modes. The inevitable error in observing the smallest scales of motion must then contaminate larger scales and finally destroy the accuracy of any predicition. The weather prediction can therefore be seen as an unstable problem in the sense that small initial differences have large final effects. Although the problem as such is <u>deterministic</u>, it is, for practical reasons <u>non-deterministic</u> since the initial state can never be perfectly known. Theoretically, the weather prediction has therefore much in common with more general non-deterministic problems, such as economical and social systems. Atmospheric prediction models may, therefore, also serve as a useful prototype for better understanding of a more general class of problems, where the dynamical laws are not yet too well understood. Although the basic physical laws governing the atmosphere have been known since the last century, no real progress took place before the advent of computers. The first integration of a simple two-dimensional atmospheric model was done by Charney, Fjörtoft and von Neumann (1950). Since then there has been a rapid development of successively more realistic models following the very fast development of computers. These models have been used for a large range of atmospheric problems from short range weather prediction to the simulation of the climate of the earth.

The problem of predicting the atmosphere from time scales from a few days to a few weeks is particularly challenging and important. It is regarded as perhaps the most difficult prediction problem because we have to rely on accurate treatment of the atmospheric observations as well as on accurate modelling. In the appreciation of the importance of this problem, the Western European countries decided to jointly set up and finance a special European Centre dedicated to medium range prediction, the European Centre for Medium Range Weather Forecasts (ECMWF). ECMWF was established in 1975. It started to make daily 7-day forecasts from August 1979. The Centre's medium range forecasts are now the best available.

2. THE PHYSICAL AND MATHEMATICAL BASIS FOR NUMERICAL MODELS

The behaviour of the atmosphere is governed by fundamental physical laws and their boundary conditions. The macrostructure of these laws has been known for over a century. What has been lacking and to some extent is still lacking, is the understanding of the interaction of the macroscales (greater than a few hundred kilometers) with processes of much smaller dimensions such as radiative transfer, turbulence fluxes, cloud and precipitation processes.

The physical laws governing the atmosphere are:

The gas law

$$p = \rho RT \tag{1}$$

The continuity equation for dry air

$$\frac{d\rho_d}{dt} = -\rho_d \nabla \cdot \underline{v} \tag{2}$$

The continuity equation for moist air

$$\frac{d\rho_w}{dt} = -\rho_w \nabla \cdot \underline{v} + S \tag{3}$$

The first law of thermodynamics

$$c_p \frac{dT}{dt} - R\frac{dp}{dt} = Q \tag{4}$$

The equation of motion

$$\frac{d\underline{v}}{dt} = \frac{1}{p} \nabla p - g - 2\underline{\Omega} x \underline{v} + \underline{F} \tag{5}$$

For a definition of symbols and expressions see Table I. The physical parameters entering these laws are: $\underline{v}$, the three-dimensiona velocity vector relative to the earth; T, the temperature; p, the pressure; ρ_d, the density of air; the ρ_w the density of water vapour. Some atmospheric models have additional conservation laws for liquid water (cloud water) and for ozone.

The equations (1) to (5) constitute a closed system which can be solved at all future times from a given initial state and with the necessary prescribed boundary conditions. However, the equations still contain some unspecified source and sink terms and it is necessary to provide a second set of expressions where we can specify these in terms of known physical quantities and/or in the basic parameters $\underline{v}$, T, p, ρ_d and ρ_w.

In the case where the source and sink terms, $\underline{F}$, Q and S are zero the system is energetically closed and can only describe adiabatic processes. It is also reversible and can be integrated backward in time as well as forward. The diabatic term, $\underline{F}$ represents dissipation of momentum and Q and S represent sources and sinks for heat and water vapour.

The complete atmospheric equations in (1) to (5) have not so far been used for operational forecasting. They are very general and represent in principle all scales of motion from the microscale to the largest planetary scale.

For practical purposes, we can only resolve the scale of motion which can be analysed by standard data and it has therefore become common practise to simplify the equations by omitting small terms and filtering out unwanted motion. The approximate equations are derived by considering the temporal and spatial scales of interest which are from about 1 hour and 100 km respectively.

With these assumptions, which are not only dictated by the data distribution but also by computational considerations, the basic equations (1) to (5) can be simplified. It is found for instance that the vertical equation of motion can be reduced to a diagnostic relation where the vertical pressure force is balanced by the gravitational force:

$$\frac{1}{p}\frac{\partial p}{\partial z} + g = 0 \qquad (6)$$

The hydrostatic equation (6) leaves us without a prognostic equation for the vertical motion, w, but the adoption of (6) will make it possible to determine w diagnostically from the remaining equations. An additional consequence of (6) is that other approximations must be made to guarantee that the resulting system, in the absence of sources and sinks, conserves energy and momentum. These approximations lead to that w is eliminated from the expression of kinetic energy and the radial distance is replaced by the average radius of the earth. The resulting equations are generally called the "primitive equations" by the meteorological community. An interesting consequence arising from the adoption of the hydrostatic relation is the removal of vertical travelling sound waves.

Observational and computation restrictions have made it necessary to confine the atmospheric model to the description of phenomena larger than a certain given scale. Present computers put this limit around 100 km in the horizontal and around one kilometre in the vertical. The dimension of such a volume is a measure of the *computational resolution.* What happens on scales smaller than that of the volume is known as *subgrid scale processes* and the simplified prescription of how they are related to the macroscale is known as *parameterization.* Finally, the empirical constants as well as the dependent variables of the macroscale that enter these relations are known as *parameters*.

The physical processes which go into the description of $\underline{F}$, Q and S consist of a manifold of subgrid processes which must be parameterized. These processes do represent a considerable fraction of the numerical calculation to be carried out in each grid point.

3. NUMERICAL ASPECTS

We will next consider the numerical and computational aspects of atmospheric modelling. If A is the integration domain then the number of grid points in a horizontal area is $A/(\Delta s)^2$ where Δs is the average horizontal grid size. For K vertical levels the total number of grid points is $KA/(\Delta s)^2$. If there are n variables per point, then the total number of variables at any one time is $nKA/(\Delta s)^2$ which is a measure of the number of degrees of freedom in the model. The maximum time increment Δt, to guarantee stability depends on the integration scheme. If we, for instance, use an explicit integration scheme the condition for computational stability is given by an expression of the kind

$$\Delta t \;\mathrm{o}\; \frac{\Delta s}{\sqrt{2}(c+U_{max})} \tag{7}$$

where $c \sim 300$ m sec^{-1} is the speed of the fastest gravity waves and U_{max} is the maximum horizontal wind speed.

Since we have eliminated vertically propagating sound waves by using the hydrostatic relation the vertical condition is less severe than the horizontal one. However, with more economical integration techniques such as semi-implicit Kwizak and Robert (1971) a longer time step can be used. Hereby gravity waves are treated implicitly, while the slower Rossby waves are treated explicitly. For a model using a semi-implicit scheme and a staggered grid a relation of the kind

$$\Delta t \;\mathrm{o}\; \frac{\Delta s}{U_{max}} \tag{8}$$

will hold.

For a total integration time T at least the following number of time steps are needed

$$\frac{T(U_{max})}{\Delta s}$$

Finally, if we need N number of operations per variable/time step the following number of arithmetic operations are needed to make a forecast with the length of time T for an area A:

$$\frac{A\ K\ nT\ (U_{max})\ N}{(\Delta s)^3} \qquad (9)$$

4. ATMOSPHERIC PREDICTABILITY AND PRESENT PREDICTIVE SKILL

Atmospheric models are used both for weather prediction and for climate simulation. One can, in principle, as for the ECMWF model, use the same model for both these applications. Following the continuing improvements in meteorological observations, computers and associated progress in modelling technique, remarkable achievements have taken place during the last 20 years and useful predictive skill has been extended from a few days in the 1960s to 5 to 7 days at present. In the short time scale detailed simulations of intense vortices such as Genoa cyclogenesis, polar lows and tropical hurricanes have been possible.

As mentioned previously, numerical weather prediction can never be exact due to the inevitable errors in the determination of the initial state coupled with an inherent tendency for errors to grow. This error growth is not an artifact of a numerical model, but a consequence of the non-linearity and instability of the dynamics of the atmosphere. Considerable research is taking place to assess the predictability of the atmosphere, both from data and by numerical experiments. It has been found that the doubling time for small errors is of the order of two to three days and successively decreasing as the error is approaching some asymptotic value. For practical purposes this growth seems to limit the prediction of the day to day weather to something in the order of two to three weeks. However, the error growth is much lower for the largest scales of motion and it may be possible to predict large scale anomalies well beyond this time.

5. CONCLUSIONS

It is obvious that the improvement of supercomputers is crucial for a further development of the meteorological models and hence our ability to make better weather forecasts. The speed of computers has approximately increased by a factor of 10 every five to seven years and we may therefore expect computers in the Giga flops range towards the end of this decade if we are bold enough to extrapolate.

It appears that in order to achieve performances of several Giga flops and beyond, we may have to turn to multiprocessing systems. The meteorological forecasting problem has a simple logical structure and it is straightforward, at least in principle, to program an atmospheric model for a multi-processing system, due to the fact that we are essentially carrying out the same kind of calculations in each grid point. The Centre's new prediction model is simultaneously being executed on two processors of the Cray X-MP and, as will be described in following lectures, the code for the ECMWF model has been designed in such a way that the calculation can be executed simultaneously on an even number of processors. This is not the first time this has happened in meteorology. The US Navy Weather Service in Monterey uses a CDC Multiprocessor system more than 15 years ago. However, it did not start there. Lewis F. Richardson, who was the first to integrate a numerical model more than 60 years ago and long before the existence of any computers, outlined in his absolutely amazing book 'Weather Prediction by Numerical Processes' (Richardson 1922), a futuristic idea of a "weather forecasting factory" consisting of 64000 human "multiprocessors". Fig. 1. Perhaps therefore one day when we can build the 64000 multiprocessing system, we may solve the weather forecasting problem.

Table I Symbols and expressions used in the article

$\underline{v}$	=	(u,v,w) - 3-dimensional velocity vector relative to the earth
T	=	temperature
p	=	pressure
ρ_d	=	density of dry air
ρ_w	=	density of water vapour
ρ	=	density of moist air. $\rho = \rho_d + \rho_w$
$\underline{F}$	=	dissipation of momentum
S	=	water vapour source/sink term
Q	=	diabatic heating
R	=	gas constant for moist air
c_p	=	specific heat at constant pressure
g_a	=	acceleration due to gravity
$\underline{\Omega}$	=	earth angular velocity
$\underline{R}$	=	radial co-ordinate measure from the centre of the earth
g	=	$g_a - \underline{\Omega}\times\underline{\Omega}\times\underline{R}$ (local gravity acceleration)
x,y,z	=	spatial co-ordinate
t	=	time
$\frac{d}{dt}$	=	individual derivative = $\frac{\partial}{\partial t} + \underline{v}\cdot\underline{\nabla}$
$\underline{\nabla}$	=	horizontal gradient operator

Table II

Estimated computational requirements for some numerical models. The independent parameters can be reduced to 4/level plus surface pressure by a combination of the continuity equation (2) and the hydrostatic equation (6). Version I is an estimate of the present situation. Version II is a projection of the requirements around 1990.

A	=	5.10^{14} m ^{2}	(area of the globe)
K	=	20	number of vertical levels
n	=	4	number of independent parameters
U_{max}	=	150 ms $^{-1}$	estimated maximum wind speed
N	=	number of operations/gridpoints/parameters	
Δs	=	horizontal grid length	
T	=	Total integration time	
M	=	Total number of operations	

(a) Short-range forecast, limited area model (1/20 globe)

	I	II
N	~ 250	400
Δs	= 50 km	25 km
T	= 2 days	2 days
M	= 10^{11}	= 10^{12}

(b) Medium-range forecast, global

	I	II
N	~ 1000	1500
Δs	= 150 km	75 km
T	= 10 days	1 month
M	= 5.10^{12}	= 2.10^{14}

(c) Climate prediction/simulation, global

	I	II
N	~ 2000	3000
Δs	= 300 km	150 km
T	= 10 years	100 years
M	= 5.10^{14}	= 6.10^{16}

Figure 1 Weather Forecasting by Numerical methods as envisaged in 1922 by L.F. Richardson

"Imagine a large hall like a theatre, except that the circles and galleries go right round through the space usually occupied by the stage. The walls of this chamber are painted to form a map of the globe. The ceiling represents the north polar regions, the tropics in the upper circle, and the antarctic in the pit. A myriad computers are at work upon the weather of the part of the map where each sits, but each computer attends only to one equation or part of an equation. The work of each region is co-ordinated by an official of higher rank. Numerous little "night signs" display the instantaneous values so that neighbouring computers can read them. From the floor of the pit a tall pillar rises to half the height of the hall. It carries a large pulpit on its top. In this sits the man in charge of the whole theatre. One of his duties is to maintain a uniform speed of progress in all parts of the globe. In this respect he is like the conductor of an orchestra in which the instruments are slide-rules and calculating machines. But instead of waving a baton he turns a beam of rosy light upon any region that is running ahead of the rest, and a beam of blue light upon those who are behindhand.

Four senior clerks in the central pulpit are collecting the future weather as fast as it is being computed, and despatching it by pneumatic carrier to a quiet room. There it will be coded and telephoned to the radio transmitting station".
Picture credit: A. Lannerback, Dagens Nyheter, Stockholm.

References:

Charney, J., R. Fjörtoft and J. von Neumann, 1950: Numerical integration of the barotropic vorticity equation. Tellus 2, 4, 237-254.

Kwizak, M. and A. Robert, 1971: A semi-implicit scheme for grid point atmospheric models of the primitive equations. Monthly Weather Review, 99, 1, 32-36.

Richardson, L.F. 1922: Weather prediction by numerical process. Cambridge University Press 236 pp.

The Current State of Meteorological Modelling

D. M. BURRIDGE

European Centre for Medium-Range Weather Forecasts, Shinfield Park, Reading, Berkshire RG2 9AX, United Kingdom

1. INTRODUCTION

Numerical weather prediction involves the solution of large systems of coupled partial differential equations which describe atmospheric motions. A brief description of the physical and mathematical basis of numerical models of the atmosphere is given in the paper by Bengtsson in this volume. For our purposes the governing equations may be written formally as

$$\frac{\partial X}{\partial t} = \underline{D}(X) + \underline{P}(X) \tag{1}$$

where X is any model variable .... wind, temperature, humidity etc. ... The problem is 3-dimensional, with the dependent variable X being a function of spatial co-ordinates x, y and z and a function of time t. The term $\underline{D}$ represents the "dynamics" - advection, pressure forces etc - and the term $\underline{P}$, normally referred to as the "physics", is a source/sink term which for large-scale models of the atmosphere describes physical processes such as the evaporation and condensation of water, solar heating, infra-red cooling, and frictional drag at the earth's surface. For fine-scale models of thunderstorms a description of cloud physics processes is required; this adds new variables to the problem - cloud water drop-size spectra and for each drop-size an electrical charge spectrum. With present computing power, severe compromises must still be made for such models. In contrast to the "dynamics" embodied in the term $\underline{D}$, the "physical" processes, $\underline{P}$, are intermittent and are often represented by on/off processes. The programming or implementation for these processes largely effects the efficiency of vectorization on computers such as the CRAY-1 and the CYBER-205, with different strategies being required for different computers. For large-scale models such problems have been overcome by carrying out redundant calculations. The effect on parallel processing is discussed in the paper by Dent in this volume.

In the rest of this paper I shall describe the basic numerical techniques in current use - Section 2 - and in Section 3 future requirements for global weather forecasting models.

2. NUMERICAL SCHEMES AND THEIR IMPLEMENTATION

Three different techniques are in general used for numerical models of the atmosphere - finite differences, finite element and spectral. With each of these techniques a 3-dimensional grid of points is required. At the time of writing typical horizontal grid spacings for global models are 200 km on 15 levels (in the vertical) ranging from the earth's surface to heights of 25 km- these resolutions requiring approximately 200*100*15 = 300,000 grid points. For the vertical representation in numerical models of the atmosphere the finite difference technique predominates though there have been some successful implementations of the finite element technique. The next three sub-sections 2.1 to 2.3, describe each of the techniques and their implementation. The fourth sub-section, 2.4, discusses the implementation of time-stepping algorithms.

2.1 Finite Differences

In the early 1970s, almost all models of the atmosphere were based on finite difference techniques, and, of the three main approaches, finite difference methods are perhaps the most readily understood and the most easy to implement. Model variables are defined at grid points and nowhere else; the derivatives in $\underline{D}(X)$ being approximated by differences. The simplest difference technique requires 9 points (i, j; i:1-200, j:1-100) in the horizontal at each of 3 levels (k; k:1-15) in the vertical. The "physics" requires, for an efficient implementation, the presence of all dependent variables for a particular vertical column. (15 levels of information for a 15-level model). The memory sizes of the present generation of supercomputers will, normally, only allow a fraction of the model's grid point description be retained in memory at any one time. The normal approach is to retain in memory vertical slices of the data for a few values of j - for our model problem data arrays for memory resident variables would be DIMENSIONED (200,15); typical vector lengths for vector calculations being 200. The full description is retained on a secondary memory - normally a high speed disc system driven by asynchronous i/o drivers. However whereas such an approach is very efficient on CRAY computers, indeed there is little overhead if the secondary storage is a CRAY-SSD system, the optimal approach for the CYBER-205 is to design a "memory-resident" model such as that developed by the British Meteorological Office; this permits the use of very long vectors. Obviously a "memory-resident" model limits flexibility particularly with regard to resolution increases.

2.2 Finite Element Schemes

The essence of finite element schemes is to expand the model's dependent variables in terms of a set of basis functions, each of which is a low-order polynomial of compact support (i.e. each basis function is non-zero only over a small sub-domain or element of the full region); a familiar basis set being the linear-spline or "hat-functions"

$$e_i(x) = \begin{cases} \dfrac{x-x_{i-1}}{x_i-x_{i-1}} & x_{i-1} \leq x \leq x_i \\ \dfrac{x_{i+1}-x}{x_{i+1}-x_i} & x_i \leq x \leq x_{i+1} \\ 0 & \text{otherwise} \end{cases}$$

Insertion of this representation into the model's equations leads to a set of equations of the form

$$\underline{m}\,\frac{\partial X}{\partial t} = \underline{d}(X) + p(X) \qquad (2)$$

where $\underline{m}$ and $\underline{d}$ are large sparse matrices and X is a vector of amplitudes. To all intents and purposes the algebraic form of the right hand-side of this equation is similar to that obtained with finite difference treatments. The main difference, computationally, between finite element and finite difference is the requirement to invert the sparse matrix $\underline{m}$. Apart from the purely computational cost, this inversion normally presents considerable technical problems. This arises because, in general, the structure of $\underline{m}$ requires the whole of the right hand side of equation 2 in memory for the inversion. However, schemes can be designed in which $\underline{m}$ can be written as a product of matrices each of which is relatively simple to invert. For example with a bi-linear spline description for the horizontal representation, $\underline{m}$ can be written as the product $\underline{m}_x * \underline{m}_y$, where $\underline{m}_x$ and $\underline{m}_y$ correspond to the x and y discretizations respectively; both matrices being narrow banded. The matrix $\underline{m}_x$ is normally inverted by Fourier analysis and synthesis and the matrix $\underline{m}_y$ by a L/U decomposition involving a forward elimination step and back substitution step. In this case the inversion of $\underline{m}$ is similar to the problem of carrying out Fourier analysis and re-synthesis in two dimensions, for which there exists many efficient "out of memory" implementations.

2.3 Spectral Methods

The last decade has seen the rapid development of spectral methods for the horizontal description in large-scale models of the atmosphere. As with finite element approach dependent variables are represented by expansions in terms of a set of basis functions - normally orthogonal polynomials - however the description is "global" rather than local. For global models, the basis functions are normally surface harmonics (Legendre functions for the North/South representation and Fourier series for the East/West representation). The most efficient implementation of the spectral technique requires two related representations - the spectral description and a grid point description. Horizontal derivatives are first calculated exactly at each grid point using the spectral representation. The second stage is to calculate $\underline{D}(X)$, using the exact values for derivatives; and $\underline{P}(X)$ at grid points. A time-stepping procedure determines new values of the dependent variables at grid-points from which new values of the spectral representation are calculated. The implementation of a spectral model on the CRAY-XMP22 is discussed in the paper by Dent in this volume.

In contrast with finite difference techniques, spectral methods are very non-local; the complete spectral representation is needed to calculate values at a single grid point and each grid point value is required to calculate individual amplitudes in the spectral representation. With the present memory sizes on super computers this requires, for an efficient implementation, at least a two-pass structure for the calculation. For the ECMWF spectral model the complete spectral representation is retained in central memory and is used in the first pass to calculate primary variables and their derivatives at grid points; these grid point values are stored on an external I/O device in vertical slices. In the second pass these data are re-input into memory and a new 3 dimensional spectral description is computed - this includes all dynamical and physics calculations. The I/O requirements are similar to those for finite difference and finite element techniques. Indeed, there is a great deal of similarity between all the implementations. Essentially, they "look like" grid point techniques and the same basic data structures can be used.

The spectral model that ECMWF plans to introduce into operations in the spring of 1985 will require 16 mw of storage (storage medium CRAY-SSD) for primary variables and a little less than the 2 mw of the CRAY-XMP22's central memory.

2.4 Time-stepping algorithms

The procedures outlined in the preceding three sub-sections are used to calculate the right hand-side of equation

$$\frac{\partial X}{\partial t} = \underline{D}(X) + \underline{P}(X)$$

The forecasting problem is, given initial values of X=X(t) at time t, to determine values of X at successive times using a time-stepping procedure. The simplest schemes are explicit, and a commonly used scheme is

$$\frac{X(t+\Delta t) - X(t-\Delta t)}{2\Delta t} = \underline{D}(X(t)) + \underline{P}(X(t-\Delta t)) \qquad (3)$$

This scheme requires complete model states at two time-levels (t and t-Δt) in order to calculate values at the new time-level t+Δt. Explicit schemes such as (3) are easy to implement and do not impose severe programming problems.

Despite their simplicity, explicit schemes can carry a serious disadvantage for meteorological models. This arises because the governing equations describe fast time-scale phenomena - gravity waves with horizontal phase speeds of 300 ms^{-1} - in addition to the slow time-scale weather events. The earliest numerical weather forecasting models were based on sets of equations which "filtered-out" fast time-scale phenomena. The presence of the fast time-scales require short time-steps - small Δt - with explicit schemes. Implicit schemes can be used, and can be designed to be unconditionally stable - allowing the use of very large time-steps; a typical example being

$$\frac{X(t+\Delta t) - X(t-\Delta t)}{2\Delta t} = \underline{D}\left\{\frac{(X(t+\Delta t) + X(t-\Delta t)}{2}\right\} + \underline{P}\left\{\frac{(X(t-\Delta t)) + X(t+\Delta t)}{2}\right\}$$

However, since $\underline{D}$ and $\underline{P}$ are a very complicated non-linear expression, the determination of $X(t+\Delta t)$ requires a prohibitively expensive inversion procedure, which for operational weather forecasting would be impractical. The compromise is to treat only a linear part of $\underline{D}$ implicitly, that is use a semi-implicit scheme. A typical semi-implicit scheme is

$$\frac{X(t+\Delta t) - X(t-\Delta t)}{2\Delta t} = \tfrac{1}{2}\left\{\underline{D}_o(X(t+\Delta t) + \underline{D}_o(X(t-\Delta t))\right\} + \underline{D}'(X(t)) + \underline{P}(X(t-\Delta t))$$

where $\underline{D}' = \underline{D}-\underline{D}_o$

Since $\underline{D}_o$ is a linear operator the inversion process is simple; usually $\underline{D}_o$ can be reduced by simple algebraic manipulation to a Laplacian.

The application of semi-implicit techniques to spectral models results in particularly efficient schemes, since the basis functions are normally chosen to be the eigenfunctions of the Laplacian and in this case the inversion process requires only a multiplication. Because they are not fully implicit, semi-implicit schemes are in general conditionally stable, and typically we require

$$\Delta t < \alpha\left(\frac{h}{U_{max}}\right)$$

where h is a horizontal grid length, U_{max} the maximum horizontal wind and α a constant of order 1.

Recently, Lagrangian schemes have been developed in order to allow further increases in the time-step. The essential features of Lagrangian schemes are illustrated in Fig. 1 below (Fig. 1 is a space-time, x-t, diagram for a 1-dimensional problem.)

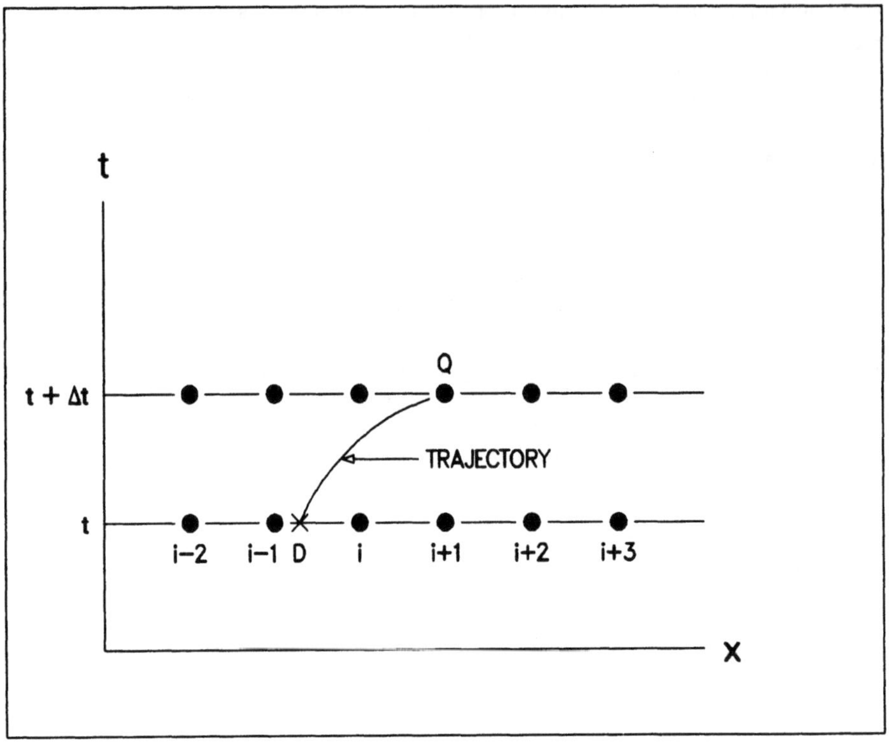

Fig. 1 Space-time x-t diagram for
a Lagrangian scheme

In order to determine the values of the dependent variables at $Q(x=x_i, t=t+\Delta t)$ a "departure" point D is required - the curve DQ being a trajectory; values at Q are identical to those at D. Since D is not normally a grid-point the values at D have to be determined by interpolation from surrounding grid point values; the accuracy of this interpolation determining the accuracy of the scheme. In numerical weather prediction, Lagrangian techniques are used for the advection processes - the slow time-scale phenomena - and when combined with semi-implicit treatments of the fast time-scales allow the use of large time-steps.

3. FUTURE DEVELOPMENTS

The schemes outlined in Section 2 will undoubtedly continue to provide the basic numerical schemes of future forecasting models. However, the approach to modelling the dynamics may require a different use of the basic equations to provide a better description of dynamical processes and a more widespread use of implicit methods. By the end of the decade we shall see a dramatic increase in the resolution of operational global forecasting models. It can be expected that the operational use of a global model with 600*300 grid points on 25 levels will be justified by the research and development currently being carried out by many research groups. Such a model, if based on a spectral representation, will require about 100 mw of storage; this requirement increasing by about 20 mw for each additional dependent variable added to the model. For such a model to be operationally feasible for medium range forecasting - forecasting up to 10 days ahead - the minimum requirements is for a machine with fifty (50) times the throughput of a CRAY-1S. Undoubtedly this power will only be provided by significant increases in the number of processors and large increases in central memory.

Machines with large amounts of central are being planned - successors to the current CRAY and CYBER machines for example - and will, I am sure be built. For the programmer the two most important characteristics are the presence or absence of shared memory - that is shared between processors - and the multi-tasking support. About 30% of the code in the present generation of "out-of-memory" models is used to manage the I/O and the central memory for primary and temporary variables, and it is to be hoped that the next generation of large - memory supercomputers will have efficient mechanisms (hardware/software) to manage shared memory and/or a memory hierarchy.

Turning finally to multi-processing, I feel optimistic that parallel processing, particularly for weather forecasting models, overheads will reduce through algorithm re-design; the scheme described by Dent in this volume being a good first start with a large-scale code.

Aspects of Using Multiprocessors for Meteorological Modelling

G.-R. Hoffmann[1], P. N. Swarztrauber[2], and R. A. Sweet[3]

1. INTRODUCTION

Since the introduction of computers for scientific research, meteorological modelling has always been one of the first disciplines to use the most advanced computers available at any given time. Today, meteorological institutions throughout the world operate class VI supercomputers like the CYBER 205, CRAY-1, or CRAY X-MP. Average processing speeds in excess of 200 Mflops (10^6 floating operations per second) are achievable for global weather prediction models with a horizontal resolution of about 100 km in grid-point space. As the results of the operational meteorological models are only useful during a very short time--the forecast for today is history tomorrow--the demands for greater processing power of computers are obvious, especially when one considers that a doubling in horizontal resolution extends the forecast range but in turn increases the computing requirements by the factor $8 = 2^3$. The next logical step envisaged by meteorologists for the end of the decade is to double the model resolution without changing the elapsed time for the computation. Therefore, a computer capable of delivering around 1600 Mflops on average and in excess of 3000 Mflops as peak rate is required. This estimate of the computing speed assumes that the complexity of the calculations does not increase as well.

Judging from the current trend in technology and assuming that no major breakthrough in physics like the step from valves to transistors occurs, it can be safely postulated that any future computer capable of achieving a performance in excess of 2.5 Gflops (10^9 floating operations per second) will employ multiple vector processors. Depending on the architecture used, the number of processors may vary from as few as possibly four to as many as thousands. However, currently known projects to build computers of the required performance seem to assume that the use of a large number of processors is still extremely difficult and only in the research phase, while the commercially undertaken studies for class VII computers only use between four and 256 processors.

---

[1] European Centre for Medium Range Weather Forecasts (ECMWF), Shinfield Park, Reading, Berks., RG2 9AX, England

[2] National Center for Atmospheric Research (NCAR), Boulder, CO. 80307, USA

[3] National Bureau of Standards (NBS), Boulder, CO. 80303, USA

In the following, the case of a very large number of processors will not be discussed: instead, emphasis will be given to three commercial developments which are based on currently available class VI machines, i.e., the CDC CYBER 205, the CRAY X-MP, and the Denelcor HEP-1. Rather detailed descriptions of the architecture of these machines may be found in many places, including [5,6], and they are therefore not repeated here. Reference will be made to special features only when we consider it necessary for explaining the expected structures of the follow-on products. However, we would like to stress that all qualitative or quantitative statements about the successors to the three computers discussed here are based on extrapolation only and certainly do not imply any commitment from the manufacturers. We also wish to note that we do not intend the exclusion of any class VII supercomputer project to reflect in any way on the validity of that project.

After discussing some general aspects of using multiprocessor systems and investigating the expected architecture of three future systems, we will consider algorithmic structures, and highlight possible problem areas. We will give the solution of a simple fluid dynamic model, and describe the experience of implementing it on the CRAY-1, the CYBER 205, and the HEP-1. We will detail the changes required by using multiprocessor configurations on the CRAY X-MP/2200 and the HEP-1. We will also show the influence of multiprocessor architecture for some algorithms commonly used in weather modelling.

2. ASPECTS OF MULTIPROCESSING

In the following, we make some remarks applicable to all multiprocessing environments. We emphasis aspects which are usually not relevant for SISD or SIMD architectures, but which affect users of multiprocessor systems (MIMD).

2.1 Parallelisation of Code

If a program runs for T units of time on a single processor and is modified to use p processors in parallel ($p > 1$), then the percentage S of the total time T can be found during which the program will use only one processor. Using this figure, the theoretical maximum speedup SP of the program for using p processors can be expressed as:

$$SP = \frac{1}{S + \frac{(1-S)}{p}}$$

Fig. 1 shows the curves for p = 16, 12, 8, and 4.

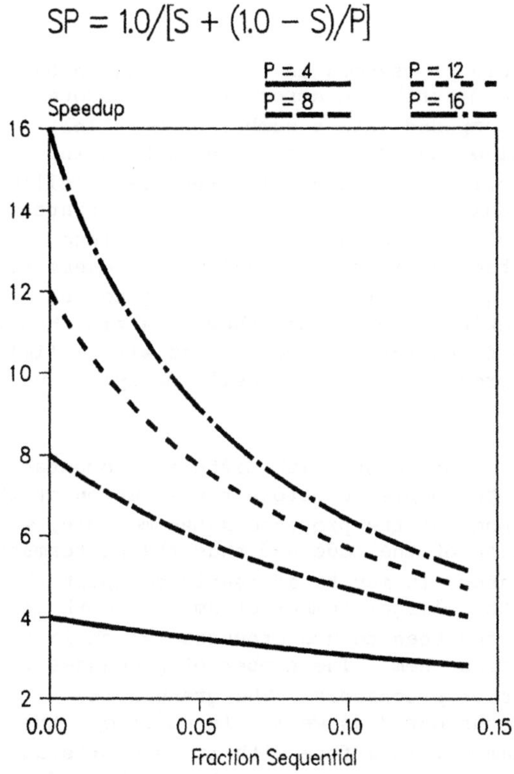

Fig. 1: Speedup when using p processors for varying percentages of of sequential code (p = 16, 12, 8, 4)

In order to achieve an economically feasible CPU utilisation of at least 75%, S must be less than 0.022; i.e., the sequential code must take less than 2.2% of the total time T. If overheads are included in the calculation, the figure has to further decrease in order to achieve the required CPU utilisation; hence it becomes clear that either the code has to contain very few sequential passages or the configuration has to have sufficient resources to sustain a reasonable multiprogramming factor. For operational weather forecasting the turnaround time for a program plays the central role, and hence, the demand for code parallelisation becomes paramount. Indeed it may be easier to develop parallel algorithms for the sequential passages than to develop faster scalar processors.

2.2 Program Decomposition

A program which uses multiple processors in parallel has to be split into processes which may be distributed across the processors. How a sequential program is decomposed into a number of parallel processes depends on the algorithm used and on the scheduling technique chosen. The algorithms and their inherent parallelism are described later. For scheduling, there exist two basic strategies: one assumes that the work of the program has been split into separate tasks which have been assigned statically to processes, while the other requires a queue of work units waiting to be processed and processes picking up work when they have finished with their previous tasks. The first strategy works best with a limited number of processes and a problem which can easily be split into identical work units.

If there is a large number of processes with different computational requirements that must all be completed before the solution of the problem can proceed, then some of the processors become idle, which decreases the parallelisation of the code and thus the performance of the system. In these cases, it may be advisable to split the solution of the problem into a larger number of smaller work units to put these into a queue, and then to use processes which in a loop dequeue work units and process them. The number of processes in this case equals the number of processors. The granularity discussed below has to be considered as well. In addition, the decomposition of the program may interfere with its vectorising features; in particular, the average vector length may be reduced. In these circumstances, a detailed analysis of the effects of the decomposition is required and may lead to the choice of a different algorithm which allows a more suitable partitioning of the data.

2.3 Code granularity

In this paper we define the granularity G of a process as the time, in machine cycles, during which the process can continue without external or internal interruption. Usually an interruption is caused by the need to wait for another process to finish. There will be some overhead when the process begins and when it reaches the end of its uninterrupted spell. If we call the overhead in machine cycles at the beginning O_B and at the end O_E, if we assume $S = 0.02$, and if we take into consideration that O_B and O_E are, in most cases, spent in sequential processing mode, then it follows from the discussion in Section 2.1 that $G > 49(O_B + O_E)$; i.e., G must be about two orders of magnitude greater than $(O_B + O_E)$. If operating system calls are involved in either O_B or O_E, then the size of $O_B + O_E$ is easily $O(10^3)$, which implies that G must be $O(10^5)$.

2.4 Memory requirements

Each process active in a processor must have sufficient memory available to keep it busy during its granularity G. Subject to the algorithms used, this condition may imply that some hundreds of thousands of memory cells are to be accessible without interruption. As weather models seem to be very data- intensive, we assume in the following that a process of granularity G will use O(G) memory cells during its computation. If the memory available to each process is not big enough to hold all the data for the problem to be solved, then sufficient space must be available for a read, write and active process. In this case, however, the granularity of the active process has to increase substantially because the time in machine cycles for a read/write process involving 10^5 memory cells is $O(10^5)$ to $O(10^7)$ depending on the storage medium used, and this time influences the size of $(O_B + O_E)$. Taking into consideration that the data requirements for high-resolution weather models are on the order of 10^8 memory cells, then we can deduce that the minimum memory configuration accessible per process should be on the order of 10^6 to 10^8. In addition, one can easily see that the read/write process must be able to transfer one memory cell per machine cycle because of the granularity consideration. As future supercomputers will have cycle times between 1 and 4 nanoseconds with 64 bits per memory cell, the I/0 device has to support transfer rates of 16 to 64 Gbits per second. This requirement can currently only be met with solid-state memory devices. The memory bandwidth in this case must be sufficient to allow at least three simultaneous accesses to memory per machine cycle and per processor.

2.5 Synchronisation tools

The tools provided by the manufacturer of a multiprocessor supercomputer for the mutual exclusion and synchronisation of programs must fit into the framework of program granularity. The method of implementing these tools directly influences the granularity of the programs using them, because O_B and O_E are determined by the time it takes to synchronise two tasks or to use locks. Which particular method, e.g., semaphores, monitors, etc., is chosen by a manufacturer to allow synchronisation by user programs affects programming ease for the user but does not have any other impact.

2.6 Process communication

It is mainly the algorithm which determines how often and in which way the processes of a program running in a multiprocessor environment have to communicate. As is shown later, there often exists a wide spectrum of algorithms with different attributes to solve a particular problem. The required granularity of the processes and the memory architecture of the target machine determine the right choice.

For current configurations with relatively few processors this problem is only occasionally evident as memory bank conflicts or processor synchronisation. However, as the number of processors increases, the communication time between processors or between processors and memory could dominate the computational time. This remains a concern even for computations that are 100% parallel.

It is clear that future computers can be designed with a very large number of processors. What is not so clear is how to interconnect and control the processors. The goal of the interconnection scheme is to minimize communication time. For a particular problem one can design a special-purpose computer with interconnections that match the computational communication paths. However, communication algorithms and processor interconnection are not well defined for a general-purpose computer.

Future general-purpose multiprocessors should be configured in such a way as to guarantee parallel nonblocking communication. Without this capability a multiprocessor is in effect reduced to a serial computer for problems in which any part of the communication is blocked.

3 FUTURE SYSTEMS

A number of computer manufacturers are currently engaged in designing supercomputers with performance in excess of 1 Gflops. In the following, we attempt to examine the proposed architecture of three of these machines. The already available class VI machines of the same manufacturers provide the basis for the expected structures.

3.1 CYBER 205 Successor

As a spin-off company of Control Data Corporation, ETA Systems has got the brief to develop a successor to the CYBER 205. This machine is expected to become available in 1986 and is called ETA^{10}. Its performance will lie in the region of 10 Gflops and will be achieved by liquid-nitrogen-cooled multiple processors. Each of the processors will be similar to the CYBER 205; in particular, the dual pipeline structure and the paged memory are expected to be continued. All processors will have local memories, but will also share a big global memory and some kind of synchronisation buffer. The expected architecture of the machine is depicted in Fig. 2.

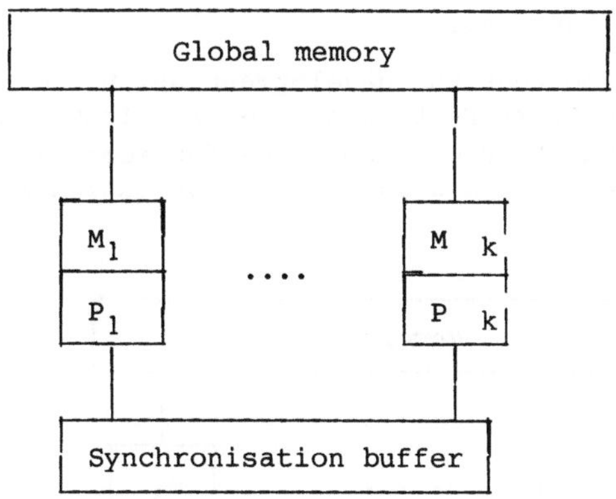

Fig. 2: Expected structure of CYBER 205 successor with k ($k \geq 1$) processors

We can be assume that the memory sizes both--local and global--will be sufficiently large to fulfill the requirements outlined in Section 2.4 above. However, because of the memory hierarchy, read and write processes will have to be provided, possibly by using built-in paging routines. The optimum way to use such an architecture seems to be the following:

(i) Acquire exclusive access to a set of pages in global memory which require changing.

(ii) Move data to local memory.

(iii) Process data.

(iv) Move processed data back to global memory.

(v) Continue with (i) until algorithm finishes.

Of course, steps (ii) to (iv) can be carried out in parallel if sufficient local memory is available. Synchronisation will be required for step (i) only, but may be quite time-consuming if the order of access to pages is not well planned in advance and the termination of a number of processors has to be awaited. In addition, the choice of the most suitable algorithm will have to be considered very carefully, because data changed within one process should occupy as few pages as possible in order to minimise the set of pages being exclusively kept by any one process. Because of the parallel read/write processes, the granularity of the active process will have to be rather large. This, however, will probably be advisable anyway, because the best vector performance of the machine can be expected with long vectors, as is the case with the CYBER 205.

3.2 CRAY X-MP successor

It can be expected that the development which led to a four-processor CRAY X-MP after progressing from a CRAY-1 to a two-processor CRAY X-MP will continue for some time in the future. The architecture of such a multiprocessor CRAY X-MP can be seen in Fig. 3.

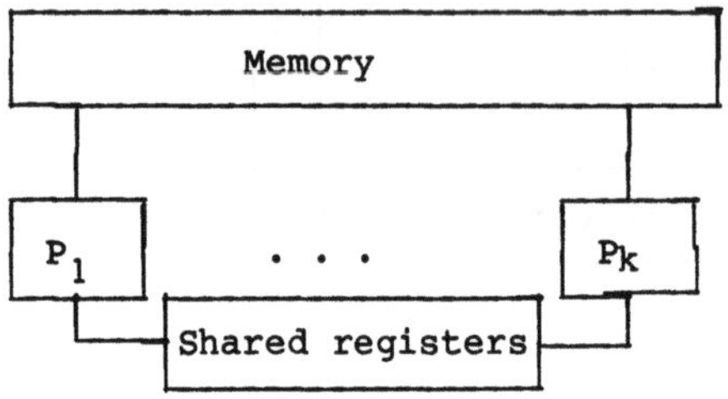

Fig. 3: Expected structure of CRAY X-MP successor with k (k > 1) processors

The main difference from the CYBER 205 successor will be the absence of local memory if one does not count the CRAY machine registers as such. Furthermore, the current addressing scheme of the CRAY prevents very large memories, which necessitates the addition of a large solid-state storage device (SSD) to the configuration. The SSD acts as a fast I/O device with a transfer speed meeting the requirements outlined in Section 2.4 above, but nevertheless requiring read/write processes. In addition, k processors will simultaneously access the global memory within vector instructions. The demands on the bandwidth of the memory and the requirement to avoid memory bank conflicts will be major problems. It is conceivable that the memory access patterns of different processes will have to be coordinated in order to avoid excessive memory wait times. A detailed study of the memory structure of the CRAY X-MP can found in [3].

Another area of concern will be the absence of any individual memory protection for the processes of one program. All the memory will be accessible to any process and the only safeguard against improper access will be faultless programming. The reason for this rather unsatisfactory behaviour of the CRAY X-MP is the fact that only one pair of base address and field length is used for all processes of one program. Experience has already shown that debugging a multiprocessor version of such a program becomes an extremely tedious task.

The granularity of the processes can vary widely without affecting performance because it is possible either to use high-level synchronisation tools or to access the shared registers directly, resulting in a very short synchronisation time.

3.3 HEP-1 successor

The main development of the HEP-1 system can be expected to be an increase in performance of each processing element (PEM). This may be achieved by using more advanced technology and possibly by adding some kind of vector capabilities. The structure of the new system can be seen in Fig. 4.

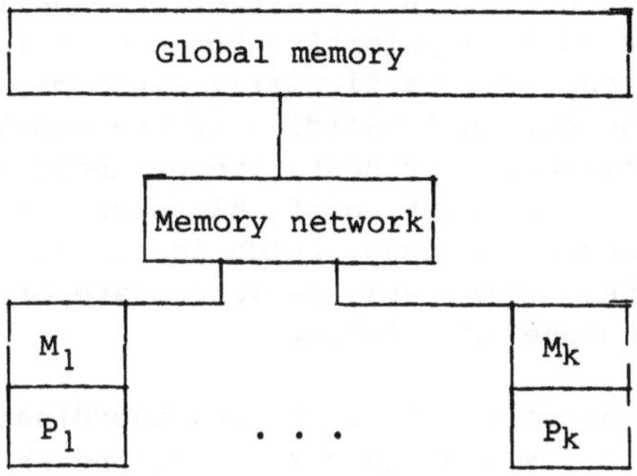

Fig. 4: Expected structure of HEP-1 successor with $k (k > 1)$ PEMs

One of the major differences between the HEP architecture and those of the previously discussed systems is the existence of an instruction pipeline within each PEM which contains instructions for all the active processes. Each PEM, therefore, is able to be looked upon as a multiprocessor system of its own. From the example discussed later it follows that currently about 20 processes are required to use a single PEM efficiently for a weather model. The local memory of a PEM has therefore to be large enough to allow sufficient active processes. For the successor of the HEP-1, the number of active processes per PEM may even be larger because the accesses to global memory may become a bottleneck if the speed of the memory network is not increased in line with the speed of the individual PEMs, especially if vector capabilities are added. We expect that in excess of 200 separate processes will be required to achieve the envisaged performance of the HEP system for weather modelling applications. Since the synchronisation tools are easy to use and applicable without large overheads, processes of relatively small granularity may be used. However, the vector capabilities may enforce a lower limit on the granularity. Nevertheless, the large number of active processes will probably require the self-scheduling programming technique outlined in Section 2.2. The algorithms for implementation on the HEP system will have to support such a scheme without too much overhead.

4 THE SHALLOW-WATER MODEL

In this section we describe a simple atmospheric dynamics model based on what are known as the shallow-water equations. The model and the computational methods are given in detail in [8]. The solutions to these equations are intended to provide rough estimates of the performance of more complex atmospheric models and to identify areas in which future computer designs might be improved in order to enhance performance. The shallow-water equations are accepted as a primitive but useful model of the dynamics of the atmosphere, particularly since we are concerned, not so much with the physical validity of the model, as with having the computations somewhat representative of those found in atmospheric models. The solution of this simple model adapts well to current supercomputer designs, which is not too surprising since the designers of these computers are well aware of the importance of computers to atmospheric science.

The equations will be solved in Cartesian coordinates on the rectancular domain $a \leqslant x \leqslant b$ and $c \leqslant y \leqslant d$. The shallow-water equations without the Coriolis term can be written in the form

$$\frac{\partial u}{\partial t} - \zeta v + \frac{\partial H}{\partial x} = 0 \tag{4.1}$$

$$\frac{\partial v}{\partial t} + \zeta u + \frac{\partial H}{\partial y} = 0 \tag{4.2}$$

$$\frac{\partial P}{\partial t} + \frac{\partial}{\partial x}(Pu) + \frac{\partial}{\partial y}(Pu) = 0 \tag{4.3}$$

where u and v are velocities in the x and y directions, respectively; P is density or pressure; and ζ is vorticity, given by

$$\zeta = \frac{\partial v}{\partial x} - \frac{\partial u}{\partial y} \tag{4.4}$$

A quantity H which is related to the height field is defined as

$$H = P + \tfrac{1}{2}(u^2+v^2) \tag{4.5}$$

If, in addition, we define the mass fluxes

$$U = Pu \tag{4.6}$$

$$V = Pv \tag{4.7}$$

and potential velocity

$$Z = \frac{\zeta}{P} \tag{4.8}$$

then (4.1), (4.2), and (4.3) take the form

$$\frac{\partial u}{\partial t} - ZV + \frac{\partial H}{\partial x} = 0 \qquad (4.9)$$

$$\frac{\partial v}{\partial t} + ZU + \frac{\partial H}{\partial y} = 0 \qquad (4.10)$$

$$\frac{\partial P}{\partial t} + \frac{\partial U}{\partial x} + \frac{\partial V}{\partial y} = 0 \qquad (4.11)$$

In order to facilitate presentation and implementation, we specify periodic boundary conditions, in which the variables u, v, U, V, and P satisfy $f(x+b,y) = f(x+a,y)$ and $f(x,y+d) = f(x,y+c)$. We wish to obtain an approximate solution to these equations on the rectangle $a<x<b$ and $c<y<d$. To this end we select integers M and N and define the grid

$$x_i = i\Delta x + a \qquad i = 0, \tfrac{1}{2}, 1, \ldots M+1$$

$$y_j = j\Delta y + b \qquad j = 0, \tfrac{1}{2}, 1, \ldots N+1$$

where $\Delta x=(b-a)/(M+1)$ and $\Delta y=(d-c)/(N+1)$. A staggered grid is used, with the positions of the dependent variables given in Fig.5 below.

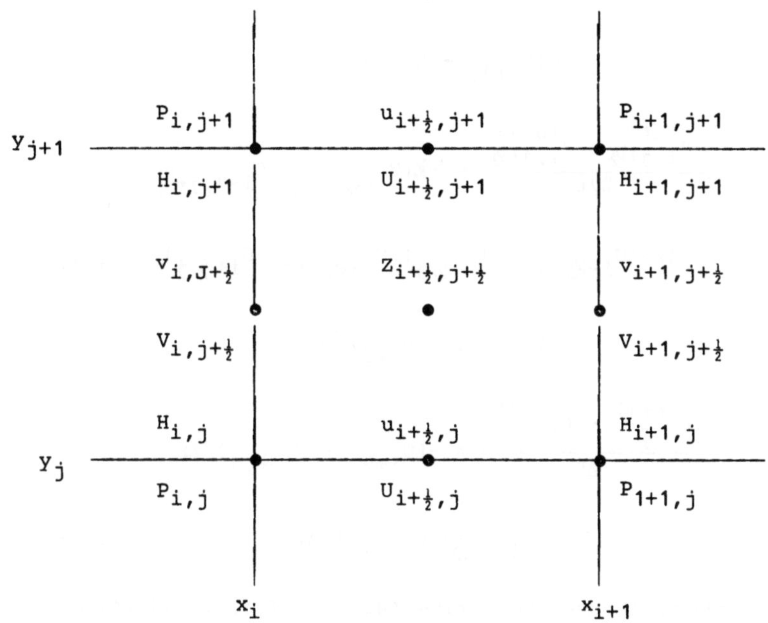

Figure 5

The variables P and H have integer subscripts, Z has half subscripts, u has half and integer subscripts, and v has integer and half subscripts. In order to obtain an approximate solution of (4.9), (4.10) and (4.11) in the form of a tabulation, the dependent variables are required to satisfy the following finite-difference approximations of (4.5) through (4.11):

$$U_{i+\frac{1}{2},j} = \tfrac{1}{2}(P_{i+1,j} + P_{i,j})u_{i+\frac{1}{2},j} \tag{4.12}$$

$$V_{i,j+\frac{1}{2}} = \tfrac{1}{2}(P_{i,j+1} + P_{i,j})v_{i,} \tag{4.13}$$

$$Z_{i+\frac{1}{2},j+\frac{1}{2}} = \frac{\left[(v_{i+1,j+\frac{1}{2}} - v_{i,j+\frac{1}{2}})/\Delta x - (u_{i+\frac{1}{2},j+1} - u_{i+\frac{1}{2},j})/\Delta y\right]}{1/4(P_{i,j} + P_{i+1,j} + P_{i+1,j+1} + P_{i,j+1})} \tag{4.14}$$

$$H_{i,j} = P_{i,j} + \tfrac{1}{2}\left[\frac{(u^2_{i+\frac{1}{2},j} + u^2_{i-\frac{1}{2},j})}{2} + \frac{v^2_{i,j+\frac{1}{2}} + v^2_{i}}{2}\right. \tag{4.15}$$

$$\frac{u^{(n+1)}_{i+\frac{1}{2},j} - u^{(n-1)}_{i+\frac{1}{2},j}}{2\Delta t} = \tfrac{1}{2}(Z_{i+\frac{1}{2},j+\frac{1}{2}} \overset{v^{y}}{+} Z_{i+\frac{1}{2},j-\frac{1}{2}})$$

$$1/4(V_{i+1,j+\frac{1}{2}} + V_{i,j+\frac{1}{2}} + V_{i+1,j-\frac{1}{2}} + V_{i,j-\frac{1}{2}}) \tag{4.16}$$

$$-(H_{i+1,j} - H_{i,j})/$$

$$\frac{v^{(n+1)}_{i,j+\frac{1}{2}} - v^{(n-1)}_{i,j+\frac{1}{2}}}{2\Delta t} = -\tfrac{1}{2}(Z_{i+\frac{1}{2},j+\frac{1}{2}} + Z_{i-\frac{1}{2},j+\frac{1}{2}})x$$

$$1/4(U_{i-\frac{1}{2},j+1} + U_{i-\frac{1}{2},j} + U_{i+\frac{1}{2},j+1} + U_{i+\frac{1}{2},j}) \tag{4.17}$$

$$-(H_{i,j+1} - H_{i,j})/\Delta y$$

$$\frac{P^{(n+1)}_{i,j} - P^{(n-1)}_{i,j}}{2\Delta t} = -(U_{i+\frac{1}{2},j} - U_{i-\frac{1}{2},j})/\Delta x$$

$$-(V_{i,j+\frac{1}{2}} - V_{i,j-\frac{1}{2}})/\Delta y \tag{4.18}$$

The leapfrog scheme is used to derive (4.16), (4.17), and (4.18). The super-scripts on the left side of these equations designate the time level. All quantities on the right side are evaluated at $t_n = n\Delta t$ where Δt must satisfy the stability criterion $H^{\frac{1}{2}} \frac{\Delta t}{\Delta x} < 1$.

The initial velocity fields u(x,y,0) and v(x,y,0) will be selected so that the balance conditions

$$\text{div } \mathbf{v} \qquad \frac{\partial u}{\partial x} + \frac{\partial v}{\partial y} = 0 \tag{4.19}$$

and

$$\frac{\partial}{\partial t} \text{div } \mathbf{v} = 0 \tag{4.20}$$

are satisfied. To this end we define a stream function

$$\Psi = A \sin \frac{2\pi x}{b-a} \sin \frac{2\pi y}{d-c} \tag{4.21}$$

and set

$$u(x,y,0) = -\frac{\partial \Psi}{\partial y} \text{ and } v(x,y,0) = \frac{\partial \Psi}{\partial x} \tag{4.22}$$

It is clear that these initial values will satisfy (4.19).

From (4.9) and (4.10)

$$\frac{\partial}{\partial t} \text{div } \mathbf{v} = \frac{\partial}{\partial x}(ZV) - \frac{\partial}{\partial y}(ZV) - \nabla^2 H \tag{4.23}$$

Substituting (4.4) through (4.8) into (4.23), we obtain

$$\frac{\partial}{\partial t} \text{div } \mathbf{v} = 2\left(\frac{\partial u}{\partial x}\frac{\partial v}{\partial y} - \frac{\partial u}{\partial y}\frac{\partial v}{\partial x}\right) - \nabla^2 P \tag{4.24}$$

In order to satisfy the balance equation (4.20), we set (4.24) equal to zero and solve for P. Substituting (4.20), (4.21), and (4.22) into (4.24), we obtain

$$\nabla^2 P = -\frac{16\pi^4 A^2}{(b-a)^2(d-c)^2}\left(\cos \frac{4\pi x}{b-a} + \cos \frac{4\pi y}{d-c}\right) \tag{4.25}$$

Integrating we obtain

$$P = \frac{A^2}{4}\left[\left(\frac{2\pi}{d-c}\right)^2 \cos \frac{4\pi x}{b-a} + \left(\frac{2\pi}{b-a}\right)^2 \cos \frac{4\pi y}{d-c}\right] + P_0 \tag{4.26}$$

where P_0 is any constant. Equation (4.26) is used to define the initial values of P at t=0. In practice, the finite-difference approximation of Poisson's equation is solved in order to initialise P.

The discrete forms of (4.21) and (4.22) are

$$\Psi_{i+\frac{1}{2},j+\frac{1}{2}} = A \sin \frac{2\pi x_{i+\frac{1}{2}}}{b-a} \sin \frac{2\pi y_{j+\frac{1}{2}}}{d-c} \tag{4.27}$$

$$u_{i+\frac{1}{2},j} = -\frac{(\Psi_{i+\frac{1}{2},j+\frac{1}{2}} - \Psi_{i+\frac{1}{2},j-\frac{1}{2}})}{\Delta y} \tag{4.28}$$

$$v_{i,j+\frac{1}{2}} = \frac{(\Psi_{i+\frac{1}{2},j+\frac{1}{2}} - \Psi_{i-\frac{1}{2},j+\frac{1}{2}})}{\Delta x} \tag{4.29}$$

The leapfrog time differencing implied by (4.16), (4.17), and (4.18) requires values of the dependent variables u, v, and P at two levels, namely, $t_0 = 0$ and $t_1 = \Delta t$. The values at t_0 are provided by the initial data in (4.26) through (4.29). The values at t_1 are computed using one step of Euler's method

$$f^{(1)} = f^{(0)} + \frac{\partial f^{(0)}}{\partial t} \Delta t \qquad (4.30)$$

where $\partial f^{(0)}/\partial t$ for f = u, v, and P, is computed from (4.16), (4.17), and (4.18) evaluated at t_0. Finally, to handle the weak instability in the leapfrog scheme, we use the following time filter for f = u, v, and P.

$$F^{(n)} = f^{(n)} + \alpha(f^{(n+1)} - 2f^{(n)} + F^{(n-1)}) \qquad (4.31)$$

where α is the filter parameter.

The filtered values $F^{(n)}$ enter the computation in (4.16), (4.17), and (4.18) at time level t_{n-1}.

5 IMPLEMENTATION AND PERFORMANCE OF THE MODEL

In this section we examine the implementation and performance of the shallow water model on several supercomputers. The performance of a super-computer is difficult to evaluate because it depends on many factors. In practice, maximum computational rates are rarely achieved, and a gap, which appears to be widening, exists between expected and possible performance. The extent of this gap depends both on the problem and on the user's ability to adapt the computation to a particular architecture. More responsibility has been placed on the user for the performance of the computer, which may be one to two orders of magnitude below maximum unless the computation is formulated in a way that takes advantage of the architectural features.

The computations in the model are all parallelisable or distributable, which resulted in relatively high computational rates on all of the computers. The remainder of this section is divided into four parts in which the results for each of the computers are presented. In addition to performance figures, each part includes a discussion about programming the model, which includes any modifications that were necessary in order to take advantage of the particular architecture.

5.1 The CRAY-1

The program for solving the shallow-water equations is listed in Appendix A. It was first written for the CRAY-1 and subsequently adapted to the other computers. The first version of the program contained about 120 lines of code and took about two hours to write using a screen editor. This version then took about four hours to debug. The final (second) version in Appendix A contains 242 lines of code and required an additional eight hours for both code development and debugging.

The program contains three main double loops that account for most of the computing time. The quantities $U_{i+\frac{1}{2}j}$, $V_{ij+\frac{1}{2}}$, $Z_{i+\frac{1}{2}j+\frac{1}{2}}$, and H_{ij} that are defined in (4.12) through (4.15) are computed in loops 100. Their FORTRAN names are CU(I+1,J), CV(I,J+1), Z(I+1,J+1), and H(I,J), respectively. The half subscripts $i+\frac{1}{2}$ or $j+\frac{1}{2}$ are converted to the FORTRAN subscripts I+1 or J+1. In general, for fractional indices, the FORTRAN subscripts are generated by adding one-half.

The quantities $U^{(n-1)}_{i+\frac{1}{2},j}$, $V^{(n+1)}_{i,j+\frac{1}{2}}$, and $P^{(n+1)}_{i,j}$ that are defined in(4.16, (4.17), and (4.18) are computed in loops 200. Their FORTRAN names are UNEW(I+1,J), VNEW(I,J+1), and PNEW(I,J), respectively. Time filtering and updating occur in loops 300. The variables at the three different time levels t_{n-1}, t_n, and t_{n+1} are stored in three separate two dimensional arrays. For example $U^{(n-1)}_{i+\frac{1}{2},j}$, $U^{(n)}_{i+\frac{1}{2},j}$, and $U^{(n+1)}_{i+\frac{1}{2},j}$ are stored in the arrays UOLD(I+1,J), U(I+1,J), and UNEW(I+1,J), respectively.

Problem and program parameters are initialized beginning with Δt = DT = 90 seconds. The variable TDT is set equal to Δt on the first pass where Euler's method is used to compute the values of u, v, and P at $t = t_1$ as discussed near (4.30). The variable TDT is then set equal to $2\Delta t$ on all subsequent cycles. Most of the parameters that are defined at the beginning of the program are self-explanatory, with the possible exception of ITMAX, which is the maximum number of time cycles, and MPRINT, which is the number of cycles between prints. The variable EL is set equal to both b - a and d - c and hence the domain is square.

The initial stream function $\psi_{i+\frac{1}{2},j+\frac{1}{2}}$ and the balanced $P_{di,j}$, as defined in (4.27) and (4.26), are computed in loops 50. Their FORTRAN equivalents are PSI(I+1,J+1) and P(I,J). The initial velocities, as defined in (4.28) and (4.29), are computed in loops 60. Periodic boundary conditions are applied throughout the code in loops 70, 75, 86, 110, 115, 210, 215, 320, and 325. Output is produced beginning with the cycle number and model time at the statement WRITE(6,350) NCYCLE,PTIME. Computational times in seconds are computed for loops 100, 200, and 300 and stored in the variables T100, T200, and T300, respectively. The megaflops for each of these loops are computed based on 24, 26, and 15 floating operations per loop and stored in the variables MFS100, MFS200, and MFS300, respectively. The computer times and megaflops are listed in Table 1. These results were obtained using the CRAY FORTRAN complier CFT 1.13.

TABLE 1
CRAY-1 performance on the shallow water model for a 64 x 64 grid

loop	time (ms)	Mflops
100	1.405	70.0
200	1.670	63.8
300	1.282	47.9

5.2 The CYBER-205

The Control Data Corporation CYBER 205 [1], unlike the CRAY-1, has no intermediate storage through which the vector arithmetic units receive operands and return results. Instead, the vector operands and results are streamed directly to and from main memory. Furthermore, on the CYBER 205 the elements of a vector must occupy consecutive locations in memory. In order to guarantee conflict-free memory access to any vectors during an operation a sophisticated microcoded algorithm is executed before each vector instruction. Consequently, there are rather large start-up times for vector operations. To reduce this cost one must arrange the calculations in such a way as to create the longest possible vectors.

For example, operations on modest-sized (on the order of 100 x 100) arrays may be done column- or row-wise on the CRAY-1 at near peak efficiency. On the CYBER 205 however, column operations alone (using vectors of length 100) yield a throughput on the order of 50% of the maximum, whereas if the operations can be performed over the entire array (working on vectors of length 10,000), the throughput will rise to about 99% of the maximum. All numbers related to speed or efficiency are based on the timing formula for vector addition and multiplication which are the fastest-executing. Other vector operations give slower speeds.

In order to increase vector lengths, bit (logical) vectors have been defined in the CYBER 205 extension to FORTRAN. Bit vectors may be used to control the storage of elements of result vectors and are absolutely required for efficiency when one is working over a subarray that is large compared to the array which contains it. Continuing the example above, suppose that one wants to perform a calculation on a 99 x 99 sub-array of a 100 x 100 array. It is no longer possible to perform one vector calculation over the entire subarray because the elements of the vector are no longer contiguously located in memory. Without bit vectors the calculation would have to be performed over vectors of length 99 (at 50% efficiency). Defining a bit vector that causes every 100th result to not be stored will result in the ability to perform operations on vectors of length 99 x 100 = 9,900 (at 99% efficiency).

A two-pipe CYBER 205 has an asymptotic megaflop rating of 100 for memory-to-memory vector operations on full 64-bit words. However, it is possible to direct the output of one vector arithmetic unit into an input port of another vector arithmetic unit. With this linking process it is possible to double the asymptotic megaflop rating to 200.

The FORTRAN program SHALOW, described in the previous section and listed in Appendix A, was run on a two-pipe CYBER 205 located at Colorado State University in Fort Collins, Colorado. There were two versions run on the machine: (i) the original FORTRAN program listed in Appendix A, and (ii) a modified version in which the principal do loops were recoded using the special vector extensions to FORTRAN developed by CDC [2]. The modified version is listed in Appendix B.

For the first version, no changes were made to the original code. The vectorization and optimization options specified to the FORTRAN 2.1.5 compiler were O=UOV, which directed the compiler to fully optimize all scalar code and vectorize as much of it as it could automatically. Examination of the main time-stepping loop (statements 99 through 400) revealed that the compiler vectorized the inner portion of the double loops 100, 200, 300, and 400 and the single loops 115, 215, and 325. The performance for the three main loops is given in Table 2.

For the modified version of the code all of the loops mentioned in the previous paragraph were recoded in special explicit vector notation. The double loops 100 and 200 could not be replaced by single vector operations because they only worked over the submatrices I = 1,2,...,M, J = 1,2...,N of the M+1 X N+1 matrices. A bit vector was defined that allowed those double loops to be replaced by single vector operations over the contiguous subarrays I = 1,2,...,M+1, J = 1,2,...,N. The resulting vector lengths were 4159 (=N*(M+1)-1) so we expect the vector portion of the code to be running at about 97% of its peak rate. The double loops 300 and 400 were replaced by vector operations over the entire M+1 X N+1 arrays. The conversion effort required about eight hours to study, code, and debug. The results of the customisation for the CYBER 205 are given in Table 3.

Comparison of the results in Tables 2 and 3 vividly demonstrate the benefit derived by operating on long vectors on the CYBER 205. An unfortunate side effect is the nonportability of the resulting FORTRAN program.

TABLE 2

CYBER 205 performance on the standard FORTRAN version of the shallow-water model for a 64 x 64 grid.

loop	time (ms)	Mflops
100	2.66	37.0
200	2.53	42.1
300	1.60	38.4

TABLE 3

CYBER 205 performance on the customized FORTRAN version of the shallow-water model for a 64 x 64 grid.

loop	time (ms)	Mflops
100	1.08	92.6
200	0.861	126
300	0.526	119

5.3 The Denelcor HEP-1

Shallowp, a parallel version of the shallow-water equations allowing for an arbitrary number of parallel processes, written from the original CRAY-1 version, is listed in Appendix C. In the parallel version each of the primary loops 100, 200, and 300 were put into subroutines, as were the periodic continuation loops. Each of these routines is called in turn from a subroutine called PARSUB. Each loop routine has two inputs: iproc = the process number and nproc = total number of processes. In each case the outer loop is split across the processes by the technique called prescheduling. The loop has the form:

```
          DO 100 I = iproc, N, nproc
              .
              .
              .
      100 CONTINUE
```

For example, if there were two processes the first would perform all odd rows and the second would perform all even rows. The value of nproc is input at run time. In PARSUB a call to barrier is made between each of the various routines for the purpose of synchronising the processes. As each process enters the barrier it is terminated, until the last process enters, at which time all processes are recreated and processing resumes. Using this technique a larger problem could be run on a four-to eight-PEM system with 128-256 parallel processes without modifying the program.

These changes required about 40 minutes of design and editing time with a full screen editor. The program ran on the first trial, but no correct answers were available at the time for comparisons. The program has since been debugged.

The program was run on System-0 at Denelcor's home offices in Aurora, Colorado, under the company's UNIX^{TM}-based operating system HEP/UPX. System-0 is a minimum 1-PEM configuration. Since the Fortran 77 optimizer was unavailable, we felt it only fair for the purposes of comparison that some simple optimisation should be performed by hand. David Snelling, analyst for Denelcor, adapted the compiler output in such a way as to simulate the loop invariant removal function of the optimiser. No other optimisation was done. The Mflop rates for loops 100, 200, and 300 are 3.9, 3.5, and 3.9, respectively.

System-0 at Denelcor was configured with an Eulerian switch, in which each message visits all ports on the network rather than taking the shortest route. The memory latency characteristics created by this routing simulate those of an eight to ten-PEM system, and hence 20 processes were required to achieve full machine utilisation instead of the 13 processes normally required on a single PEM.

5.4 The CRAY X-MP/2

The CRAY X-MP/2 has two vector processors, each of which is similar to the CRAY-1, except for the following major differences:

(i) The clock period has been reduced from 12.5 to 9.5 nanoseconds.
(ii) The ports to memory have been increased to four per processor, which permits two vector loads, a store, and an I/O transfer to occur simultaneously.
(iii) The solid-state disk (SSD), which is available, can be written at a rate of about one word per CPU cycle.

The two CPUs operate in a multitask mode. The user generates tasks which are then scheduled on the processors by the operating system. The user generates a task either as a main program or by executing a CALL TSKSTART statement. In effect, this calls a subroutine which can be executed at the same time as the main program. However, unless the computer is dedicated to a particular user it is likely that the two processors will be executing the programs of two different users.

At certain times within the user's program it may be necessary to wait for a set of tasks to be completed before starting new ones. This can either be achieved by a CALL TSKWAIT statement or by synchronisation using the event statements, CALL EVPOST, CALL EVWAIT, and CALL EVCLEAR. The subroutine EVPOST posts an event, EVWAIT waits for an event to occur in the other processor, and EVCLEAR clears the event, which can then be reused. A detailed description of the use of these subprograms is contained in the multitasking guide [4] available from Cray Research. An overview of multitasking is given in [7].

Two multiprocessor programs for the sample model were written and run on a dedicated CRAY X-MP/2. The first program is listed in Appendix D and uses TSKSTART and TSKWAIT subroutines in order to synchronise the processors. Six tasks are defined for each time step, two for each of the three main loops, namely, 100, 200, and 300. These loops were extracted from the main program and made into three subroutines. Each of the three subroutines was called with different parameters which define the region of computation for each processor.

For example, loop 100 has been replaced by the statements

```
      CALL TSKSTART(I100A,L100,M,1,N/2)

      CALL TSKSTART(I100B,L100,M,N/2+1,N)

      CALL TSKWAIT(I100A)

      CALL TSKWAIT(I100B)
```

The first argument, I100A, is called the task array and is used by the system when a task is defined. The second argument, L100, is the subroutine that performs the task, and the remaining arguments are used by the subroutine L100. The first TSKSTART statement initiates the computations in loop 100 that correspond to the values J=1,N/2. The second TSKSTART statement initiates the computations that correspond to the remaining values J=N/2+1,N. The tasks are synchronised prior to the periodic continuation by the CALL TSKWAIT statements. This procedure is repeated for loops 200 and 300. The performance of this program is presented in Table 5. For the purpose of comparison, the performance of the single-processor code in Appendix A on a single CRAY X-MP/2 processor is listed in Table 4.

TABLE 4

Performance on single-processor CRAY X-MP of shallow-water model on a 64 x 64 grid

loop	time (ms)	mflops
100	0.963	102
200	1.10	97
300	0.653	94

TABLE 5

CRAY X-MP/2 performance on "tasked" version of shallow water model in Appendix D on a 64 x 64 grid

loop	time (ms)	Mflops	speedup
100	0.559	176	1.73
200	0.616	173	1.78
300	0.407	151	1.61

The second multiprocessor code is listed in Appendix E. It is synchronised using events. The program was developed by duplicating loops 100, 200, and 300 and including them in a subroutine called TASK, which is called only once. The loops are synchronised with those in the main program by using EVPOST, EVWAIT, and EVCLEAR subroutines.

The performance of the "events" program is given in Table 6. As suggested by Larson [7] and others, its performance is superior to the "tasks" program in Table 5. The reason is simply that the TSKSTART subprogram is time-consuming compared to the events program. Of course the results in Tables 5 and 6 would be closer if the grid size were increased.

TABLE 6

CRAY X-MP/2 performance on "events" version of the shallow-water model in Appendix E on a 64 x 64 grid

loop	time (ms)	mflops	speedup
100	0.504	195	1.91
200	0.563	189	1.95
300	0.351	175	1.86

The program in Appendix D that uses tasks was both easier to write and easier to debug. The total time required was about eight hours using a line editor. The program in Appendix E took about the same time, but that was only because the author (Swarztrauber) had developed more knowledge about both the editor and multitasking.

6 THE DISTRIBUTED SOLUTION OF POISSON'S EQUATION

In this section we describe the solution of Poisson's equation in a distributed processing environment. Although Poisson's equation does not appear in the sample model, it is nevertheless solved in a number of atmospheric applications and is therefore deserving of some discussion. As we will see, the distributed solution of Poisson's equation is not quite as straightforward as the solution of the sample model. On the other hand, the effort that is required to solve Poisson's equation is likely more representative of the effort that is required to distribute the computations in a more realistic and complex model. We will discuss the solution of Poisson's equation on a computer with p processors, where p is a relatively small number. We will also assume that the processors share memory and that each processor is pipelined. These two assumptions significantly influence the method of solution. A different approach can be taken if memory is not shared [10].

We wish to solve Poisson's equation:

$$\Delta u = g \tag{6.1}$$

on a rectangle $a < x < b$ and $c < y < d$ and subject to an appropriate boundary condition. We assume that $u = 0$ is specified on the boundary; however, the results are applicable to any of the standard boundary conditions [11,12]. More precisely, we will obtain an approximate solution of (6.1) in the form of a tabulation. Given the integers M and N, a grid (x_i, y_j) is defined by $x_i = a + i\Delta x$ and $y_j = c + j\Delta y$, where $\Delta x = (b-a)/(M+1)$ and $\Delta y = (d-c)/(N+1)$. We seek a tabulation $u_{i,j}$ that approximates the exact solution $u(x_i, y_j)$ at each of the grid points. To this end we require $u_{i,j}$ to satisfy the following finite-difference approximation of (6.1).

$$\frac{u_{i+1,j} - 2u_{i,j} + u_{i-1,j}}{\Delta x^2} + \frac{u_{i,j+1} - 2u_{i,j} + u_{i,j-1}}{\Delta y^2} = g_{i,j} \qquad (6.2)$$

where $g_{i,j} = g(x_i, y_j)$. A convenient notation is obtained by introducing the vectors

$$u_j^T = (u_{1,j}, u_{2,j}, \ldots, u_{M,j}) \text{ and } g_j^T = \Delta y^2 (g_{1,j}, g_{2,j}, \ldots, g_{M,j}).$$

Then(6.2) can be written:

$$u_{j-1} + Au_j + u_{j+1} = g_j \quad j=1,\ldots,N \qquad (6.3)$$

where

$$A = \begin{bmatrix} -2\alpha & 1 & & & & \\ 1 & -2\alpha & 1 & & & \\ & 1 & \cdot & & & \\ & & & \cdot & & \\ & & & \cdot & 1 & \\ & & & 1 & -2\alpha & 1 \\ & & & & 1 & -2\alpha \end{bmatrix} \qquad (6.4)$$

and $\alpha = \left(\frac{\Delta y}{\Delta x}\right)^2 + 1$. With no loss of generality we can assume that $\Delta x = \Delta y$.

There are a number of methods for solving the system (6.3), which may be equally attractive, particularly in a time-dependent setting. In this paper we investigate the Fourier method, which is known to take only 5 to 10% of the total computer time in current atmospheric models.

Let Q be the orthogonal matrix whose columns are the eigenvectors of A; then

$$q_{i,j} = \sqrt{\frac{2}{M+1}} \sin ij\frac{\pi}{M+1} \qquad (6.5)$$

and $Q^TAQ = \text{diag}(\lambda_1,\ldots,\lambda_M)$, where

$$\lambda_k = 2\left[\sin^2 k\frac{\pi}{2(M+1)} - \frac{(\Delta y)^2}{(\Delta x)}\right] \qquad k=1,\ldots,M \qquad (6.6)$$

If we define $\hat{u}_j = Q^T u_j$ and compute $g_j = Q^T g_j$, then, substituting into (6.2), we obtain

$$\hat{u}_{j-1} + Q^T AQ\hat{u}_j + \hat{u}_{j+1} = g_j$$

Let $\hat{u}_{k,j}$ and $g_{k,j}$ denote the kth component of $\hat{u}_j$ and g_j, respectively; then for each k=1,...M+1 we obtain a tridiagonal system of order N.

$$\hat{u}_{k,j-1} + \lambda_k \hat{u}_{k,j} + \hat{u}_{k,j+1} = g_{k,j} \quad j=1,\ldots N \qquad (6.8)$$

The Fourier method can now be summarized.

(i) Compute $g_i = Q^T g_j$ using the FFT.

ii) Solve the tridiagonal systems (6.8) for $\hat{u}_{k,j}$ and hence $\hat{u}_j$.

(iii) Compute the solution $u_j = Q\hat{u}_j$ again using the FFT.

Note that the Fourier transforms are computed in the x (or i) direction, whereas the tridiagonal systems are solved in the y (or j) direction, as depicted in Figure 6.

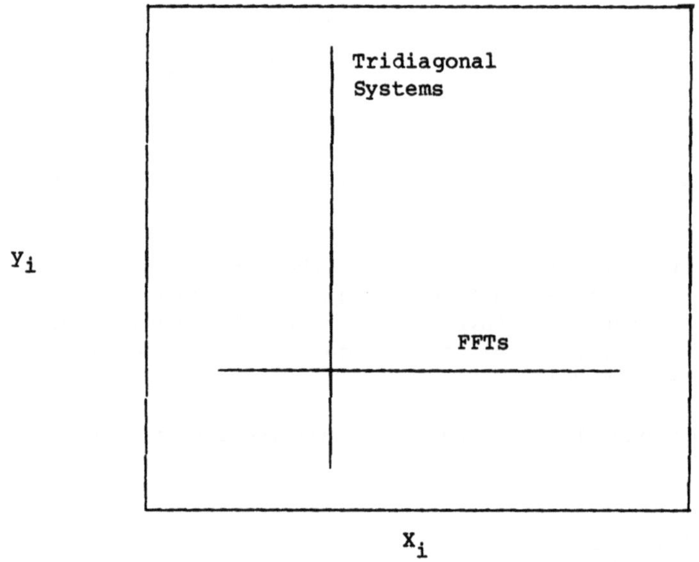

Figure 6

We proceed now to discuss the implementation of the Fourier method on a vector multiprocessor. We assume that the memory is shared ad that the computer has p processors. Consider first the FFTs of N sequences of length M, which is required in steps (i) and (iii) of the Fourier method. Four methods are presented below. The fourth method will be discussed in detail in the next section.

(i) Method I for FFTs

The simplest approach is to assign N/p FFTs to each processor and to perform the FFTs individually. The minimum vector length is $\sqrt{M/2}$, which can be relatively small.

(ii) Method II for FFTs

Again N/p FFTs are assigned to each processor. However, each operation in the FFT is applied to all N/p sequences. All vectors have length N/p, which can also be relatively short. This is an example of how distribution and vectorisation can be at odds. As the number of processors increases, the vectors become shorter.

(iii) Method III for FFTs

Apply each operation in a distributed version of the FFT to all N sequences. All vectors have length N. This approach requires a distributed version of the FFT which is discussed in the next section. This approach is relatively efficient, since the FFT distributes quite well; i.e., the number of operations in the distributed FFT is the same as in the non-distributed FFT.

(iv) Method IV for FFTs

All N sequence of length M are transformed as a single sequence of length MN using the "truncated" FFT. The minimum vector length is $\sqrt{MN/p}$ and the maximum vector length is MN/p. This approach is discussed in detail in the next section.

Next we consider the efficient solution of the tridiagonal systems. The asymptotic operation count for solving the tridiagonal systems is O(MN), which is less than O(MNlogM) for the FFTs. However, it is more difficult to vectorise the solution of the tridiagonal systems, and in practice it takes a significant part of the total computing time.

(i) Method I for Solving the Tridiagonal Systems

The simplest approach is to assign M/p tridiagonal systems to each processor. The systems can then be solved individually using the tridiagonal version of Gauss elimination. The difficulty with this approach is that the Gauss algorithm does not vectorise.

(ii) Method II for Solving the Tridiagonal Systems
In each processor, each operation in the Gauss algorithm is applied to all M/P systems. All vector lengths are M/P; however, a significant amount of additional storage is required for the LU decompositions.

(iii) Method III for Solving the Tridiagonal Systems
Each operation in a distributed version of the Gauss algorithm is applied to all tridiagonal systems. All vector lengths are M. This method also requires a substantial amount of storage. Another problem is that the Gauss algorithm does not distribute well. The number of operations in the distributed algorithm is greater than in the non-distributed algorithm [9].

(iv) Method IV for Solving the Tridiagonal Systems
Each operation in a distributed version of the cyclic reduction algorithm is applied to all tridiagonal systems, with the result that the vector lengths are all equal to M. The cyclic reduction algorithm distributes well and requires only a modest amount of additional storage.

7 COMPUTING THE FFT ON A VECTOR MULTIPROCESSOR

In the previous section an overview of the Fourier method for solving Poisson's equation was presented. At the end of the section, vector method IV was presented in which N sequences of length M are transformed as a single sequence of length MN using a distributed version of the truncated FFT. In this section we discuss this method in greater detail. A survey of vector methods for the FFT is given in [13].

This section is divided into the following parts:

- 7.1 A review of the Cooley-Tukey FFT
- 7.2 The loop inversion method for maintaining long vectors
- 7.3 The "truncated" FFT for multiple transforms
- 7.4 Eliminating the sort phase in the FFT
- 7.5 The distributed FFT on the Cray X-MP/2

7.1 A review of the Cooley-Tukey FFT

The presentation in this section involves modifications of the computation and storage allocations that are internal to the FFT. Therefore, in order to develop notation and to describe the results, it is necessary to review the FFT algorithm for computing the discrete Fourier transform. Given the complex sequence x_m, then we wish to compute the discrete Fourier transform:

$$X_k = \frac{1}{M} \sum_{m=0}^{M-1} x_m e^{-ikm\frac{2\pi}{M}} \qquad (7.1)$$

This computation requires M^2 complex multiplications and M(M-1) complex additions. We begin our discussion of the FFT with a description of the splitting algorithm which is fundamental to the FFT and which computes the X_k with about half the number of operations required by (7.1).

If M is even then the sum on the right side of (7.1) can be split into two sums in which the subscripts on x_m are even and odd, respectively. It is customary to ignore the scale factor of M^{-1}, in which case:

$$X_k = \sum_{m=0}^{M/2-1} x_{2m} e^{-ik2m\frac{2\pi}{M}} + \sum_{m=0}^{M/2-1} x_{2m+1} e^{-ik(2m+1)\frac{2\pi}{M}} \tag{7.2}$$

If we define the new transforms

$$Y_k = \sum_{m=0}^{M/2-1} x_{2m} e^{-ik2m\frac{2\pi}{M/2}} \tag{7.3}$$

$$Z_k = \sum_{m=0}^{M/2-1} x_{2m+1} e^{-ik2m\frac{2\pi}{M/2}} \tag{7.4}$$

then (7.2) takes the form:

$$X_k = Y_k + e^{-ik\frac{2\pi}{M}} Z_k \tag{7.5}$$

From (7.3) and (7.4) it can be determined that Y_k and Z_k are periodic with period M/2. That is, $Y_{k+M/2} = Y_k$ and $Z_{k+M/2} = Z_k$. Also, since

$$e^{-i(k+M/2)\frac{2\pi}{M}} = -e^{-ik\frac{2\pi}{M}} \tag{7.6}$$

$$X_{k+N/2} = Y_k + e^{-ik\frac{2\pi}{M}} Z_k \tag{7.7}$$

the splitting algorithm can now be given:

(i) For k=1,...M/2 compute Y_k and Z_k from (7.3) and (7.4)

(ii) Compute X_k from Y_k and Z_k using (7.5) and (7.7)

The computation of Y_k and Z_k requires $M^2/2$ multiplications and M(M-1)/2 additions. The computation of X_k from (7.5) and (7.7) is proportional to M which can be ignored for sizeable M. Therefore, the splitting algorithm requires about half the time taken by (7.1).

The FFT is now a simple extension of the splitting algorithm. Since (7.3) and (7.4) have the same form as (7.1) but with M replaced by M/2, the splitting algorithm can also be used to compute Y_k and Z_k. If $N=2^{\alpha}$for some integer α then the splitting can be continued α times, which results in the FFT. A sample FFT is shown in Table 7 for the case $M=2^3=8$. The original sequence is in column 1 and subsequent splittings are in columns 2, 3 and 4. From (7.1) we note that the transform of a sequence of length 1 is just the element itself and therefore column 4 also contains the Fourier transforms of the sequences of length 1.

Columns 5, 6 and 7 are computed using the splitting formulas (7.5) and (7.7) with M = 2, 4 and 8, respectively. Columns 1 through 4 are called the sort phase of the FFT, and columns 5, 6, and 7 are called the combine phase. In Section 7.4 it is shown that sorting can be eliminated. This is important since sorting is expensive on a vector processor.

7.2 The loop inversion method for maintaining long vectors

One of the difficulties associated with computing the FFT on a vector processor is that the vectors get smaller as the computation proceeds. This is evident in Table 7 where the lengths of the sequences are halved from one column to the next until column 4, after which the lengths of the transforms double from column to column. The short vectors in the centre columns result in the inefficient use of a vector processor. This is also evident in the expository FORTRAN programs, given in Figure 7, for the combine phase of the Cooley-Tukey FFT. The sort phase is not presented since, as we show later, it can be eliminated.

```
      SUBROUTINE CTCMBN (IS,M,C)
C
C     COMBINE PHASE FOR THE COOLEY-TUKEY FFT
C
      COMPLEX C(1)
      N = 2**M
      DO 100 L=1,M
      LS = 2**(L-1)
      NS = N/(LS+LS)
      CALl CTCMB1 (IS,LS,NS,C)
100   CONTINUE
      RETURN
      END
      SUBROUTINE CTCMB1 (IS,LS,NS,C)
      COMPLEX OMEGA,OMEGK,WYK,C(LS,2,NS)
      ANGLE=FLOAT(IS)*4.*ATAN(1.)/FLOAT(LS)
      OMEGA=CMPLX(COS(ANGLE),SIN(ANGLE))
      DO 200 J=1,NS
      OMEGK=1.
      DO 200 I=1,LS
      WYK=OMEGK*C(I,2,J)
      C(I,1,J0=C(I,1,J)+WYK
      C(I,2,J)=C(I,2,J)-WYK
      OMEGK=OMEGA*OMEGK
200   CONTINUE
      RETURN
      END
```

Figure 7

In the program, the length of the original sequence is N and M is $\log_2 N$. IS=-1 produces a forward transform and IS=1 produces a backward transform. Loop 100 with index L computes the columns in the combine phase of the FFT. In Table 7, columns 5, 6, and 7 are computed with L = 1, 2, and 3, respectively. The variable 2LS is the length of the sequences in the Lth column, and NS contains the number of sequences. For example, when computing column 6 in Table 7, LS=2 and NS=2. The splitting formulas (7.5) and (7.7) are evaluated in loop 200.

For L=1 the inner loop has length LS=1 which significantly degrades the performance of a vector processor. This problem could be eliminated by using the Pease FFT, in which all vectors have length N/2. However, the Pease algorithm requires $N\log_2 N$ additional storage locations, which makes it unattractive for large N. A more attractive approach is to lengthen the vectors by simply inverting the order of the loop. That is, we replace loop 200 in Figure 7 by the FORTRAN code in Figure 8.

```
      IF (LS.LT.NS)  GO TO 300
      DO 200 J=1,NS
      OMEGK=1.
      DO 200 I=1,LS
      WYK=OMEGK*C(I,2,J)
      C(I,1,J)=C(I,1,J)+WYK
      C(I,2,J)=C(I,2,J)-WYK
      OMEGK=OMEGA*OMEGK
200   CONTINUE
      RETURN
300   OMEGK=1.
      DO 500 I=1,LS
      DO 400 J=1,NS
      WYK=OMEGK*C(I,2,J)
      C(I,1,J)=C(I,1,J)+WYK
      C(I,2,J)=C(I,2,J)-WYK
400   CONTINUE
      OMEGK=OMEGA*OMEGK
500   CONTINUE
```

Figure 8

Loop 200 is duplicated but with the order of the loops inverted and renumbered as loops 500 and 400. The IF statement ensures that longest vectors correspond to the inner loop. This straightforward modification can reduce computing time by 50% depending on N. The vector lengths still vary as the computation proceeds but this minimum length is now N/2. In practice the crossover point may not be LS = NS but rather a function of the architecture. For example, in loop 400 the data are not accessed with a stride of 1, which may create memory conflicts or require a scatter- gather operation.

7.3 The "truncated" FFT for multiple transforms

In this part we will continue the search for long vectors which is particularly important for a distributed FFT. Consider now the problem in which N sequences of length M are to be transformed. In Table 7 the two sequences in column 6 are Y_k and Z_k, defined in (7.3) and (7.4) which are the Fourier transforms on the two sequences X_{2m} and X_{2m+1}in column 2. More generally, the sequences in columns 5, 6, and 7 are the Fourier transforms of the sequences in columns 3, 2, and 1, respectively. If we choose to ignore columns 1 and 7 and assume that the two sequences in column 2 are given, then these transforms Y_k and Z_k are computed in column 6. Therefore columns 2 through 6 correspond to a multiple transform in which M=4 and N=2. Similarly, columns 3, 4, and 5 correspond to a multiple transform problem in which M=2 and N=4.

This result generalises to arbitrary M and N, and the resulting algorithm is called the truncated FFT. It is attractive on a vector processor since the minimum vector length is now $\sqrt{MN/2}$. A FORTRAN program for the truncated FFT is given in the next part. However, it is based on an algorithm for the FFT that does not require sorting.

7.4 Eliminating the sort phase in the FFT

Fortunately, sorting can be eliminated, which results in a significant improvement on a vector processor. There are several variants of the Cooley-Tukey FFT which differ only in the way that the intermediate computations are stored. These variants, together with the Cooley-Tukey FFT are discussed in detail in [13]. Of particular interest are the Stockham FFTs because they do not have a sort phase. The intermediate computations in the combine phase are stored in such a manner as to ensure that the final transform is ordered.

An expository FORTRAN program for the Stockham algorithm is given in Fig. 9. It differs from the Cooley-Tukey FFT in Fig. 7, only in the order of the subscripts in loop 200 and in its use of a work array. Unlike the Cooley-Tukey FFT, the Stockham FFT cannot be computed in place.

```
      SUBROUTINE STOCK(IS,M,C,WORK)
C
C     THE STOCKHAM AUTO-SORT FFT
C
      COMPLEX C(1), WORK(1)
      N = 2**M
      DO 100 L=1,M
      LS = 2**(L-1)
      NS = N/(LS+LS)
      CALL STOCK1(IS,LS,NS,C,WORK)
      DO 100 I=1,N
      C(I)=WORK(I)
  100 CONTINUE
      RETURN
      END
      SUBROUTINE STOCK1(IS,LS,NS,C,CH)
      COMPLEX OMEGA,OMEGK,WYK,C(NS,2,LS),CH(NS,LS,2)
      ANGLE=FLOAT(IS)*4.*ATAN(1.)/FLOAT(LS)
      OMEGA=CMPLX(COS(ANGLE),SIN(ANGLE))
      DO 200 J=1,NS
      OMEGK=1.
      DO 200 I=1,LS
      WYK=OMEGK*C(J,2,I)
      CH(J,I,1)=C(J,1,I)+WYK
      CH(J,I,2)=C(J,1,I)-WYK
      OMEGK=OMEGA*OMEGK
  200 CONTINUE
      RETURN
      END
```

Figure 9

The Stockham FFT is particularly suited for use with the truncated FFT. If the N sequences are stored in an array C(I,J), where I=1,...,M and J=1,...,N and C is dimensioned as C(M,N), then the truncated FFT computes the transforms in the J direction. That is, M transforms of length N are computed. This is particularly useful when the transforms are being performed as part of a solution to Poisson's equation. Since the FFTs are performed in the J direction, the solution of the tridiagonal systems can then be performed in the I direction where the vector elements are stored consecutively. An expository FORTRAN program for the truncated FFT is given in Figure 10.

```
      SUBROUTINE MSTOCK(IS,LI,LJ,C,WORK)
C
C     THE TRUNCATED STOCKHAM AUTOSORT FFT FOR MULTIPLE TRANSFORMS
C
C     DEFINE M=2**LI AND N=2**LJ THEN M TRANSFORMS OF LENGTH N
C     ARE COMPUTED.  THE TRANSFORMS ARE COMPUTED IN THE DIRECTION
C     OF THE FIRST INDEX OF THE TWO DIMENSIONAL ARRAY C.
C     THE FIRST DIMENSION OF C MUST BE EQUAL TO M.  THAT IS,
C     THE SEQUENCES MUST BE STORED CONSECUTIVELY.
C
C     IS = -1 FORWARD TRANSFORM
C     IS =  1 BACKWARD TRANSFORM
C
      COMPLEX C(1), WORK(1)
      N = 2**(LI+LJ)
      DO 100 L=1,LJ
      LS = 2**(L-1)
      NS = N/(LS+LS)
      CALL STOCKH(IS,LS,NS,C,WORK)
      DO 100 I=1,N
      C(I)=WORK(I)
100   CONTINUE
      RETURN
      END
      SUBROUTINE STOCKH(IS,LS,NS,C,CH)
      COMPLEX OMEGA,OMEGK,WYK,C(NS,2,LS),CH(NS,LS,2)
      ANGLE=FLOAT(IS)*4.*ATAN(1.)/FLOAT(LS)
      OMEGA=CMPLX(COS(ANGLE),SIN(ANGLE))
      DO 200 J=1,NS
      OMEGK=1.
      DO 200 I=1,LS
      WYK=OMEGK*C(J,2,I)
      CH(J,I,1)=C(J,1,I)+WYK
      CH(J,I,2)=C(J,1,I)-WYK
      OMEGK=OMEGA*OMEGK
200   CONTINUE
      RETURN
      END
```

Figure 10

7.5 The distributed FFT on the Cray X-MP/2

With the results presented above, the distribution of the FFT on a vector multiprocessor is straightforward. From Table 7 it would seem reasonable to partition the computations between the upper and lower halves, that is, to assign the computations in the upper and lower halves of Table 7 to processors 0 and 1, respectively. The processors would be synchronised at only one point, namely, between columns 6 and 7.

However, longer vectors are possible if the computations are distributed in the FORTRAN program for the truncated Stockham FFT given in Fig. 9. Although the processors must be synchronised between each column (or value of L) the long vectors should result in a net increase in speed. A FORTRAN program for the distributed computation is given in Appendix F.

This program differs from the single processor program in the following respects.

(i) The computations in loops 200 on the rectangle J=1, NS; I=1,LS are distributed between the processors. The half that corresponds to J=1,NS/2 is computed in subroutine STOCKH, and the other half for J=NS/2+1, NS, is computed in subroutine STOCKG. One processor executes subroutine MSTOCK, which calls STOCKH. The other processor executes subroutine TASK, which calls STOCKG.

(ii) The processors are synchronised using events following the computations in loops 200 in STOCKH and STOCKG, respectively.

(iii) To avoid the need for additional synchronisation points, the transfer from the WORK array to the C array has been eliminated by duplicating the CALL STOCKH and CALL STOCKG statements, with the order of the arguments C and WORK reversed.

It is important to note that the program is strictly expository. The loops 200 in STOCKH and STOCKG will not vectorise because of the statement OMEGK=OMEGA*OMEGK. Also, the loops 200 should be inverted as discussed in Section 7.2. The computations for the inverted loops would then be distributed on the index I. That is, the computations for I=1,LS/2 and I=LS/2+1, LS would be assigned to processors 0 and 1, respectively.

In addition, the CALL TSKSTART statement should be moved to the program that calls subroutine MSTOCK or to an initialising program. In practice the trigonometric computations for OMEGK are computed only once in a separate subroutine that is called the initialising program. The reason for this is that the trigonometric computations take a substantial amount of time, and they can be used repeatedly for subsequent transforms. The situation is similar for the CALL TSKSTART statement, which can require up to 4 milliseconds.

The task is started in the initialising program, and its computations are synchronised using the statements CALL EVPOST, CALL EVWAIT, and CALL EVCLEAR, whose combined time is about 4 microseconds. The use of these statements is illustrated in Appendix F.

The distribution of the FFT for a larger number of processors is somewhat more complicated because the number of processors is greater than the range NS on the index J. For example, when L=M then NS=2. If four processors are available, then the inner loop must also be distributed, which will result in shorter vectors. However, LS achieves its maximum value N/2 when L=M, so the length of the inner loop on I is N/4 which is still a relatively long vector and hence any reduction in performance should be minimal.

8 ACKNOWLEDGEMENT

The authors gratefully acknowledge the valuable assistance afforded them by colleagues and staff at Cray Research Inc., Control Data Corporation, Denelcor Inc. and ETA Corporation. In particular, the contributions made by D. Dent (ECMWF) and J. Larson (Cray Research Inc.) deserve mention.

REFERENCES

(1) CDC CYBER 200 Model 205 Computer System Hardware Reference Manual, Rev. A, Pub. No. 60256020, Control Data Corp., Minneapolis, Minnesota, 1981.

(2) CDC CYBER 200 FORTRAN Version 2 Reference Manual, Rev. B., Pub. No. 60485000, Control Data Corp., Minneapolis, Minnesota, 1981.

(3) T. Cheung and J.E. Smith, An analysis of the CRAY X-MP memory system, Proceedings of the 1984 International Conference on Parallel Processing, Bellaire, Michigan, IEEE Computer Society, Los Angeles, California 1984.

(4) Cray Research Multitasking User Guide, CRAY Computer Systems Technical Note No. SN-0222, Cray Research, Inc., Mendota Heights, Minnesota, 1984.

(5) R. Hockney and C. Jesshope, Parallel Computers, Adam Hilger Ltd., Bristol, England, 1981.

(6) K. Hwang and F.A. Briggs, Computer Architecture and Parallel Processing, McGraw-Hill, New York, 1984.

(7) J. Larson, "Multitasking on the CRAY X-MP/2 Multiprocessor", Computer, 17 (1984), pp. 62-69.

(8) R. Sadourney, The dynamics of finite difference models of the shallow water equations, J. Atmos. Sciences, 32 (1975), pp. 680-689

(9) A.H. Sameh and D. Kuck, On stable parallel linear system solvers, JACM, 25 (1978), pp. 80-91.

(10) A.H. Sameh, A fast Poisson solver for multiprocessors, Elliptic Problem Solvers II, Academic Press Inc., New York, 1984.

(11) Paul N. Swarztrauber, The methods of cyclic reduction, Fourier analysis and the FACR algorithm for the discrete solution of Poisson's equation on a rectangle, SIAM Rev., 19 (1977), pp.490-501.

(12) -------------------, Fast Poisson Solvers,MAA Studies in Numerical Analysis, Vol.24, Mathematical Association of America, 1984.

(13) -------------------, FFT algorithms for vector computers, Parallel Computing, 1(1984), pp.45-63.

APPENDIX A

Single processor CRAY-1 version of shallow water model

```
      PROGRAM SHALOW
C
C     BENCHMARK WEATHER PREDICTION PROGRAM FOR COMPARING THE
C     PERFORMANCE OF CURRENT SUPERCOMPUTERS. THE MODEL IS
C     BASED OF THE PAPER - THE DYNAMICS OF FINITE-DIFFERENCE
C     MODELS OF THE SHALLOW-WATER EQUATIONS, BY ROBERT SADOURNY
C     J. ATM. SCIENCES, VOL 32, NO 4, APRIL 1975.
C
C     CODE BY PAL N. SWARZTRAUBER, NATIONAL CENTER FOR
C     ATMOSPHERIC RESEARCH, BOULDER, CO, OCTOBER 1984.
C
      DIMENSION U(65,65),V(65,65),P(65,65),UNEW(65,65),VNEW(65,65),
     1          PNEW(65,65),UOLD(65,65),VOLD(65,65),POLD(65,65),
     2          CU(65,65),CV(65,65),Z(65,65),H(65,65),PSI(65,65)
      REAL MFS100,MFS200,MFS300
C
C     NOTE BELOW THAT TWO DELTA T (TDT) IS SET TO DT ON THE FIRST
C     CYCLE AFTER WHICH IT IS RESET TO DT+DT
C
      DT = 90.
      TDT = DT
C
      DX = 1.E5
      DY = 1.E5
      A = 1.E6
      ALPHA = .001
      ITMAX = 1200
      MPRINT = 120
      M = 64
      N = 64
      MP1 = M+1
      NP1 = N+1
      EL = FLOAT(N)*DX
      PI = 4.*ATAN(1.)
      TPI = PI+PI
      DI = TPI/FLOAT(M)
      DJ = TPI/FLOAT(N)
      PCF = PI*PI*A*A/(EL*EL)
C
C     INITIAL VALUES OF THE STREAM FUNCTION AND P
C
      DO 50 J=1,NP1
      DO 50 I=1,MP1
      PSI(I,J) = A*SIN((FLOAT(I)-.5)*DI)*SIN((FLOAT(J)-.5)*DJ)
      P(I,J) = PCF*(COS(2.*FLOAT(I-1)*DI)
     1                 +COS(2.*FLOAT(J-1)*DJ))+50000.
   50 CONTINUE
C
C     INITIALIZE VELOCITIES
C
      DO 60 J=1,N
      DO 60 I=1,M
      U(I+1,J) = -(PSI(I+1,J+1)-PSI(I+1,J))/DY
```

```
      V(I,J+1) = (PSI(I+1,J+1)-PSI(I,J+1))/DX
   60 CONTINUE
C
C     PERIODIC CONTINUATION
      DO 70 J=1,N
      U(1,J) = U(M+1,J)
      V(M+1,J+1) = V(1,J+1)
   70 CONTINUE
      DO 75 I=1,M
      U(I+1,N+1) = U(I+1,1)
      V(I,1) = V(I,N+1)
   75 CONTINUE
      U(1,N+1) = U(M+1,1)
      V(M+1,1) = V(1,N+1)
      DO 86 J=1,NP1
      DO 86 I=1,MP1
      UOLD(I,J) = U(I,J)
      VOLD(I,J) = V(I,J)
      POLD(I,J) = P(I,J)
   86 CONTINUE
C
C     PRINT INITIAL VALUES
C
      WRITE(6,390) N,M,DX,DY,DT,ALPHA
  390 FORMAT(*1NUMBER OF POINTS IN THE X DIRECTION* I8/
     1       * NUMBER OF POINTS IN THE Y DIRECTION* I8/
     2       * GRID SPACING IN THE X DIRECTION    * F8.0/
     3       * GRID SPACING IN THE Y DIRECTION    * F8.0/
     4       * TIME STEP                          * F8.0/
     5       * TIME FILTER PARAMETER              * F8.3)
      MNMIN = MIN0(M,N)
      WRITE(6,391) (POLD(I,I),I=1,MNMIN)
  391 FORMAT(/* INITIAL DIAGONAL ELEMENTS OF P * //(8E15.6))
      WRITE(6,392) (UOLD(I,I),I=1,MNMIN)
  392 FORMAT(/* INITIAL DIAGONAL ELEMENTS OF U * //(8E15.5))
      WRITE(6,393) (VOLD(I,I),I=1,MNMIN)
  393 FORMAT(/* INITIAL DIAGONAL ELEMENTS OF V * //(8E15.6))
      TSTART = 9.5E-9*RTC(DUM)
      T300 = 1.
      TCYCH = 0.
      TIME = 0.
      NCYCLE = 0
   90 NCYCLE = NCYCLE + 1
      TCYC = 9.5E-9*RTC(DUM)
C
C     COMPUTE CAPITAL U, CAPATIL V, Z AND H
C
      FSDX = 4./DX
      FSDY = 4./DY
      T100 = 9.5E-9*RTC(DUM)
      DO 100 J=1,N
      DO 100 I=1,M
      CU(I+1,J) = .5*(P(I+1,J)+P(I,J))*U(I+1,J)
      CV(I,J+1) = .5*(P(I,J+1)+P(I,J))*V(I,J+1)
```

```
      Z(I+1,J+1) = (FSDX*(V(I+1,J+1)-V(I,J+1))-FSDY*(U(I+1,J+1)
     1          -U(I+1,J)))/(P(I,J)+P(I+1,J)+P(I+1,J+1)+P(I,J+1))
      H(I,J) = P(I,J)+.25*(U(I+1,J)*U(I+1,J)+U(I,J)*U(I,J)
     1                +V(I,J+1)*V(I,J+1)+V(I,J)*V(I,J))
  100 CONTINUE
      T100 = 9.5E-9*RTC(DUM)-T100
C
C     PERIODIC CONTINUATION
C
      DO 110 J=1,N
      CU(1,J) = CU(M+1,J)
      CV(M+1,J+1) = CV(1,J+1)
      Z(1,J+1) = Z(M+1,J+1)
      H(M+1,J) = H(1,J)
  110 CONTINUE
      DO 115 I=1,M
      CU(I+1,N+1) = CU(I+1,1)
      CV(I,1) = CV(I,N+1)
      Z(I+1,1) = Z(I+1,N+1)
      H(I,N+1) = H(I,1)
  115 CONTINUE
      CU(1,N+1) = CU(M+1,1)
      CV(M+1,1) = CV(1,N+1)
      Z(1,1) = Z(M+1,N+1)
      H(M+1,N+1) = H(1,1)
C
C     COMPUTE NEW VALUES U,V AND P
C
      TDTS8 = TDT/8.
      TDTSDX = TDT/DX
      TDTSDY = TDT/DY
      T200 = 9.5E-9*RTC(DUM)
      DO 200 J=1,N
      DO 200 I=1,M
      UNEW(I+1,J) = UOLD(I+1,J)+
     1       TDTS8*(Z(I+1,J+1)+Z(I+1,J))*(CV(I+1,J+1)+CV(I,J+1)+CV(I,J)

     2       +CV(I+1,J))-TDTSDX*(H(I+1,J)-H(I,J))
      VNEW(I,J+1) = VOLD(I,J+1)-TDTS8*(Z(I+1,J+1)+Z(I,J+1))
     1       *(CU(I+1,J+1)+CU(I,J+1)+CU(I,J)+CU(I+1,J))
     2       -TDTSDY*(H(I,J+1)-H(I,J))
      PNEW(I,J) = POLD(I,J)-TDTSDX*(CU(I+1,J)-CU(I,J))
     1       -TDTSDY*(CV(I,J+1)-CV(I,J))
  200 CONTINUE
      T200 = 9.5E-9*RTC(DUM)-T200
C
C     PERIODIC CONTINUATION
C
      DO 210 J=1,N
      UNEW(1,J) = UNEW(M+1,J)
      VNEW(M+1,J+1) = VNEW(1,J+1)
      PNEW(M+1,J) = PNEW(1,J)
  210 CONTINUE
      DO 215 I=1,M
      UNEW(I+1,N+1) = UNEW(I+1,1)
```

```
      VNEW(I,1) = VNEW(I,N+1)
      PNEW(I,N+1) = PNEW(I,1)
  215 CONTINUE
      UNEW(1,N+1) = UNEW(M+1,1)
      VNEW(M+1,1) = VNEW(1,N+1)
      PNEW(M+1,N+1) = PNEW(1,1)
      IF(NCYCLE .GT. ITMAX) CALL EXIT
      TIME = TIME + DT
      TCYC = 9.5E-9*RTC(DUM) - TCYC
      IF(MOD(NCYCLE,MPRINT) .NE. 0) GO TO 370
      PTIME = TIME/3600.
      WRITE(6,350) NCYCLE,PTIME
  350 FORMAT(//* CYCLE NUMBER*I5* MODEL TIME IN  HOURS* F6.2)
      WRITE(6,355) (PNEW(I,I),I=1,MNMIN)
  355 FORMAT(/*DIAGONAL ELEMENTS OF P * //(8E15.6))
      WRITE(6,360) (UNEW(I,I),I=1,MNMIN)
  360 FORMAT(/* DIAGONAL ELEMENTS OF U * //(8E15.6))
      WRITE(6,365) (VNEW(I,I),I=1,MNMIN)
  365 FORMAT(/* DIAGONAL ELEMENTS OF V * //(8E15.6))
      MFS100 = 24.*M*N/T100/1.E6
      MFS200 = 26.*M*N/T200/1.E6
      MFS300 = 15.*M*N/T300/1.E6
      CTIME = 9.5E-9*RTC(DUM)-TSTART
      TCYC = CTIME/FLOAT(NCYCLE)
      WRITE(6,375)
      NCYCLE,CTIME,TCYCH,T100,MFS100,T200,MFS200,T300,
     1              MFS300
  375 FORMAT(* CYCLE NUMBER*I5* TOTAL COMPTER TIME* E15.6
     1         * TIME PER CYCLE* E15.6 /
     2         * TIME AND MEGAFLOPS FOR LOOP 100 * E15.6,F6.1/
     3         * TIME AND MEGAFLOPS FOR LOOP 200 * E15.6,F6.1/
     4         * TIME AND MEGAFLOPS FOR LOOP 300 * E15.6,F6.1/ )
  370 IF(NCYCLE .LE. 1) GO TO 310
      CTCYC = 9.5E-9*RTC(DUM)
      T300 = 9.5E-9*RTC(DUM)
      DO 300 J=1,N
      DO 300 I=1,M
      UOLD(I,J) = U(I,J)+ALPHA*(UNEW(I,J)-2.*U(I,J)+UOLD(I,J))
      VOLD(I,J) = V(I,J)+ALPHA*(VNEW(I,J)-2.*V(I,J)+VOLD(I,J))
      POLD(I,J) = P(I,J)+ALPHA*(PNEW(I,J)-2.*P(I,J)+POLD(I,J))
      U(I,J) = UNEW(I,J)
      V(I,J) = VNEW(I,J)
      P(I,J) = PNEW(I,J)
  300 CONTINUE
      T300 = 9.5E-9*RTC(DUM)-T300
C
C     PERIODIC CONTINUATION
C
      DO 320 J=1,N
      UOLD(M+1,J) = UOLD(1,J)
      VOLD(M+1,J) = VOLD(1,J)
      POLD(M+1,J) = POLD(1,J)
      U(M+1,J) = U(1,J)
      V(M+1,J) = V(1,J)
      P(M+1,J) = P(1,J)
```

```
320 CONTINUE
    DO 325 I=1,M
    UOLD(I,N+1) = UOLD(I,1)
    VOLD(I,N+1) = VOLD(I,1)
    POLD(I,N+1) = POLD(I,1)
    U(I,N+1) = U(I,1)
    V(I,N+1) = V(I,1)
    P(I,N+1) = P(I,1)
325 CONTINUE
    UOLD(M+1,N+1) = UOLD(1,1)
    VOLD(M+1,N+1) = VOLD(1,1)
    POLD(M+1,N+1) = POLD(1,1)
    U(M+1,N+1) = U(1,1)
    V(M+1,N+1) = V(1,1)
    P(M+1,N+1) = P(1,1)
    TCYC = TCYC + 9.5E-9*RTC(DUM) - CTCYC
    TCYCH = TCYC
    GO TO 90
310 TDT = TDT+TDT
    DO 400 J=1,NP1
    DO 400 I=1,MP1
    UOLD(I,J) = U(I,J)
    VOLD(I,J) = V(I,J)
    POLD(I,J) = P(I,J)
    U(I,J) = UNEW(I,J)
    V(I,J) = VNEW(I,J)
    P(I,J) = PNEW(I,J)
400 CONTINUE
    GO TO 90
    END
```

APPENDIX B

Cyber 205 version of shallow water model

```
      PROGRAM SHALOW(INPUT,OUTPUT,TAPE5=INPUT,TAPE6=OUTPUT)
C
C     BENCHMARK WEATHER PREDICTION PROGRAM FOR COMPARING  THE
C     PERFORMANCE OF CURRENT SUPERCOMPUTERS. THE MODEL IS
C     BASED ON THE PAPER - THE DYNAMICS OF FINITE-DIFFERENCE
C     MODELS OF THE SHALLOW-WATER EQUATIONS, BY ROBERT SADOURNY
C     J. ATM. SCIENCES, VOL 32, NO 4, APRIL 1975.
C
C     CODE BY PAUL N. SWARZTRAUBER, NATIONAL CENTER FOR
C     ATMOSPHERIC RESEARCH, BOULDER, CO, OCTOBER 1984.
C
C     CONVERSION TO CYBER 205 VECTOR FORTRAN BY ROLAND SWEET,
C     NATIONAL BUREAU OF STANDARDS, BOULDER, CO. OCTOBER 1984.
C
      DIMENSION U(65,65),V(65,65),P(65,65),UNEW(65,65),VNEW(65,65),
     1          PNEW(65,65),UOLD(65,65),VOLD(65,65),POLD(65,65),
     2          CU(65,65),CV(65,65),Z(65,65),H(65,65),PSI(65,65)
      REAL MFS100,MFS200,MFS300
C
C     CYBER 205 DECLARATION OF BIT VECTOR FOR CONTROLLED STORE
C     OPERATION
C
C     BIT INTER((M+1)*N-1)
C
      BIT INTER(4159)
C
C     NOTE BELOW THAT TWO DELTA T (TDT) IS SET TO DT ON THE FIRST
C     CYCLE AFTER WHICH IT IS RESET TO DT+DT.
C
      DT = 90.
      TDT = DT
C
      DX = 1.E5
      DY = 1.E5
      A = 1.E6
      ALPHA = .001
      ITMAX = 1200
      MPRINT = 120
      M = 64
      N = 64
      MP1 = M+1
      NP1 = N+1
C
C     DEFINE CYBER 205 BIT VECTOR TO STORE ONLY IN THE SUB-ARRAY
C     I=1,2,...,M, J=1,2,...,N OF THE ARRAY DEFINED FOR
C     I=1,2,...,M,M+1, J=1,2,...,N,N+1.
C
      LV = MP1*N-1
      INTER(1;LV) = Q8VMK0(M,MP1;INTER(1;LV))
      L = MP1*NP1
C
```

```
      EL = FLOAT(N)*DX
      PI = 4.ATAN(1.)
      TPI = PI+PI
      DI = TPI/FLOAT(M)
      DJ = TPI/FLOAT(N)
      PCF = PI*PI*A*A/(EL*EL)
C
C     INITIAL VALUES OF THE STREAM FUNCTION AND P
C
      DO 50 J=1,NP1
      DO 50 I=1,MP1
      PSI(I,J) = A*SIN((FLOAT(I)-.5)*DI)*SIN((FLOAT(J)-.5)*DJ)
      P(I,J) = PCF*(COS(2.*FLOAT(I-1)*DI)
     1                +COS(2.*FLOAT(J-1)*DJ))+50000.
   50 CONTINUE
C
C     INITIALIZE VELOCITIES
C
      DO 60 J=1,N
      DO 60 I=1,M
      U(I+1,J) = -(PSI(I+1,J+1)-PSI(I+1,J))/DY
      V(I,J+1) =  (PSI(I+1,J+1)-PSI(I,J+1))/DX
   60 CONTINUE
C
C     PERIODIC CONTINUATION
C
      DO 70 J=1,N
      U(1,J) = U(M+1,J)
      V(M+1,J+1) = V(1,J+1)
   70 CONTINUE
      DO 75 I=1,M
      U(I+1,N+1) = U(I+1,1)
      V(I,1) = V(I,N+1)
   75 CONTINUE
      U(1,N+1) = U(M+1,1)
      V(M+1,1) = V(1,N+1)
      DO 86 J=1,NP1
      DO 86 I=1,MP1
      UOLD(I,J) = U(I,J)
      VOLD(I,J) = V(I,J)
      POLD(I,J) = P(I,J)
   86 CONTINUE
C
C     PRINT INITIAL VALUES
C
      WRITE(6,390) N,M,DX,DY,DT,ALPHA
  390 FORMAT(*1NUMBER OF POINTS IN THE X DIRECTION* I8/
     1       * NUMBER OF POINTS IN THE Y DIRECTION* I8/
     2       * GRID SPACING IN THE X DIRECTION    * F8.0/
     3       * GRID SPACING IN THE Y DIRECTION    * F8.0/
     4       * TIME STEP                          * F8.0/
     5       * TIME FILTER PARAMTER               * F8.3)
      MNMIN = MIN0(M,N)
      WRITE(6,391) (POLD(I,I),I=1,MNMIN)
```

```
  391 FORMAT(/* INITIAL DIAGONAL ELEMENTS OF P * //(4E25.13))
      WRITE(6,392) (UOLD(I,I),I=1,MNMIN)
  392 FORMAT(/* INITIAL DIAGONAL ELEMENTS OF U * //(4E25.13))
      WRITE(6.393) (VOLD(I,I),I=1,MNMIN)
  393 FORMAT(/* INITIAL DIAGONAL ELEMENTS OF V * //(4E25.13))
      TSTART = SECOND(DUM)
      T300 = 1.
      TIME = 0.
      NCYCLE = 0
   90 NCYCLE = NCYCLE + 1
C
C     COMPUTE CAPITAL U, CAPITAL V, Z AND H
C
      FSDX = 4./DX
      FSDY = 4./DY
      T100 = SECOND(DUM)
C
C     CYBER 205 VERSION OF DO 100 LOOP
C
      WHERE (INTER(1;LV))
         CU(2,1;LV) = .5*(P(2,1;LV)+P(1,1;LV))*U(2,1;LV)
         CV(1,2;LV) = .5*(P(1,2;
         Z(2,2;LV)  = (FSDX*(V(2,2;LV)-V(1,2;LV))-
     *                 FSDY*(U(2,2;LV)-U(2,1;LV)))/
     *                (P(1,1;LV)+P(1,2;LV)+P(2,2;LV)+P(2,1;LV))
         H(1,1;LV)  = P(1,1;LV)+.25*(U(2,1;LV)*U(2,1;LV)
     *                +U(1,1;LV)*U(1,1;LV)+V(1,2;LV)*V(1,2;LV)
                     +V(1,1;LV)*V(1,1;LV))
      END WHERE
C
      T100 = SECOND(DUM)-T100
C
C     PERIODIC CONTINUATION
C
      DO 110 J=1,N
      CU(1,J) = CU(M+1,J)
      CV(M+1,J+1) = CV(1,J+1)
      Z(1,J+1) = Z(M+1,J+1)
      H(M+1,J) = H(1,J)
  110 CONTINUE
C
C     CYBER 205 VERSION OF DO 115 LOOP
C
      CU(2,NP1;M) = CU(2,1;M)
      CV(1,1;M) = CV(1,NP1;M)
      Z(2,1;M) = Z(2,NP1;M)
      H(1,NP1;) = H(1,1;M)
C
      CU(1,N+1) = CU(M+1,1)
      CV(M+1,1) = CV(1,N+1)
      Z(1,1) = Z(M+1,N+1)
      H(M+1,N+1) = H(1,1)
C
```

```
C     COMPUTE NEW VALUES U,V AND P
C
      TDTS8 = TDT/8.
      TDTSDX = TDT/DX
      TDTSDY = TDT/DY
      T200 = SECOND(DUM)
C
C     CYBER 205 VERSION OF DO 200 LOOP
C
      WHERE (INTER(1;LV))
         UNEW(2,1;LV) = UOLD(2,1;LV)+TDTS8*(Z(2,2;LV)+Z(2,1;LV))*
     *   (CV(2,2;LV)+CV(1,2;LV)+CV(1,1;LV)+CV(2,1;LV))
     *                  -TDTSDX*(H(2,1;LV)-H(1,1;LV))
         VNEW(1,2;LV) = VOLD(1,2;LV)-TDTS8*(Z(2,2;LV)+Z(1,2;LV))*
     *   (CU(2,2;LV)+CU(2,1;LV)+CU(1,1;LV)+CU(1,2;LV))
                        -TDTSDY*(H(1,2;LV)-H(1,1;LV))
         PNEW(1,1;LV) = POLD(1,1;LV)-TDTSDX*(CU(2,1;LV)-CU(1,1;LV))
     *                  -TDTSDY*(CV(1,2;LV)-CV(1,1;LV))
      END WHERE
C
      T200 = SECOND(DUM)-T200
C
C
C     PERIODIC CONTINUATION
C
      DO 210 J=1,N
      UNEW(1,J) = UNEW(M+1,J)
      VNEW(M+1,J+1) = VNEW(1,J+1)
      PNEW(M+1,J) = PNEW(1,J)
  210 CONTINUE
C
C     CYBER 205 VERSION OF DO 215 LOOP
C
      UNEW(2,NP1;M) = UNEW(2,1;M)
      VNEW(1,1;M) = VNEW(1,NP1;M)
      PNEW(1,NP1;M) = PNEW(1,1;M)
C
      UNEW(1,N+1) = UNEW(M+1,1)
      VNEW(M+1,1) = VNEW(1,N+1)
      PNEW(M+1,N+1) = PNEW(1,1)
      IF(NCYCLE .GT. ITMAX) STOP
      TIME = TIME + DT
      IF(MOD(NCYCLE,MPRINT) .NE. 0) GO TO 370
      PTIME = TIME/3600.
      WRITE(6,350) NCYCLE,PTIME
  350 FORMAT(//* CYCLE NUMBER*I5* MODEL TIME IN HOURS* F6.2)
      WRITE(6,355) (PNEW(I,I),I=1,MNMIN)
  355 FORMAT(/* DIAGONAL ELEMENTS OF P * //(4E25.13))
      WRITE(6,360) (UNEW(I,I),I=1,MNMIN)
  360 FORMAT(/* DIAGONAL ELEMENTS OF U * //(4E25.13))
      WRITE(6,365) (VNEW(I,I),I=1,MNMIN)
```

```
  365 FORMAT(/* DIAGONAL ELEMENTS OF V * //(4E25.13))
      MFS100 = 24.*((M+1)*N-1)/T100/1.E6
      MFS200 = 26.*((M+1)*N-1)/T200/1.E6
      MFS300 = 15.*((M+1)*N-1)/T300/1.E6
      CTIME = SECOND(DUM)-TSTART
      TCYC = CTIME/FLOAT(NCYCLE)
      WRITE(6,375) NCYCLE,CTIME,TCYC,T100,MFS100,T200,MFS200,T300,MFS300
  375 FORMAT(* CYCLE NUMBER*I5* TOTAL COMPUTER TIME* E15.6
     1       * TIME PER CYCLE* E15.6 /
     2       * TIME AND MEGAFLOPS FOR LOOP 100 * E15.6,F6.1/
     3       * TIME AND MEGAFLOPS FOR LOOP 200 * E15.6,F6.1/
     4       * TIME AND MEGAFLOPS FOR LOOP 300 * E15.6,F6.1/ )
C
C     TIME SMOOTHING AND UPDATE FOR NEXT CYCLE
C
  370 IF(NCYCLE .LE. 1) GO TO 310
      T300 = SECOND(DUM)
C
C     CYBER 205 VERSION OF DO 300 LOOP
C
      UOLD(1,1;L) = U(1,1;L)+ALPHA*(UNEW(1,1;L)-2.*U(1,1;L)+UOLD(1,1;L))
      VOLD(1,1;L) = V(1,1;L)+ALPHA*(VNEW(1,1;L)-2.*V(1,1;L)+VOLD(1,1;L))
      POLD(1,1;L) = P(1,1;L)+ALPHA*(PNEW(1,1;L)-2.*P(1,1;L)+POLD(1,1;L))
      U(1,1;L) = UNEW(1,1;L)
      V(1,1;L) = VNEW(1,1;L)
      P(1,1;L) = PNEW(1,1;L)
C
      T300 = SECOND(DUM)-T300
C
C     PERIODIC CONTINUATION
      DO 320 J=1,N
      UOLD(M+1,J) = UOLD(1,J)
      VOLD(M+1,J) = VOLD(1,J)
      POLD(M+1,J) = POLD(1,J)
      U(M+1,J) = U(1,J)
      V(M+1,J) = V(1,J)
      P(M+1,J) = P(1,J)
  320 CONTINUE
C
C     CYBER 205 VERSION OF DO 325 LOOP
C
      UOLD(1,NP1;M) = UOLD(1,1;M)
      VOLD(1,NP1;M) = VOLD(1,1;M)
      POLD(1,NP1;M) = POLD(1,1;M)
      U(1,NP1;M) = U(1,1;M)
      V(1,NP1;M) = V(1,1;M)
      P(1,NP1;M) = P(1,1;M)
C
      UOLD(M+1,N+1) = UOLD(1,1)
      VOLD(M+1,N+1) = VOLD(1,1)
      POLD(M+1,N+1) = POLD(1,1)
      U(M+1,N+1) = U(1,1)
      V(M+1,N+1) = V(1,1)
      P(M+1,N+1) = P(1,1)
      GO TO 90
```

```
  310 TDT = TDT+TDT
C
C     CYBER 205 VERSION OF DO 400 LOOP
C
      UOLD(1,1;L)L = U(1,1;L)
      VOLD(1,1;L)L = V(1,1;L)
      POLD(1,1;L)L = P(1,1;L)
      U(1,1;L) = UNEW(1,1;L)
      V(1,1;L) = VNEW(1,1;L)
      P(1,1;L) = PNEW(1,1;L)
C
      GO TO 90
      END
```

APPENDIX C

HEP.1 version of shallow water model

```
      PROGRAM SHALLOWP
C
C     BENCHMARK WEATHER PREDICTION PROGRAM FOR COMPARING THE
C     PERFORMANCE OF CURRENT SUPERCOMPUTERS. THE MODEL IS
C     BASED ON THE PAPER - THE DYNAMICS OF FINITE-DIFFERENCE
C     MODELS OF THE SHALLOW-WATER EQUATIONS, BY ROBERT SADDURNY
C     J. ATM. SCIENCES VOL. 32,NO. 4 APRIL 1975.
C
C     CODE BY DAVID SNELLING, DENELCOR, DENVER, CO OCTOBER 1984.
C
C     PROGRAM PERFORMS NDIM**2 * (29+25) + (NDIM+1)**2 * 15 FLOATING
C     POINT OPERATIONS PER TIME STEP.
C
C     ON THE CRAY-1 AT NCAR FOR NDIM=64
C             RUN TIME PER TIME STEP = 0.00525 SECONDS.
C             MFLOPS                 = 54.2
C             MACHINE CYCLE TIME (MS) * MFLOPS = 0.677 (EFFICIENT)
C
C
      PARAMETER (NDIM = 64)
      PARAMETER (NDP1 = NDIM + 1)

C
      COMMON /DATA/ U(NDP1,NDP1),V(NDP1,NDP1),P(NDP1,NDP1),
     1UNEW(NDP1,NDP1),VNEW(NDP1,NDP1),PNEW(NDP1,NDP1),
     2UOLD(NDP1,NDP1),VOLD(NDP1,NDP1),POLD(NDP1,NDP1),
     3CU(NDP1,NDP1),CV(NDP1,NDP1),Z(NDP1,NDP1),H(NDP1,NDP1),
     4PSI(NDP1,NDP1)
C
      COMMON /PARA/ A,TDT,DX,DY,ALPHA,M,N,MP1,NP1,PI,TPI,DI,DJ
     1   ,ISTART,ITMAX,MPRINT,EL,PCF
C
      COMMON /SYNC/ KOUT,NPROC,DONE,IBARR(52),FINISH
C
      EXTERNAL PARSUB
C
C
```

```
C             INITIALIZE THE RUNTIME PARAMETERS.
C
      A        = 1.0 E 6
      TDT      = 90.0
      DX       = 1.0 E 5
      DY       = 1.0 E 5
      ALPHA    = 0.001
      ITMAX    = 1200
      M        = NDIM
      N        = NDIM
      MP1      = M + 1
      NP1      = N + 1
      PI       = 4.0 * ATAN(1.0)
      TPI      = 2.0 * PI
      DI       = TPI/FLOAT(M)
      DJ       = TPI/FLOAT(N)
      EL       = FLOAT (N) * DX
      PCF      = PI*PI*A*A/(EL*EL)
C
      WRITE (6,77)
   77 FORMAT (' ENTER NPROC AND MPRINT (32  120).')
      READ (5,1001) NPROC,MPRINT
 1001 FORMAT (I2,I5)
      WRITE (6,78) NPROC
   78 FORMAT(///,'NPROC = ',I5,///)
C
C             INITIALIZE DATA FOR PROBLEM.
C
      CALL INIT
C
      CALL PRINTIT (6)
C
      CALL INTIME
      CALL CLOCK (ISTART)
C
      CALL SETE (DONE)
      CALL SETE (KOUNT)
      IBARR (1) = NPROC
      IBARR (2) = 0
      FINISH = 0
C
      DO 10 I=1,NPROC-1
      CALL AWRITE (KOUNT,I)
      CALL CREATE (PARSUB)
   10 CONTINUE
      CALL AWRITE (KOUNT,NPROC)
      CALL PARSUB
C
      CALL WAITF (DONE)
C
      CALL OUTIME (6)
      STOP
      END
      SUBROUTINE PARSUB
```

```
C
      PARAMETER (NDIM = 64)
      PARAMETER (NDP1 = NDIM + 1)
C
      COMMON /DATA/ U(NDP1,NDP1),V(NDP1,NDP1),P(NDP1,NDP1),
     1UNEW(NDP1,NDP1),VNEW(NDP1,NDP1),PNEW(NDP1,NDP1),
     2UOLD(NDP1,NDP1),VOLD(NDP1,NDP1),POLD(NDP1,NDP1),
     3CU(NDP1,NDP1),CV(NDP1,NDP1),Z(NDP1,NDP1),H(NDP1,NDP1),
     4PSI(NDP1,NDP1)
C
      COMMON /PARA/ A,TDT,DX,DY,ALPHA,M,N,MP1,NP1,PI,TPI,DI,DJ
     1   ,ISTART,ITMAX,MPRINT,EL,PCF
C
      COMMON /SYNC/ KOUNT,NPROC,DONE,IBARR(52),FINISH
C
C
      IPROC = IAREAD(KOUNT)
      NCYCLE = 0
C
C             TOP OF LOOP.
C
   90 CONTINUE

      NCYCLE = NCYCLE + 1
C
C              COMPUTE CAPITAL U, CAPITAL V, Z, AND H.
C
      CALL CAPUV (IPROC,NPROC)
C
      CALL BARRIER (IBARR)
C
C              PERIODIC CONTINUATION.
C
      CALL CONCAPS (IPROC,NPROC)
C
      CALL BARRIER (IBARR)
C
C              COMPUTE NEW U, V, AND P.
C
      CALL NEWUVP (IPROC,NPROC)
C
      CALL BARRIER (IBARR)
C
C              PERIODIC CONTINUATION.
C
      CALL CONNEW(IPROC,NPROC)
C
      IF (IPROC .EQ. 1) THEN
C
C             TEST FOR EXIT AND PRINT INTERMEDIATE RESULTS.
```

```
C
      IF (NCYCLE .GT. ITMAX) FINISH = 1
      CALL CLOCK (ITIME)
      TIME = 1.0E-7 * FLOAT(ITIME-ISTART)
      TYC = TIME/FLOAT(NCYCLE)
      IF (MOD(NCYCLE,MPRINT) .EQ. 0) WRITE (6,1000) NCYCLE,TIME,TCYC
 1000 FORMAT (' NCYCLE = ',I5,' TOTAL TIME =',E12.4,
     1  ' TIME PER CYCLE =',E12.4)
      IF (FINISH .EQ. 1) CALL AWRITE (DONE,1.)
      ENDIF
C
C              TIME SMOOTHING AND UPDATE FOR NEXT CYCLE.
C
      CALL BARRIER (IBARR)
C
      IF (FINISH .EQ. 1) GOTO 999
C
      IF (NCYCLE .GT. 1) THEN
              CALL SMOOTH (IPROC,NPROC)
              CALL BARRIER (IBARR)
              CALL CONSMO (IPROC,NPROC)
      ELSE
              IF (IPROC .EQ. 1) TDT = 2.0 * TDT
              DO 400 J=IPROC,NP1,NPROC
              DO 400 I=1,MP1
              UOLD(I,J) = U(I,J)
              VOLD(I,J) = V(I,J)
              POLD(I,J) = P(I,J)
              U(I,J) = UNEW(I,J)
              V(I,J) = VNEW(I,J)
              P(I,J) = PNEW(I,J)
  400         CONTINUE
      ENDIF
C
      IF (MOD(NCYCLE,MPRINT) .EG. 0) THEN
              CALL BARRIER (IBARR)
              IF (IPROC .EQ. 1) CALL PRINTIT(6)
      ENDIF
      CALL BARRIER (IBARR)
C
      GOTO 90
  999 CONTINUE
      RETURN
      END
C
      SUBROUTINE INIT
C
      PARAMETER (NDIM = 64)
      PARAMETER (NDP1 = NDIM + 1)
C
      COMMON /DATA/ U(NDP1,NDP1),V(NDP1,NDP1),P(NDP1,NDP1),

     1UNEW(NDP1,NDP1),VNEW(NDP1,NDP1),PNEW(NDP1,NDP1),
     2UOLD(NDP1,NDP1),VOLD(NDP1,NDP1),POLD(NDP1,NDP1),
     3CU(NDP1,NDP1),CV(NDP1,NDP1),Z(NDP1,NDP1),H(NDP1,NDP1),
     4PSI(NDP1,NDP1)
```

```
C
      COMMON /PARA/ A,TDT,DX,DY,ALPHA,M,N,MP1,NP1,PI,TPI,DI,DJ
     1   ,ISTART,ITMAX,MPRINT,EL,PCF
C
C             INITIALIZE VALUES OF THE STREAM FUNCTION.
C
      DO 50 J=1,NP1
      DO 50 I=1,MP1
      PSI(I,J) = A*SIN((FLOAT(I)-.5)*DI) * SIN((FLOAT(J)-.5)*DJ)
      P(I,J)=PCF*(COS(2.0*FLOAT(I-1)*DI)+COS(2.0*FLOAT(J-1)*DJ))
     1                  + 50000.0
   50 CONTINUE
C
C             INITIALIZE VELOCITIES.
C
      DO 60 J=1,N
      DO 60 I=1,M
      U(I+1,J) = -(PSI(I+1,J+1)-PSI(I+1,J))/DY
      V(I,J+1) = (PSI(I+1,J+1)-PSI(I,J+1))/DX
   60 CONTINUE
C
C             PERIODIC CONTINUATION.
C
      DO 70 J=1,N
      U(1,J) = U(M+1,J)
      V(M+1,J+1) = V(1,J+1)
   70 CONTINUE
C
      DO 80 I=1,M
      U(I+1,N+1) = U(I+1,1)
      V(I,1) = V(I,N+1)
   80 CONTINUE
C
      U(1,N+1) = U(M+1,1)
      V(M+1,1) = V(1,N+1)
C
C             INITIALIZE OLD VALUES.
C
      DO 86 J=1,NP1
      DO 86 I=1,MP1
      UOLD(I,J) = U(I,J)
      VOLD(I,J) = V(I,J)
      POLD(I,J) = P(I,J)
   86 CONTINUE
C
      RETURN
      END
      SUBROUTINE CAPUV (IPROC,NPROC)
C
      PARAMETER (NDIM = 64)
      PARAMETER (NDP1 = NDIM + 1)
C
      COMMON /DATA/ U(NDP1,NDP1),V(NDP1,NDP1),P(NDP1,NDP1),
     1UNEW(NDP1,NDP1),VNEW(NDP1,NDP1),PNEW(NDP1,NDP1),
     2UOLD(NDP1,NDP1),VOLD(NDP1,NDP1),POLD(NDP1,NDP1),
     3CU(NDP1,NDP1),CV(NDP1,NDP1),Z(NDP1,NDP1),H(NDP1,NDP1),
     4PSI(NDP1,NDP1)
```

```
C
      COMMON /PARA/ A,TDT,DX,DY,ALPHA,M,N,MP1,NP1,PI,TPI,DI,DJ
     1  ,ISTART,ITMAX,MPRINT,EL,PCF
C

      DO 100 J=IPROC,N,NPROC
      DO 100 I=1,M
      CU(I+1,J) = 0.5 * (P(I+1,J)+P(I,J))*U(I+1,J)
      CV(I,J+1) = 0.5 * (P(I,J+1)+P(I,J))*V(I,J+1)
      Z(I+1,J+1) = 4.0 * ((V(I+1,J+1)-V(I,J+1))/DX-(U(I+1,J+1)
     1    -U(I+1,J)).DY) / (P(I,J)+P(I,J+1)+P(I+1,J)+P(I+1,J+1))
      H(I,J) = P(I,J)+0.25*(U(I+1,J)**2+U(I,J)**2+V(I,J+1)**2+V(I,J)**2)
  100 CONTINUE
C
      RETURN
      END

      SUBROUTINE CONCAPS (IPROC,NPROC)
C
      PARAMETER (NDIM = 64)
      PARAMETER (NDP1 = NDIM + 1)
C
      COMMON /DATA/ U(NDP1,NDP1),V(NDP1,NDP1),P(NDP1,NDP1),
     1UNEW(NDP1,NDP1),VNEW(NDP1,NDP1),PNEW(NDP1,NDP1),
     2UOLD(NDP1,NDP1),VOLD(NDP1,NDP1),POLD(NDP1,NDP1),
     3CU(NDP1,NDP1),CV(NDP1,NDP1),Z(NDP1,NDP1),H(NDP1,NDP1),
     4PSI(NDP1,NDP1)
C
      COMMON /PARA/ A,TDT,DX,DY,ALPHA,M,N,MP1,NP1,PI,TPI,DI,DJ
     1  ,ISTART,ITMAX,MPRINT,EL,PCF
C

      DO 110 J=IPROC,N,NPROC
      CU(1,J) = CU(M+1,J)
      CV(M+1,J+1) = CV(1,J+1)
      Z(1,J+1) = Z(M+1,J+1)
      H(M+1,J) = H(1,J)
  110 CONTINUE
      DO 115 I=IPROC,M,NPROC
      CU(I+1,N+1) = CU(I+1,1)
      CV(I,1) = CV(I,N+1)
      Z(I+1,1) = Z(I+1,N+1)
      H(I,N+1) = H(I,1)
  115 CONTINUE

C
      IF (IPROC .EQ.1) THEN
              CU(1,N+1) = CU(M+1,1)
              CV(M+1,1) = CV(1,N+1)
              Z(1,1) = Z(M+1,N+1)
              H(M+1,N+1) = H(1,1)
      ENDIF
```

```
C
      RETURN
      END
      SUBROUTINE NEWUVP (IPROC,NPROC)
C
      PARAMETER (NDIM = 64)
      PARAMETER (NDP1 = NDIM + 1)

C
      COMMON /DATA/ U(NDP1,NDP1),V(NDP1,NDP1),P(NDP1,NDP1),
     1UNEW(NDP1,NDP1),VNEW(NDP1,NDP1),PNEW(NDP1,NDP1),
     2UOLD(NDP1,NDP1),VOLD(NDP1,NDP1),POLD(NDP1,NDP1),
     3CU(NDP1,NDP1),CV(NDP1,NDP1),Z(NDP1,NDP1),H(NDP1,NDP1),
     4PSI(NDP1,NDP1)
C
      COMMON /PARA/ A,TDT,DX,DY,ALPHA,M,N,MP1,NP1,PI,TPI,DI,DJ
     1   ,ISTART,ITMAX,MPRINT,EL,PCF
C
      DO 200 J=IPROC,N,NPROC
      DO 200 I=1,M
      UNEW(I+1,J) = UOLD(I+1,J)+TDT*
     1(0.125*(Z(I+1,J+1)+Z(I+1,J))*(CV(I+1,J+1)+CV(I,J+1)+CV(I,J)
     2+CV(I+1,J))-(H(I+1,J)-H(I,J))/DX)
      VNEW(I,J+1) = VOLD(I,J+1)-TDT*
     1(0.125*(Z(I+1,J+1,)+Z(I,J+1))*(CU(I+1,J+1)+CU(I,J+1(+CU(I,J)

     2+CU(I+1,J))-((H(I,J+1)-H(I,J))/DY)
      PNEW(I,J) = POLD(I,J)-TDT*
     1((CU(I+1,J)-CU(I,J))/DX + (CV(I,J+1)-CV(I,J))/DY)
  200 CONTINUE
C
      RETURN
      END
      SUBROUTINE CONNEW (IPROC,NPROC)
C
      PARAMETER (NDIM = 64)
      PARAMETER (NDP1 = NDIM + 1)
C
      COMMON /DATA/ U(NDP1,NDP1),V(NDP1,NDP1),P(NDP1,NDP1),
     1UNEW(NDP1,NDP1),VNEW(NDP1,NDP1),PNEW(NDP1,NDP1),
     2UOLD(NDP1,NDP1),VOLD(NDP1,NDP1),POLD(NDP1,NDP1),
     3CU(NDP1,NDP1),CV(NDP1,NDP1),Z(NDP1,NDP1),H(NDP1,NDP1),
     4PSI(NDP1,NDP1)
C
      COMMON /PARA/ A,TDT,DX,DY,ALPHA,M,N,MP1,NP1,PI,TPI,DI,DJ
     1   ,ISTART,ITMAX,MPRINT,EL,PCF
C

      DO 210 J=IPROC,N,NPROC
      UNEW(1,J) = UNEW(M+1,J)
      VNEW(M+1,J+1) = VNEW(1,J+1)
      PNEW(M+1,J) = PNEW(1,J)
```

```
  210 CONTINUE
      DO 215 I=IPROC,M,NPROC
      UNEW(I+1,N+1) = UNEW(I+1,1)
      VNEW(I,1) = VNEW(I,N+1)
      PNEW(I,N+1) = PNEW(I,1)
  215 CONTINUE
C
      IF (IPROC .EQ. 1) THEN
              UNEW(1,N+1) = UNEW(M+1,1)
              VNEW(M+1,1) = VNEW(1,N+1)
              PNEW(M+1,N+1) = PNEW(1,1)
      ENDIF
C

      RETURN
      END
      SUBROUTINE SMOOTH (IPROC,NPROC)
C
      PARAMETER (NDIM = 64)
      PARAMETER (NDP1 = NDIM + 1)
C
      COMMON /DATA/ U(NDP1,NDP1),V(NDP1,NDP1),P(NDP1,NDP1),
     1UNEW(NDP1,NDP1),VNEW(NDP1,NDP1),PNEW(NDP1,NDP1),
     2UOLD(NDP1,NDP1),VOLD(NDP1,NDP1),POLD(NDP1,NDP1),
     3CU(NDP1,NDP1),CV(NDP1,NDP1),Z(NDP1,NDP1),H(NDP1,NDP1),
     4PSI(NDP1,NDP1)
C
      COMMON /PARA/ A,TDT,DX,DY,ALPHA,M,N,MP1,NP1,PI,TPI,DI,DJ
     1   ,ISTART,ITMAX,MPRINT,EL,PCF
C

      DO 300 J=IPROC,N,NPROC
      DO 300 I=1,M
      UOLD(I,J) = U(I,J) + ALPHA*(UNEW(I,J)-2.0*U(I,J)+UOLD(I,J))
      VOLD(I,J) = V(I,J) + ALPHA*(VNEW(I,J)-2.0*V(I,J)+VOLD(I,J))
      POLD(I,J) = P(I,J) + ALPHA*(PNEW(I,J)-2.0*V(I,J)+POLD(I,J))
      U(I,J) = UNEW(I,J)
      V(I,J) = VNEW(I,J)
      P(I,J) = PNEW(I,J)
  300 CONTINUE
C
      RETURN
      END

      SUBROUTINE CONSMO (IPROC,NPROC)
C
      PARAMETER (NDIM = 64)
      PARAMETER (NDP1 = NDIM + 1)
C
      COMMON /DATA/ U(NDP1,NDP1),V(NDP1,NDP1),P(NDP1,NDP1),
     1UNEW(NDP1,NDP1),VNEW(NDP1,NDP1),PNEW(NDP1,NDP1),
     2UOLD(NDP1,NDP1),VOLD(NDP1,NDP1),POLD(NDP1,NDP1),
     3CU(NDP1,NDP1),CV(NDP1,NDP1),Z(NDP1,NDP1),H(NDP1,NDP1),
     4PSI(NDP1,NDP1)
```

```
C
      COMMON /PARA/ A,TDT,DX,DY,ALPHA,M,N,MP1,NP1,PI,TPI,DI,DJ
     1   ,ISTART,ITMAX,MPRINT,EL,PCF
C

      DO 100 J=IPROC,N,NPROC
      UOLD(M+1,J) = UOLD(1,J)
      VOLD(M+1,J) = VOLD(1,J)
      POLD(M+1,J) = POLD(1,J)
      U(M+1,J) = U(1,J)

      SUBROUTINE CONSMO (IPROC,NPROC)
C
      PARAMETER (NDIM = 64)
      PARAMETER (NDP1 = NDIM + 1)
C
      COMMON /DATA/ U(NDP1,NDP1),V(NDP1,NDP1),P(NDP1,NDP1),
     1UNEW(NDP1,NDP1),VNEW(NDP1,NDP1),PNEW(NDP1,NDP1),
     2UOLD(NDP1,NDP1),VOLD(NDP1,NDP1),POLD(NDP1,NDP1),
     3CU(NDP1,NDP1),CV(NDP1,NDP1),Z(NDP1,NDP1),H(NDP1,NDP1),
     4PSI(NDP1,NDP1)
C
      COMMON /PARA/ A,TDT,DX,DY,ALPHA,M,N,MP1,NP1,PI,TPI,DI,DJ
     1   ,ISTART,ITMAX,MPRINT,EL,PCF
C

      DO 100 J=IPROC,N,NPROC
      UOLD(M+1,J) = UOLD(1,J)
      VOLD(M+1,J) = VOLD(1,J)
      POLD(M+1,J) = POLD(1,J)
      U(M+1,J) = U(1,J)
      V(M+1,J) = V(1,J)
      P(M+1,J) = P(1,J)
  100 CONTINUE
      DO 200 I=IPROC,M,NPROC
      UOLD(I,N+1) = UOLD(I,1)
      VOLD(I,N+1) = VOLD(I,1)
      POLD(I,N+1) = POLD(I,1)
      U(I,N+1) = U(I,1)
      V(I,N+1) = V(I,1)
      P(I,N+1) = P(I,1)
  200 CONTINUE IF (IPROC .EQ. 1) THEN
              UOLD(M+1,N+1) = UOLD(1,1)
              VOLD(M+1,N+1) = VOLD(1,1)
              POLD(M+1,N+1) = POLD(1,1)
              U(M+1,N+1) = U(1,1)
              V(M+1,N+1) = V(1,1)
              P(M+1,N+1) = P(1,1)
      ENDIF
      RETURN
      END
C
      PARAMETER (NDIM = 64)
      PARAMETER (NDP1 = NDIM + 1)
```

```
C
      COMMON /DATA/ U(NDP1,NDP1),V(NDP1,NDP1),P(NDP1,NDP1),
     1UNEW(NDP1,NDP1),VNEW(NDP1,NDP1),PNEW(NDP1,NDP1),
     2UOLD(NDP1,NDP1),VOLD(NDP1,NDP1),POLD(NDP1,NDP1),
     3CU(NDP1,NDP1),CV(NDP1,NDP1),Z(NDP1,NDP1),H(NDP1,NDP1),
     4PSI(NDP1,NDP1)
C
      COMMON /PARA/ A,TDT,DX,DY,ALPHA,M,N,MP1,NP1,PI,TPI,DI,DJ
     1   ,ISTART,ITMAX,MPRINT,EL,PCF
C
      SUBROUTINE PRINTIT (IOUT)

      WRITE (IOUT,1000) N,M,DX,DY,TDT,ALPHA
 1000 FORMAT ( 'N = ',I3,' M = ',I3,' DX = ',F8.0,' DY = ',F8.0
     1       ,' DT = ',F8.0,' ALPHA = ',F8.3)
      WRITE (IOUT,1001) (P(I,I),I=1,N)
      WRITE (IOUT,1001) (U(I,I),I=1,N)
      WRITE (IOUT,1001) (V(I,I),I=1,N)
 1001 FORMAT (///(8E15.6))
      RETURN
      END
```

APPENDIX D

The "tasked" multiprocessor program for the CRAY X-MP 2

```
      PROGRAM SHALOW
C
C     BENCHMARK WEATHER PREDICTION PROGRAM FOR COMPARING THE
C     PERFORMANCE OF CURRENT SUPERCOMPUTERS. THE MODEL IS
C     BASED OF THE PAPER - THE DYNAMICS OF FINITE-DIFFERENCE
C     MODELS OF THE SHALLOW-WATER EQUATIONS, BY ROBERT SADOURNY
C     J. ATM. SCIENCES, VOL 32, NO 4, APRIL 1975.
C
C     CODE BY PAUL N. SWARZTRAUBER, NATIONAL CENTER FOR
C     ATMOSPHERIC RESEARCH, BOULDER, CO, NOVEMBER 1984.
C
      COMMON U(65,65),V(65,65),P(65,65),UNEW(65,65),VNEW(65,65),
     1          PNEW(65,65),UOLD(65,65),VOLD(65,65),POLD(65,65),
     2          CU(65,65),CV(65,65),Z(65,65),H(65,65),PSI(65,65)
     3         ,DT,TDT,DX,DY,ALPHA
      DIMENSION I100A(3),I100B(3),I200A(3),I200B(3),I300A(3),I300B(3)
      EXTERNAL L100,L200,L300
      REAL MFS100,MFS200,MFS300
C
C     NOTE BELOW THAT TWO DELTA T (TDT) IS SET TO DT ON THE FIRST
C     CYCLE AFTER WHICH IT IS RESET TO DT+DT
C
      I100A(1) = 3
      I100A(3) = 'I100A'
      I100B(1) = 3
      I100B(3) = 'I100B'
      I200A(1) = 3
      I200A(3) = 'I200A'
      I200B(1) = 3
      I200B(3) = 'I200B'
      I300A(1) = 3
      I300A(3) = 'I300A'
      I300B(1) = 3
      I300B(3) = 'I300B'
      DT = 90.
      TDT = DT
C
      DX = 1.E5
      DY = 1.E5
      A = 1.E6
      ALPHA = .001
      ITMAX = 1200
      MPRINT = 120
      M = 64
      N = 64
      MP1 = M+1
      NP1 = N+1
      EL = FLOAT(N)*DX
      PI = 4.*ATAN(1.)
      TPI = PI+PI
      DI = TPI/FLOAT(M)
      DJ = TPI/FLOAT(N)
      PCF = PI*PI*A*A/(EL*EL)
```

```
C
C     INITIAL VALUES OF THE STREAM FUNCTION AND P
C
      DO 50 J=1,NP1
      DO 50 I=1,MP1
      PSI(I,J) = A*SIN((FLOAT(I)-.5)*DI)*SIN((FLOAT(J)-.5)*DJ)
      P(I,J) = PCF*(COS(2.*FLOAT(I-1)*DI)
     1                 +COS(2.*FLOAT(J-1)*DJ))+50000.
   50 CONTINUE
C
C     INITIALIZE VELOCITIES
C
      DO 60 J=1,N
      DO 60 I=1,M
      U(I+1,J) = -(PSI(I+1,J+1)-PSI(I+1,J))/DY
      V(I,J+1) = (PSI(I+1,J+1)-PSI(I,J+1))/DX
   60 CONTINUE
C
C     PERIODIC CONTINUATION
C
      DO 70 J=1,N
      U(1,J) = U(M+1,J)
      V(M+1,J+1) = V(1,J+1)
   70 CONTINUE
      DO 75 I=1,M
      U(I+1,N+1) = U(I+1,1)
      V(I,1) = V(I,N+1)
   75 CONTINUE
      U(1,N+1) = U(M+1,1)
      V(M+1,1) = V(1,N+1)
      DO 86 J=1,NP1
      DO 86 I=1,MP1
      UOLD(I,J) = U(I,J)
      VOLD(I,J) = V(I,J)
      POLD(I,J) = P(I,J)
   86 CONTINUE
C
C     PRINT INITIAL VALUES
C
      WRITE(6,390) N,M,DX,DY,DT,ALPHA
  390 FORMAT(*1NUMBER OF POINTS IN THE X DIRECTION* I8/
     1       * NUMBER OF POINTS IN THE Y DIRECTION* I8/
     2       * GRID SPACING IN THE X DIRECTION    * F8.0/
     3       * GRID SPACING IN THE Y DIRECTION    * F8.0/
     4       * TIME STEP                          * F8.0/
     5       * TIME FILTER PARAMETER              * F8.3)
      MNMIN = MIN0(M,N)
      WRITE(6,391) (POLD(I,I),I=1,MNMIN)
  391 FORMAT(/* INITIAL DIAGONAL ELEMENTS OF P * //(8E15.6))
      WRITE(6,392) (UOLD(I,I),I=1,MNMIN)
  392 FORMAT(/* INITIAL DIAGONAL ELEMENTS OF U * //(8E15.5))
      WRITE(6,393) (VOLD(I,I),I=1,MNMIN)
  393 FORMAT(/* INITIAL DIAGONAL ELEMENTS OF V * //(8E15.6))
      TSTART = 9.5E-9*RTC(DUM)
```

```
      T300 = 1.
      TCYCH = 0.
      TIME = 0.
      NCYCLE = 0
   90 NCYCLE = NCYCLE + 1
      TCYC = 9.5E-9*RTC(DUM)
C
C     COMPUTE CAPITAL U, CAPITAL V, Z AND H
C
      T100 = 9.5E-9*RTC(DUM)
      CALL TSKSTART(I100A,L100,M,1,N/2)
      CALL TSKSTART(I100B,L100,M,N/2+1,N)
      CALL TSKWAIT(I100A)
      CALL TSKWAIT(I100B)
      T100 = 9.5E-9*RTC(DUM)-T100
C
C     PERIODIC CONTINUATION
C
      DO 110 J=1,N
      CU(1,J) = CU(M+1,J)
      CV(M+1,J+1) = CV(1,J+1)
      Z(1,J+1) = Z(M+1,J+1)
      H(M+1,J) = H(1,J)
  110 CONTINUE
      DO 115 I=1,M
      CU(I+1,N+1) = CU(I+1,1)
      CV(I,1) = CV(I,N+1)
      Z(I+1,1) = Z(I+1,N+1)
      H(I,N+1) = H(I,1)
  115 CONTINUE
      CU(1,N+1) = CU(M+1,1)
      CV(M+1,1) = CV(1,N+1)
      Z(1,1) = Z(M+1,N+1)
      H(M+1,N+1) = H(1,1)
C
C     COMPUTE NEW VALUES U,V AND P
C
      T200 = 9.5E-9*RTC(DUM)
      CALL TSKSTART(I200A,L200,M,1,N/2)
      CALL TSKSTART(I200B,L200,M,N/2+1,N)
      CALL TSKWAIT(I200A)
      CALL TSKWAIT(I200B)
C
      T200 = 9.5E-9*RTC(DUM)-T200
C
C     PERIODIC CONTINUATION
C
      DO 210 J=1,N
      UNEW(1,J) = UNEW(M+1,J)
      VNEW(M+1,J+1) = VNEW(1,J+1)
      PNEW(M+1,J) = PNEW(1,J)
  210 CONTINUE
      DO 215 I=1,M
      UNEW(I+1,N+1) = UNEW(I+1,1)
```

```
      VNEW(I,1) = VNEW(I,N+1)
      PNEW(I,N+1) = PNEW(I,1)
  215 CONTINUE
      UNEW(1,N+1) = UNEW(M+1,1)
      VNEW(M+1,1) = VNEW(1,N+1)
      PNEW(M+1,N+1) = PNEW(1,1)
      IF(NCYCLE .GT. ITMAX) CALL EXIT
      TIME = TIME + DT
      TCYC = 9.5E-9*RTC(DUM) - TCYC
      IF(MOD(NCYCLE,MPRINT) .NE. 0) GO TO 370
      PTIME = TIME/3600.
      WRITE(6,350) NCYCLE,PTIME
  350 FORMAT(//* CYCLE NUMBER*I5* MODEL TIME IN  HOURS* F6.2)
      WRITE(6,355) (PNEW(I,I),I=1,MNMIN)
  355 FORMAT(/* DIAGONAL ELEMENTS OF P * //(8E15.6))
      WRITE(6,360) (UNEW(I,I),I=1,MNMIN)
  360 FORMAT(/* DIAGONAL ELEMENTS OF U * //(8E15.6))
      WRITE(6,365) (VNEW(I,I),I=1,MNMIN)
  365 FORMAT(/* DIAGONAL ELEMENTS OF V * //(8E15.6))
      MFS100 = 24.*M*N/T100/1.E6
      MFS200 = 26.*M*N/T200/1.E6
      MFS300 = 15.*M*N/T300/1.E6
      CTIME = 9.5E-9*RTC(DUM)-TSTART
      TCYC = CTIME/FLOAT(NCYCLE)
      WRITE(6,375) NCYCLE,CTIME,TCYCH,T100,MFS100,T200,MFS200,T300,
     1             MFS300
  375 FORMAT(* CYCLE NUMBER*I5* TOTAL COMPUTER TIME* E15.6
     1       * TIME PER CYCLE* E15.6 /
     2       * TIME AND MEGAFLOPS FOR LOOP 100 * E15.6,F6.1/
     3       * TIME AND MEGAFLOPS FOR LOOP 200 * E15.6,F6.1/
     4       * TIME AND MEGAFLOPS FOR LOOP 300 * E15.6,F6.1/ )
  370 IF(NCYCLE .LE. 1) GO TO 310
      CTCYC = 9.5E-9*RTC(DUM)
      T300 = 9.5E-9*RTC(DUM)
      CALL TSKSTART(I300A,L300,M,1,N/2)
      CALL TSKSTART(I300B,L300,M,N/2+1,N)
      CALL TSKWAIT(I300A)
      CALL TSKWAIT(I300B)
      T300 = 9.5E-9*RTC(DUM)-T300
C
C     PERIODIC CONTINUATION
C
      DO 320 J=1,N
      UOLD(M+1,J) = UOLD(1,J)
      VOLD(M+1,J) = VOLD(1,J)
      POLD(M+1,J) = POLD(1,J)
      U(M+1,J) = U(1,J)
      V(M+1,J) = V(1,J)
      P(M+1,J) = P(1,J)
  320 CONTINUE
      DO 325 I=1,M
      UOLD(I,N+1) = UOLD(I,1)
      VOLD(I,N+1) = VOLD(I,1)
      POLD(I,N+1) = POLD(I,1)
```

```
      U(I,N+1) = U(I,1)
      V(I,N+1) = V(I,1)
      P(I,N+1) = P(I,1)
  325 CONTINUE
      UOLD(M+1,N+1) = UOLD(1,1)
      VOLD(M+1,N+1) = VOLD(1,1)
      POLD(M+1,N+1) = POLD(1,1)
      U(M+1,N+1) = U(1,1)
      V(M+1,N+1) = V(1,1)
      P(M+1,N+1) = P(1,1)
      TCYC = TCYC + 9.5E-9*RTC(DUM) - CTCYC
      TCYCH = TCYC
      GO TO 90
  310 TDT = TDT+TDT
      DO 400 J=1,NP1
      DO 400 I=1,MP1
      UOLD(I,J) = U(I,J)
      VOLD(I,J) = V(I,J)
      POLD(I,J) = P(I,J)
      U(I,J) = UNEW(I,J)
      V(I,J) = VNEW(I,J)
      P(I,J) = PNEW(I,J)
  400 CONTINUE
      GO TO 90
      END
      SUBROUTINE L100(M,NS,NF)
      COMMON U(65,65),V(65,65),P(65,65),UNEW(65,65),VNEW(65,65),
     1          PNEW(65,65),UOLD(65,65),VOLD(65,65),POLD(65,65),
     2          CU(65,65),CV(65,65),Z(65,65),H(65,65),PSI(65,65)
     3          ,DT,TDT,DX,DY,ALPHA
      FSDX = 4./DX
      FSDY = 4./DY
      DO 100 J=NS,NF
      DO 100 I=1,M
      CU(I+1,J) = .5*(P(I+1,J)+P(I,J))*U(I+1,J)
      CV(I,J+1) = .5*(P(I,J+1)+P(I,J))*V(I,J+1)
      Z(I+1,J+1) = (FSDX*(V(I+1,J+1)-V(I,J+1))-FSDY*(U(I+1,J+1)
     1          -U(I+1,J)))/(P(I,J)+P(I+1,J)+P(I+1,J+1)+P(I,J+1))
      H(I,J) = P(I,J)+.25*(U(I+1,J)*U(I+1,J)+U(I,J)*U(I,J)
     1               +V(I,J+1)*V(I,J+1)+V(I,J)*V(I,J))
  100 CONTINUE
      RETURN
      END
      SUBROUTINE L200(M,NS,NF)
      COMMON U(65,65),V(65,65),P(65,65),UNEW(65,65),VNEW(65,65),
     1          PNEW(65,65),UOLD(65,65),VOLD(65,65),POLD(65,65),
     2          CU(65,65),CV(65,65),Z(65,65),H(65,65),PSI(65,65)
     3          ,DT,TDT,DX,DY,ALPHA
      TDTS8 = TDT/8.
      TDTSDX = TDT/DX
      TDTSDY = TDT/DY
      DO 200 J=NS,NF
      DO 200 I=1,M
      UNEW(I+1,J) = UOLD(I+1,J)+
```

```
     1    TDTS8*(Z(I+1,J+1)+Z(I+1,J))*(CV(I+1,J+1)+CV(I,J+1)+CV(I,J)
     2        +CV(I+1,J))-TDTSDX*(H(I+1,J)-H(I,J))
      VNEW(I,J+1) = VOLD(I,J+1)-TDTS8*(Z(I+1,J+1)+Z(I,J+1))
     1        *(CU(I+1,J+1)+CU(I,J+1)+CU(I,J)+CU(I+1,J))
     2        -TDTSDY*(H(I,J+1)-H(I,J))
      PNEW(I,J) = POLD(I,J)-TDTSDX*(CU(I+1,J)-CU(I,J))
     1        -TDTSDY*(CV(I,J+1)-CV(I,J))
  200 CONTINUE
      RETURN
      END
      SUBROUTINE L300(M,NS,NF)
      COMMON U(65,65),V(65,65),P(65,65),UNEW(65,65),VNEW(65,65),
     1          PNEW(65,65),UOLD(65,65),VOLD(65,65),POLD(65,65),
     2          CU(65,65),CV(65,65),Z(65,65),H(65,65),PSI(65,65)
     3         ,DT,TDT,DX,DY,ALPHA
      DO 300 J=NS,NF
      DO 300 I=1,M
      UOLD(I,J) = U(I,J)+ALPHA*(UNEW(I,J)-2.*U(I,J)+UOLD(I,J))
      VOLD(I,J) = V(I,J)+ALPHA*(VNEW(I,J)-2.*V(I,J)+VOLD(I,J))
      POLD(I,J) = P(I,J)+ALPHA*(PNEW(I,J)-2.*P(I,J)+POLD(I,J))
      U(I,J) = UNEW(I,J)
      V(I,J) = VNEW(I,J)
      P(I,J) = PNEW(I,J)
  300 CONTINUE
      RETURN
      END
```

APPENDIX E

The "events" multiprocesor program for the CRAY X-MP 2

```
      PROGRAM SHALOW
C
C     BENCHMARK WEATHER PREDICTION PROGRAM FOR COMPARING THE
C     PERFORMANCE OF CURRENT SUPERCOMPUTERS. THE MODEL IS
C     BASED OF THE PAPER - THE DYNAMICS OF FINITE-DIFFERENCE
C     MODELS OF THE SHALLOW-WATER EQUATIONS, BY ROBERT SADOURNY
C     J. ATM. SCIENCES, VOL 32, NO 4, APRIL 1975.
C
C     CODE BY PAUL N. SWARZTRAUBER, NATIONAL CENTER FOR
C     ATMOSPHERIC RESEARCH, BOULDER, CO, NOVEMBER 1984
C
      COMMON U(65,65),V(65,65),P(65,65),UNEW(65,65),VNEW(65,65),
     1          PNEW(65,65),UOLD(65,65),VOLD(65,65),POLD(65,65),
     2          CU(65,65),CV(65,65),Z(65,65),H(65,65),PSI(65,65)
     3         ,DT,TDT,DX,DY,ALPHA,ITMAX
     4         ,IV1,IV2,IV3,IV4,IV5,IV6,IV7,IV8,IV9,IV10,IV11,IV12
      DIMENSION ITASK(3)
      EXTERNAL TASK
      REAL MFS100,MFS200,MFS300
      ITASK(1) = 3
      ITASK(3) = 'TASK'
      CALL EVASGN(IV1)
      CALL EVASGN(IV2)
      CALL EVASGN(IV3)
      CALL EVASGN(IV4)
      CALL EVASGN(IV5)
      CALL EVASGN(IV6)
      CALL EVASGN(IV7)
      CALL EVASGN(IV8)
      CALL EVASGN(IV9)
      CALL EVASGN(IV10)
      CALL EVASGN(IV11)
      CALL EVASGN(IV12)
C
C     NOTE BELOW THAT TWO DELTA T (TDT) IS SET TO DT ON THE FIRST
C     CYCLE AFTER WHICH IT IS RESET TO DT+DT
C
      DT = 90.
      TDT = DT
C
      DX = 1.E5
      DY = 1.E5
      A = 1.E6
      ALPHA = .001
      ITMAX = 1200
      MPRINT = 120
      M = 64
      N = 64
      MP1 = M+1
      NP1 = N+1
      NS2 = N/2
      NS2P1 = NS2+1
      EL = FLOAT(N)*DX
```

```
      PI = 4.*ATAN(1.)
      TPI = PI+PI
      DI = TPI/FLOAT(M)
      DJ = TPI/FLOAT(N)
      PCF = PI*PI*A*A/(EL*EL)
C
C     INITIAL VALUES OF THE STREAM FUNCTION AND P
C
      DO 50 J=1,NP1
      DO 50 I=1,MP1
      PSI(I,J) = A*SIN((FLOAT(I)-.5)*DI)*SIN((FLOAT(J)-.5)*DJ)
      P(I,J) = PCF*(COS(2.*FLOAT(I-1)*DI)
     1                +COS(2.*FLOAT(J-1)*DJ))+50000.
   50 CONTINUE
C
C     INITIALIZE VELOCITIES
C
      DO 60 J=1,N
      DO 60 I=1,M
      U(I+1,J) = -(PSI(I+1,J+1)-PSI(I+1,J))/DY
      V(I,J+1) = (PSI(I+1,J+1)-PSI(I,J+1))/DX
   60 CONTINUE
C
C     PERIODIC CONTINUATION
      DO 70 J=1,N
      U(1,J) = U(M+1,J)
      V(M+1,J+1) = V(1,J+1)
   70 CONTINUE
      DO 75 I=1,M
      U(I+1,N+1) = U(I+1,1)
      V(I,1) = V(I,N+1)
   75 CONTINUE
      U(1,N+1) = U(M+1,1)
      V(M+1,1) = V(1,N+1)
      DO 86 J=1,NP1
      DO 86 I=1,MP1
      UOLD(I,J) = U(I,J)
      VOLD(I,J) = V(I,J)
      POLD(I,J) = P(I,J)
   86 CONTINUE
C
C     PRINT INITIAL VALUES
C
      WRITE(6,390) N,M,DX,DY,DT,ALPHA
  390 FORMAT(*1NUMBER OF POINTS IN THE X DIRECTION* I8/
     1       * NUMBER OF POINTS IN THE Y DIRECTION* I8/
     2       * GRID SPACING IN THE X DIRECTION    * F8.0/
     3       * GRID SPACING IN THE Y DIRECTION    * F8.0/
     4       * TIME STEP                          * F8.0/
     5       * TIME FILTER PARAMETER              * F8.3)
      MNMIN = MIN0(M,N)
      WRITE(6,391) (POLD(I,I),I=1,MNMIN)
  391 FORMAT(/* INITIAL DIAGONAL ELEMENTS OF P * //(8E15.6))
      WRITE(6,392) (UOLD(I,I),I=1,MNMIN)
```

```
  392 FORMAT(/* INITIAL DIAGONAL ELEMENTS OF U * //(8E15.5))
      WRITE(6,393) (VOLD(I,I),I=1,MNMIN)
  393 FORMAT(/* INITIAL DIAGONAL ELEMENTS OF V * //(8E15.6))
      TSTART = 9.5E-9*RTC(DUM)
      T300 = 1.
      TCYCH = 0.
      TIME = 0.
      NCYCLE = 0
      CALL TSKSTART(ITASK,TASK,M,NS2P1,N)
   90 NCYCLE = NCYCLE + 1
      TCYC = 9.5E-9*RTC(DUM)
C
C     COMPUTE CAPITAL U, CAPITAL V, Z AND H
C
      T100 = 9.5E-9*RTC(DUM)
      CALL EVPOST(IV1)
      CALL EVWAIT(IV7)
      CALL EVCLEAR(IV7)
      FSDX = 4./DX
      FSDY = 4./DY
      DO 100 J=1,NS2
      DO 100 I=1,M
      CU(I+1,J) = .5*(P(I+1,J)+P(I,J))*U(I+1,J)
      CV(I,J+1) = .5*(P(I,J+1)+P(I,J))*V(I,J+1)
      Z(I+1,J+1) = (FSDX*(V(I+1,J+1)-V(I,J+1))-FSDY*(U(I+1,J+1)
     1           -U(I+1,J)))/(P(I,J)+P(I+1,J)+P(I+1,J+1)+P(I,J+1))
      H(I,J) = P(I,J)+.25*(U(I+1,J)*U(I+1,J)+U(I,J)*U(I,J)
     1                 +V(I,J+1)*V(I,J+1)+V(I,J)*V(I,J))
  100 CONTINUE
      CALL EVPOST(IV2)
      CALL EVWAIT(IV8)
      CALL EVCLEAR(IV8)
      T100 = 9.5E-9*RTC(DUM)-T100
C
C     PERIODIC CONTINUATION
C
      DO 110 J=1,N
      CU(1,J) = CU(M+1,J)
      CV(M+1,J+1) = CV(1,J+1)
      Z(1,J+1) = Z(M+1,J+1)
      H(M+1,J) = H(1,J)
  110 CONTINUE
      DO 115 I=1,M
      CU(I+1,N+1) = CU(I+1,1)
      CV(I,1) = CV(I,N+1)
      Z(I+1,1) = Z(I+1,N+1)
      H(I,N+1) = H(I,1)
  115 CONTINUE
      CU(1,N+1) = CU(M+1,1)
      CV(M+1,1) = CV(1,N+1)
      Z(1,1) = Z(M+1,N+1)
      H(M+1,N+1) = H(1,1)
C
C     COMPUTE NEW VALUES U,V AND P
```

```
C
      T200 = 9.5E-9*RTC(DUM)
      CALL EVPOST(IV3)
      CALL EVWAIT(IV9)
      CALL EVCLEAR(IV9)
      TDTS8 = TDT/8.
      TDTSDX = TDT/DX
      TDTSDY = TDT/DY
      DO 200 J=1,NS2
      DO 200 I=1,M
      UNEW(I+1,J) = UOLD(I+1,J)+
     1    TDTS8*(Z(I+1,J+1)+Z(I+1,J))*(CV(I+1,J+1)+CV(I,J+1)+CV(I,J)
     2       +CV(I+1,J))-TDTSDX*(H(I+1,J)-H(I,J))
      VNEW(I,J+1) = VOLD(I,J+1)-TDTS8*(Z(I+1,J+1)+Z(I,J+1))
     1       *(CU(I+1,J+1)+CU(I,J+1)+CU(I,J)+CU(I+1,J))
     2       -TDTSDY*(H(I,J+1)-H(I,J))
      PNEW(I,J) = POLD(I,J)-TDTSDX*(CU(I+1,J)-CU(I,J))
     1       -TDTSDY*(CV(I,J+1)-CV(I,J))
  200 CONTINUE
      CALL EVPOST(IV4)
      CALL EVWAIT(IV10)
      CALL EVCLEAR(IV10)
      T200 = 9.5E-9*RTC(DUM)-T200
C
C     PERIODIC CONTINUATION
C
      DO 210 J=1,N
      UNEW(1,J) = UNEW(M+1,J)
      VNEW(M+1,J+1) = VNEW(1,J+1)
      PNEW(M+1,J) = PNEW(1,J)
  210 CONTINUE
      DO 215 I=1,M
      UNEW(I+1,N+1) = UNEW(I+1,1)
      VNEW(I,1) = VNEW(I,N+1)
      PNEW(I,N+1) = PNEW(I,1)
  215 CONTINUE
      UNEW(1,N+1) = UNEW(M+1,1)
      VNEW(M+1,1) = VNEW(1,N+1)
      PNEW(M+1,N+1) = PNEW(1,1)
      IF(NCYCLE .GT. ITMAX) CALL EXIT
      TIME = TIME + DT
      TCYC = 9.5E-9*RTC(DUM) - TCYC
      IF(MOD(NCYCLE,MPRINT) .NE. 0) GO TO 370
      PTIME = TIME/3600.
      WRITE(6,350) NCYCLE,PTIME
  350 FORMAT(//* CYCLE NUMBER*I5* MODEL TIME IN  HOURS* F6.2)
      WRITE(6,355) (PNEW(I,I),I=1,MNMIN)
  355 FORMAT(/* DIAGONAL ELEMENTS OF P * //(8E15.6))
      WRITE(6,360) (UNEW(I,I),I=1,MNMIN)
  360 FORMAT(/* DIAGONAL ELEMENTS OF U * //(8E15.6))
      WRITE(6,365) (VNEW(I,I),I=1,MNMIN)
  365 FORMAT(/* DIAGONAL ELEMENTS OF V * //(8E15.6))
      MFS100 = 24.*M*N/T100/1.E6
      MFS200 = 26.*M*N/T200/1.E6
```

```
      MFS300 = 15.*M*N/T300/1.E6
      CTIME = 9.5E-9*RTC(DUM)-TSTART
      WRITE(6,375) NCYCLE,CTIME,TCYCH,T100,MFS100,T200,MFS200,T300,
     1            MFS300
  375 FORMAT(* CYCLE NUMBER*I5* TOTAL COMPUTER TIME* E15.6
     1       * TIME PER CYCLE* E15.6 /
     2       * TIME AND MEGAFLOPS FOR LOOP 100 * E15.6,F6.1/
     3       * TIME AND MEGAFLOPS FOR LOOP 200 * E15.6,F6.1/
     4       * TIME AND MEGAFLOPS FOR LOOP 300 * E15.6,F6.1/ )
  370 IF(NCYCLE .LE. 1) GO TO 310
      CTCYC = 9.5E-9*RTC(DUM)
      T300 = 9.5E-9*RTC(DUM)
      CALL EVPOST(IV5)
      CALL EVWAIT(IV11)
      CALL EVCLEAR(IV11)
      DO 300 J=1,NS2
      DO 300 I=1,M
      UOLD(I,J) = U(I,J)+ALPHA*(UNEW(I,J)-2.*U(I,J)+UOLD(I,J))
      VOLD(I,J) = V(I,J)+ALPHA*(VNEW(I,J)-2.*V(I,J)+VOLD(I,J))
      POLD(I,J) = P(I,J)+ALPHA*(PNEW(I,J)-2.*P(I,J)+POLD(I,J))
      U(I,J) = UNEW(I,J)
      V(I,J) = VNEW(I,J)
      P(I,J) = PNEW(I,J)
  300 CONTINUE
      CALL EVPOST(IV6)
      CALL EVWAIT(IV12)
      CALL EVCLEAR(IV12)
      T300 = 9.5E-9*RTC(DUM)-T300
C
C     PERIODIC CONTINUATION
C
      DO 320 J=1,N
      UOLD(M+1,J) = UOLD(1,J)
      VOLD(M+1,J) = VOLD(1,J)
      POLD(M+1,J) = POLD(1,J)
      U(M+1,J) = U(1,J)
      V(M+1,J) = V(1,J)
      P(M+1,J) = P(1,J)
  320 CONTINUE
      DO 325 I=1,M
      UOLD(I,N+1) = UOLD(I,1)
      VOLD(I,N+1) = VOLD(I,1)
      POLD(I,N+1) = POLD(I,1)
      U(I,N+1) = U(I,1)
      V(I,N+1) = V(I,1)
      P(I,N+1) = P(I,1)
  325 CONTINUE
      UOLD(M+1,N+1) = UOLD(1,1)
      VOLD(M+1,N+1) = VOLD(1,1)
      POLD(M+1,N+1) = POLD(1,1)
      U(M+1,N+1) = U(1,1)
      V(M+1,N+1) = V(1,1)
      P(M+1,N+1) = P(1,1)
      TCYC = TCYC + 9.5E-9*RTC(DUM)-CTCYC
```

```
      TCYCH = TCYC
      GO TO 90
  310 TDT = TDT+TDT
      DO 400 J=1,NP1
      DO 400 I=1,MP1
      UOLD(I,J) = U(I,J)
      VOLD(I,J) = V(I,J)
      POLD(I,J) = P(I,J)
      U(I,J) = UNEW(I,J)
      V(I,J) = VNEW(I,J)
      P(I,J) = PNEW(I,J)
  400 CONTINUE
      GO TO 90
      END
      SUBROUTINE TASK(M,NS,NF)
      COMMON U(65,65),V(65,65),P(65,65),UNEW(65,65),VNEW(65,65),
     1          PNEW(65,65),UOLD(65,65),VOLD(65,65),POLD(65,65),
     2          CU(65,65),CV(65,65),Z(65,65),H(65,65),PSI(65,65)
     3         ,DT,TDT,DX,DY,ALPHA,ITMAX
     4         ,IV1,IV2,IV3,IV4,IV5,IV6,IV7,IV8,IV9,IV10,IV11,IV12
      NCYCLE = 0
   90 NCYCLE = NCYCLE+1
      CALL EVPOST(IV7)
      CALL EVWAIT(IV1)
      CALL EVCLEAR(IV1)
      FSDX =4./DX
      FSDY= 4./DY
      DO 100 J=NS,NF
      DO 100 I=1,M
      CU(I+1,J) = .5*(P(I+1,J)+P(I,J))*U(I+1,J)
      CV(I,J+1) = .5*(P(I,J+1)+P(I,J))*V(I,J+1)
      Z(I+1,J+1) = (FSDX*(V(I+1,J+1)-V(I,J+1))-FSDY*(U(I+1,J+1)
     1          -U(I+1,J)))/(P(I,J)+P(I+1,J)+P(I+1,J+1)+P(I,J+1))
      H(I,J) = P(I,J)+.25*(U(I+1,J)*U(I+1,J)+U(I,J)*U(I,J)
     1                +V(I,J+1)*V(I,J+1)+V(I,J)*V(I,J))
  100 CONTINUE
      CALL EVPOST(IV8)
      CALL EVWAIT(IV2)
      CALL EVCLEAR(IV2)
      CALL EVPOST(IV9)
      CALL EVWAIT(IV3)
      CALL EVCLEAR(IV3)
      TDTS8 = TDT/8.
      TDTSDX = TDT/DX
      TDTSDY = TDT/DY
      DO 200 J=NS,NF
      DO 200 I=1,M
      UNEW(I+1,J) = UOLD(I+1,J)+
     1    TDTS8*(Z(I+1,J+1)+Z(I+1,J))*(CV(I+1,J+1)+CV(I,J+1)+CV(I,J)
     2        +CV(I+1,J))-TDTSDX*(H(I+1,J)-H(I,J))
      VNEW(I,J+1) = VOLD(I,J+1)-TDTS8*(Z(I+1,J+1)+Z(I,J+1))
     1        *(CU(I+1,J+1)+CU(I,J+1)+CU(I,J)+CU(I+1,J))
     2        -TDTSDY*(H(I,J+1)-H(I,J))
      PNEW(I,J) = POLD(I,J)-TDTSDX*(CU(I+1,J)-CU(I,J))
```

```
    1        -TDTSDY*(CV(I,J+1)-CV(I,J))
200 CONTINUE
    CALL EVPOST(IV10)
    CALL EVWAIT(IV4)
    CALL EVCLEAR(IV4)
    IF(NCYCLE.GT.ITMAX) GO TO 55
    IF(NCYCLE .LE. 1) GO TO 90
    CALL EVPOST(IV11)
    CALL EVWAIT(IV5)
    CALL EVCLEAR(IV5)
    DO 300 J=NS,NF
    DO 300 I=1,M
    UOLD(I,J) = U(I,J)+ALPHA*(UNEW(I,J)-2.*U(I,J)+UOLD(I,J))
    VOLD(I,J) = V(I,J)+ALPHA*(VNEW(I,J)-2.*V(I,J)+VOLD(I,J))
    POLD(I,J) = P(I,J)+ALPHA*(PNEW(I,J)-2.*P(I,J)+POLD(I,J))
    U(I,J) = UNEW(I,J)
    V(I,J) = VNEW(I,J)
    P(I,J) = PNEW(I,J)
300 CONTINUE
    CALL EVPOST(IV12)
    CALL EVWAIT(IV6)
    CALL EVCLEAR(IV6)
    GO TO 90
 55 RETURN
    END
```

APPENDIX F

The following program is an implementation of the "truncated" Stockham FFT for multiple sequences on the CRAY X-MP 2

```
      SUBROUTINE MSTOCK(IS,LI,LJ,C,WORK)
C
C     THE STOCKHAM AUTOSORT FFT FOR MULTIPLE TRANSFORMS
C     MULTIPROCESSOR VERSION FOR THE CRAY-XMP-2
C
C     DEFINE M=2**LI AND N=2**LJ THEN M TRANSFORMS OF LENGTH N
C     ARE COMPUTED. THE TRANSFORMS ARE COMPUTED IN THE DIRECTION
C     OF THE SECOND INDEX OF THE TWO DIMENSIONAL ARRAY C.
C     THE FIRST DIMENSION OF THE ARRAY C MUST BE EQUAL TO M.
C     THAT IS, THE SEQUENCES MUST BE STORED CONSECUTIVELY.
C
C     IS = -1  FORWARD TRANSFORM
C     IS =  1  BACKWARD TRANSFORM
C
      COMPLEX C(1),WORK(1)
      DIMENSION IVT(2),ITASK(2)
      EXTERNAL TASK
      ITASK(1) = 2
      CALL EVASGN(IVT(1))
      CALL EVASGN(IVT(2))
      CALL TSKSTART(ITASK,TASK,IS,LI,LJ,C,WORK,IVT)
      N = 2**(LJ+LI)
      L01 = 0
      DO 100 L=1,LJ
      LS=2**(L-1)
      NS = N/(LS+LS)
      L01 = 1-L01
      IF(L01 .EQ. 0) GO TO 102
      CALL STOCKH(IS,LS,NS,C,WORK)
      GO TO 103
  102 CALL STOCKH(IS,LS,NS,WORK,C)
  103 CALL EVPOST(IVT(1))
      CALL EVWAIT(IVT(2))
      CALL EVCLEAR(IVT(2))
  100 CONTINUE
      IF(L01 .EQ. 0) RETURN
      DO 101 I=1,N
      C(I) = WORK(I)
  101 CONTINUE
      RETURN
      END
      SUBROUTINE STOCKH(IS,LS,NS,C,CH)
      COMPLEX OMEGA,OMEGK,WYK,C(NS,2,LS),CH(NS,LS,2)
      ANGLE = FLOAT(IS)*4.*ATAN(1.)/FLOAT(LS)
      OMEGA = CMPLX(COS(ANGLE),SIN(ANGLE))
      NS2 = NS/2
      DO 200 J=1,NS2
      OMEGK = 1.
      DO 200 I=1,LS
      WYK = OMEGK*C(J,2,I)
      CH(J,I,1) = C(J,1,I)+WYK
```

```
      CH(J,I,2) = C(J,1,I)-WYK
      OMEGK = OMEGA*OMEGK
  200 CONTINUE
      RETURN
      END
      SUBROUTINE TASK(IS,LI,LJ,C,WORK,IVT)
      COMPLEX C(1),WORK(1)
      DIMENSION IVT(1)
      N = 2**(LJ+LI)
      L01 = 0
      DO 100 L=1,LJ
      LS=2**(L-1)
      NS = N/(LS+LS)
      L01 = 1-L01
      IF(L01 .EQ. 0) GO TO 102
      CALL STOCKG(IS,LS,NS,C,WORK)
      GO TO 103
  102 CALL STOCKG(IS,LS,NS,WORK,C)
  103 CALL EVPOST(IVT(2))
      CALL EVWAIT(IVT(1))
      CALL EVCLEAR(IVT(1))
  100 CONTINUE
      RETURN
      END
      SUBROUTINE STOCKG(IS,LS,NS,C,CH)
      COMPLEX OMEGA,OMEGK,WYK,C(NS,2,LS),CH(NS,LS,2)
      ANGLE = FLOAT(IS)*4.*ATAN(1.)/FLOAT(LS)
      OMEGA = CMPLX(COS(ANGLE),SIN(ANGLE))
      NS2P1 = NS/2+1
      DO 200 J=NS2P1,NS
      OMEGK = 1.
      DO 200 I=1,LS
      WYK = OMEGK*C(J,2,I)
      CH(J,I,1) = C(J,1,I)+WYK
      CH(J,I,2) = C(J,1,I)-WYK
      OMEGK = OMEGA*OMEGK
  200 CONTINUE
      RETURN
      END
```

Table 7

The Cooley-Tukey FFT for N=8

1	2	3	4	5	6	7
			.580			
		.580		1.034		
		.454	.454	.126		
	.580				2.096+.000i	
.580	.786.		.786		.126-.511i	3.656+0.000i
.951	.454	.786		1.062	-.028+.000i	.800-1.173i
.786	.276	.276	.276	.511	.126+.511i	-.028-0.354i
.298						-.547-0.151i
.454						.536+0.000i
.006	.951	.951	.951	.957	1.560+.000i	-.547+0.151i
.276	.298	.006		.944	.944+.008i	-.028+0.354i
.306	.006		.006		.354+.000i	.800+1.173i
	.306				.944-.008i	
		.298	.298	.603		
		.306		-.008		
			.306			

Atmospheric Modelling on a SIMD Computer

M. J. SUAREZ

Goddard Space Flight Center/611, Greenbelt, MD 20771, USA

As technical limits are approached on the speed of processors used in super-computers, we may expect that greater parallelism will be sought to enhance their power. The parallel execution of varied tasks is to some degree used in all modern mainframes. In the new class of parallel machines, however, parallelism is a central design concept, and its presence is much more apparent to the user. Efficient use of a two processor XMP, for example, to run a single, large task such as an atmospheric GCM, is a very different problem than that of running the same task on its predecessor, the CRAY-1. As the number of processors increases from two to say eight or sixteen, we cannot expect the total computing power to increase proportionally, but the difficulty of programming probably will. Furthermore, the increase in computational speed we get from additional processors will depend both on our task (and how we organise it) and on the organisation of the machine itself. It thus behooves us at these early stages in the development of parallel super-computers to consider how well our models could be made to perform on various computer organisations. At Goddard Space Flight Center we started a project to explore the possibilities of using machines like Goddard's Massively Parallel Processor (MPP) in atmospheric modelling.

In his landmark paper on computer organisation, Flynn (1972) discussed two types of parallel processors that he termed: single instruction stream/multiple data stream (SIMD) and multiple instruction stream/multiple data stream (MIMD). In a MIMD each processor can work on a different set of instructions and access different data. Problems in its design and use revolve around the assignment and coordination of tasks among the processors and the avoidance of conflicts in their access to data. Most parallel super-computers produced or planned are of this type.

The MPP is an example of the SIMD organisation. At any one time all its processors are executing the same instruction, or doing nothing. Each processor accesses different data, kept in local memory. But addressing, like any other instruction, is done with all processors in lock-step, so they all work on the same portion of the locally held data. Because of their relatively simple organisation SIMD machines can be built of a very large number of processors (the MPP consists of 16K processing elements (PEs). The rigidity of their lock-step execution of instructions, however, makes them suitable only for the most parallel of tasks.

It is clear that the wide range of computing tasks encountered in the atmospheric sciences requires that large, general purpose computers be available at major centers. It is unlikely SIMD machines will ever be able to meet this need. However, a large fraction of our computing resources is devoted to a single large task: the execution of high resolution general circulation and weather prediction models. Since these models are used repetitively for years in both research and operational forecasting, one can afford to devote considerable effort to optimising their codes. The question underlying our discussion is: Should one also consider the optimisation of the organisation of the computers on which they are run? Judging by the attention given to the question, the answer has been - no, it is preferable to devote our attention to improving and adjusting our codes to the best available general purpose computers. It is our starting premise that increased availability and use of parallelism could alter this answer. In the rest of this note we give a brief discussion of this question. We will identify - probably naively - MIMD with general purpose and SIMD with specialisation, since the latter is the simplest and most economical way of achieving a large degree of parallelism and produces a consequent decrease in generality.

The main argument against considering a specialised organisation is that it could be justified only for our largest tasks, general circulation models (GCMs). Because these attempt a comprehensive description of atmospheric processes, they involve most of the tasks of the field as a whole; consequently they could not be made sufficiently parallel to make efficient use of the SIMD specialisation. In particular these models, in addition to solving discrete versions of the equations of motion, deal with a number of sub-grid scale processes such as radiative transfer through cloudy atmospheres, condensation, and small scale turbulent and convective motions. Because existing parameterisations for these processes are highly conditional (do this if there is a cloud, otherwise do that) they will be very inefficient in a highly parallel SIMD machine. Two questions arise: How punitive are such "non-parallel" tasks? and how much may their formulation be modified to make them more amenable to parallel calculation? We will devote our attention to the first and easier one of the two, assuming (conservatively for the SIMD case) that the answer to the second is - not much.

A good starting model for the behaviour of a task as the number of processors working on it is increased is "Amdahl's assertion". It states, in effect, that every task contains some fixed fraction of critical work that while being performed requires all other subtasks idle, awaiting either access or results. If this fraction is denoted by f, then the time required to perform the task with n processors, given in units of the time required to perform it with a single, identical processor is

$$t = \frac{(1 - f)}{n} + f,$$

and its inverse, the speed-up, is

$$t^{-1} = \frac{n}{(1 - f) + fn} \quad (1)$$

For small n, speed increases linearly, while at large n it saturates at a value of g/f times the speed of the individual processors. The choices this model's assumptions offer is to either effectively eliminate critical tasks or use the complexity allowed the machine for a few fast processors. Less pessimistic results require relaxing the assumption of a fixed, single-tasking fraction.

To be definite we consider a simple heuristic model of a GCM. We assume that its tasks fall in three categories: local or nearly local unbranched calculations, such as those used in updating the equations of motion at grid points; nonlocal unbranched calculations, such as computing zonal FFT's; and local branched calculations, such as those used in subgrid scale processes.

Machines like the MPP -- a SIMD with a nearest neighbour network -- are ideal for the local unbranched tasks of grid-point models, provided the model's horizontal grid can be mapped onto the processors array and its vertical grid fits into each processing element. In this case the only overhead is in accessing a near neighbour's data. The rest of the time is spent doing arithmetic and accessing local memory. If as n increases the grid is made finer or the mapping more sparse (fewer grid points per processor); the calculations per unit time increases like n and the overhead remains fixed, so

$$t^{-1} \sim n.$$

This is the ideal case.

As an example of non-local unbranched calculations we take an FFT along one direction of the array of processors. This would be the first step in solving a Poisson equation on a sphere with a latitude longitude mapping onto the array and the transform taken in the zonal direction. If for simplicity we take a square array, and again assume the grid is made finer with larger n, the results for large n is

$$t^{-1} \sim \sqrt{n} \log n \quad . \quad (2)$$

This is less than ideal but still an attractive speed-up, and suggests that non-local operations such as the solution of elliptic equations need not be a serious impediment to SIMD calculations. We note, however, that the proportionality constant in (2) would depend on the speed with which near neighbour transfers can be made compared to the arithmetic speed of the processors.

The third case, branched local calculations, are the most punitive to the SIMD machine. Flynn (1972) analysed this case by considering a branched structure with equal probability of having to execute either side of the branch. With this assumption the number of processors active at branching level p is to execute

$$n_p = \frac{n}{2^p},$$

and beyond level $p = \log_2(n) - 1$, processors are executing sequentially. If f_p is the fraction of the calculations done at each level, the time required in a SIMD organisation is

$$t = 1 + \sum_{p=o}^{\log_2(n)-1} f_p[\frac{1}{n_p} - 1] \text{ , } n > 1. \tag{3}$$

The speed-up is then crucially dependent on the choice of fp, which depends on the nature of the task. If the branching is only a few levels deep the summation would be terminated after these terms and we would have

$$t = \sum_{p=o}^{P_1} \frac{f_p}{n_p}.$$

Two idealised cases considered by Flynn serve to illustrate possible behaviours of the speed-up. In the first, the fraction of the work at successively deeper levels of branching is assumed to decrease exponentially, just matching the decrease in the number of processors working.

$$f_p = \frac{1}{2p+1},$$

that is, half the operations are fully parallel, one quarter are at the first level, and so on. This circumvents Amdahl's assertion by making the single-tasking fraction decrease as n increases, the speed-up for large n is then

$$t^{-1} \sim \frac{n}{\log (n)} \tag{4}$$

this is probably the weakest dependence for which a large degree of parallelism should be considered. The second case is to assume that an equal fraction of the work must be done at each level. The asymptotic speed up is then

$$t^{-1} \sim \log (n) \tag{5}$$

this behaviour is called Minsky's conjecture, and a task that exhibits it is clearly unsuitable for highly parallel calculations.

Where does this leave us? First the dynamics calculations should be amenable to large parallelism, for both local and non-local processes. If the branched fraction in the physics behaves like (5) and cannot be modified, we will be restricted to using a few fast processors. If on average it behaves like (4), however, then, depending on how economically processors can be replicated, a fairly large n may be optimal.

Reference:

Flynn M.J. (1972) Some computer organisations and their effectiveness. IEEE Transactions on Computers, C:21, 948-960.

The Multitasking Spectral Model at ECMWF

DAVID DENT

European Centre for Medium-Range Weather Forecasts, Shinfield Park, Reading, Berkshire RG2 9AX, United Kingdom

1. INTRODUCTION

ECMWF has 3 major applications which could benefit from multitasking because they are both time critical and require a large proportion of main memory. To date, only the forecast has been modified to make use of more than one processor, although work is proceeding on the other applications. This paper discusses the spectral model in its state of development at the end of 1984.

2. HISTORY

The original ECMWF production forecast was made using a grid point model. Over a period of 2 years, a spectral model was developed to replace it. This went into daily production in April 1983 executing on a CRAY-1A, with spectral resolution T63.

The code is independent of resolution and can be run without recompilation using any desired data. It consists of:

96000 source lines 26000 Fortran statements

Since the resolution T63 was chosen as appropriate for a CRAY-1, a computationally more demanding resolution is possible on a CRAY-XMP. Given the available configuration, a resolution of T106 has been chosen for a comprehensive set of meteorological experiments with the target of making this model available for production use by April 1985.

To achieve an acceptable wall clock execution time, it is obviously essential to make efficient use of both central processors of the CRAY-XMP from within the application code. Hence, a multitasking version of the spectral model has been developed over a period of about 1 year. The first working version went into regular experimental use in July 1984. Since then, efforts have been made to reduce the execution time by identifying and removing inefficiencies.

3. USE OF MAIN MEMORY

The model is unusual in that it makes extensive use of a locally designed memory manager to provide easy and safe organisation of main memory for the various sub-processes within the code.

Since the model was developed and will continue to be developed by a team of scientists, it is extremely useful and productive for any member of the team to be able to allocate a portion of memory for his own particular needs, safe in the knowledge that his allocated space will not be used by any other member of the team.

Use of the manager depends on the POINTER statement, which is an extension to Fortran 77 supported by the Cray Fortran compiler. It allows an array to be dynamically dimensioned, since the array dimension may be a variable in COMMON storage and hence known only at run time. Memory is allocated from a block obtained at initialisation time from the Cray heap manager. An allocated array is identified uniquely by a character name and an integer code. Its base address may subsequently be obtained by another subroutine through use of a LOCATE routine. An UNLOC routine is available to return the space when no longer required.

```
      POINTER (IPT , DIV(ND) )

      CALL ALLOCA(IPT , LENGTH , NAME , KODE )
      CALL LOCATE(IPT , NAME , KODE )
      CALL UNLOC (NAME , KODE )
```

where: DIV is the name of the array of length ND addressed through the pointer IPT

NAME is the character name	which together identify
KODE is the integer code	the allocated space

In multitasking mode, a subroutine commonly executes in both processors simultaneously. Since the subroutine often requires array space to hold the results of calculations, this output space must be unique for each subtask. This can be achieved by using a locally dimensioned array in the usual way, but this space is obtained from the Fortran controlled stack and is released at the end of the subroutine. By using a different integer code value when allocating an array, the memory manager creates a unique space for each execution of the routine and this space remains available for other subsequent routines to locate until explicitly released.

Such a convenience to the programmer of course costs some execution time which must be weighed against the scientists productivity and the ease of code maintenance and development. The overhead in the spectral model is approximately 5% of execution time.

4. ECMWF CRAY XMP CONFIGURATION

From the point of view of the spectral model, the principal characteristics of the Cray-XMP/22 installed at ECMWF are-

2 Central Processors
2 Mwords of central memory
16 banks of memory
16 Mwords of Solid State storage device (SSD)
80 Mwords/sec memory to SSD transfer rate

5. COMPUTER RESOURCES USED BY THE SPECTRAL MODEL

At resolution T106, the single-tasking model requires:
1.5 Mwords of central memory
15.3 Mwords of SSD

There are 3 major work files:
Legendre coefficients - 950 KW - read twice each step
grid point data - 8.7 MW - read and written each step
fourier coefficients - 5.7 MW - read and written each step

total 15.3 MW - 30 MW I/O per step

Putting files on a device with such a high transfer rate to/from central memory allows I/O to be carried out synchronously without much overhead. This reduces the central memory requirements for buffer space and costs less than 4% of the elapsed time for a 10 day forecast.

6. MULTI-TASKING INTERFACE

The following facilities available in the Cray multi-tasking library are used in the model:

```
CALL TSKSTART(ctltab,routine)
CALL TSKWAIT (ctltab)
CALL LOCKON  (lock)
CALL LOCKOFF (lock)

where- 'ctltab' is a task control block
       'lock'   is a unique lock identifier
       'routine' is the name of a subroutine to be executed
```

These tools enable tasks to be started and synchronised, and critical areas of code to be protected against simultaneous execution. Event setting is also supported in the library but the current version of the model does not use this technique. It is possible to pass parameters to a 'routine' but this facility is also not used.

7. GENERAL STRUCTURE

The model is organised into 2 scans over the data as shown in figure 1. Within each scan, there is a loop over all latitude rows (160 for the T106 resolution). Between scans is a smaller area of computation associated with diffusion and semi-implicit calculations. The loop over time steps is repeated 1200 times for a 10 day forecast. However, every 16 steps, significant additional computation is performed by radiation calculations.

Within Scan 1, the following are the principal components of work-

I/O: read Fourier coefficients
I/O: read legendre coefficients
I/O: read and write grid point data
computations in Fourier space
FFT - grid point computations - FFT
semi-implicit computations
computations in Fourier space
compute direct legendre transforms

Within Scan 2, the main items are-

I/O: read legendre coefficients
I/O: write Fourier coefficients
compute inverse legendre transforms

A multi-tasking version of an application requires more main memory than its single-tasking equivalent. Given (a) the desire to maximise the resolution and (b) the shortage of main memory, it is important to select a multitasking strategy which has low memory requirements.

It turns out to be convenient and efficient in memory to split Scan 1 and perform it in 2 pairs of subtasks with a synchronising point in between. This is because each Northern row generates the symmetric part of a Fourier component while the equivalent antisymmetric part is generated by the appropriate Southern row. Both components are combined in different ways to provide contributions to the legendre transform. By computing one Northern row and one Southern row simultaneously, not only is the memory requirement minimised, but also the legendre computation is performed efficiently.

Part of the diffusion calculation is also multi-tasked and Scan 2 can be computed 2 rows at a time (see figure 2).

There remain some relatively small parts of the code which are computed in single-tasking mode.

The memory requirements for this multi-tasking strategy are 1.8 Mwords. Note that alternative strategies are of course possible. However, subtask structures which may be preferred for optimising reasons require either more central memory or additional SSD.

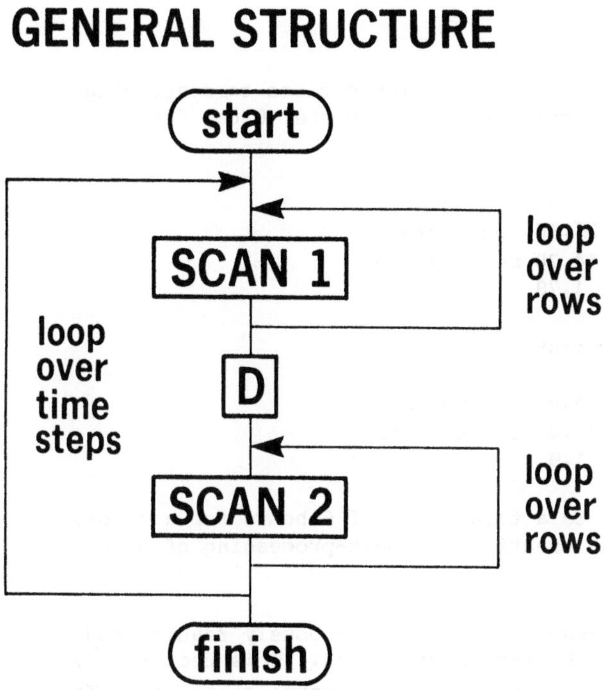

Figure 1

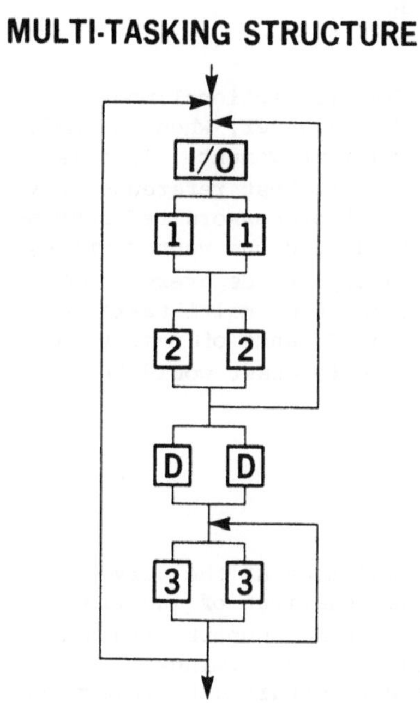

Figure 2

8. OVERALL TIMINGS

All the timings reported here are elapsed times corresponding either to a single time step or to a complete 10 day forecast.

For a normal timestep:

single tasking : 25.36 seconds/step
multi tasking : 14.28 seconds/step
speedup ratio : 1.78

For a radiation timestep:

single tasking : 75.0 seconds/step
multi tasking : 39.4 seconds/step
speedup ratio : 1.9

These times correspond to a total time of 6 hours for a 10 day forecast, including the creation and post-processing of history data.

The Cray-XMP has the capability of simultaneously reading and writing to memory from the vector registers. This feature may be switched on or off by means of a simple control statement. The above times were measured with bidirectional transfer enabled. The following times were measured when bidirectional transfer was disabled:

single tasking : 26.71 seconds/step
multi tasking : 14.21 seconds/step
speedup ratio : 1.88

Thus, for single tasking, switching on the bidirectional mode speeds up the model execution by about 5%. However, when the model is multi-tasked, there is no corresponding improvement. This is easily explained since a 16 bank memory can at best reference only 4 words per clock period. With both central processors referencing memory at the maximum rate of 2 vector reads and one vector write, the code is trying to reference a maximum of 6 words every clock period and is therefore slowed down. If the same multi-tasked model were to be run on a Cray-XMP/22 with 32 banks of memory, an estimated saving of 15 minutes for a 10 day forecast would be achieved.

9. MORE DETAILED TIMINGS

Since the above timings are very simple and made at the very highest level they tell nothing about the behaviour of individual tasks within the model. Currently, there is no support within the Cray multi-tasking library for obtaining detailed timings. Consequently, all the following timings were obtained by inserting code into the model at strategic places in order to record times as reported by the real time clock. The measurements were done in such a way as to disturb the model as little as possible. The model was run in a dedicated environment with no disturbances other than any caused by the operating system (COS X.13). Analysis of the measurements was done subsequently in a normal batch environment.

The average times taken by each of the tasks as identified in the previous section are shown in Fig. 3.

By measuring the time taken by the Cray multi-tasking library routines, it is possible to obtain estimates of the cost of starting tasks etc.

For TSKSTART , three distinctly different times are observed as follows:

40 milliseconds for one case only
0.4 milliseconds for 96% of all TSKSTARTs
0.04 milliseconds for 4% of all TSKSTARTs

The expensive start corresponds to the very first TSKSTART in the complete application, when additional memory has to be requested from the operating system for table space.

The intermediate time corresponds to the case when a 'logical CPU' has to be established (table creation etc).

The shortest time corresponds to the case when a logical CPU already exists. In this execution, the Cray multi-tasking scheduler has released the logical CPU in nearly all cases before the next task is created. The small percentage of fast TSKSTART times were all observed for TASK 2 where there is a very small time gap after completion of TASK 1. In the future it will be possible to tune the scheduler to retain the logical CPU in all cases.

The measured minimum times for other multi-tasking calls are:

TSKWAIT	0.007 milliseconds
LOCKON/LOCKOFF	0.001 milliseconds

Hence it is clear that the TSKSTART times dominate the task overheads.

The approximate total overhead cost in a 10 day forecast is:

3 * 80 * 1200 * 0.4 milliseconds
which is about 2 minutes or 0.7% of the total time

With scheduler tuning, this is likely to be reduced to 0.1%.

An obvious conclusion is that task overheads are small compared to the size of tasks which exist in the spectral model.

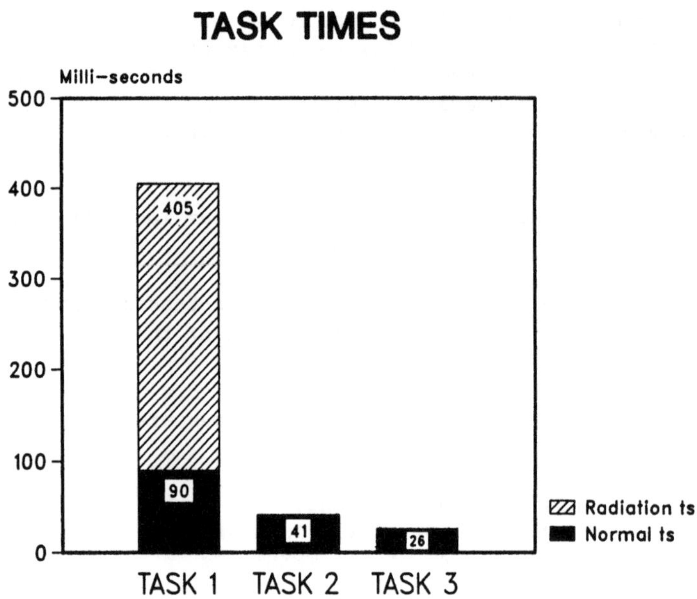

Figure 3

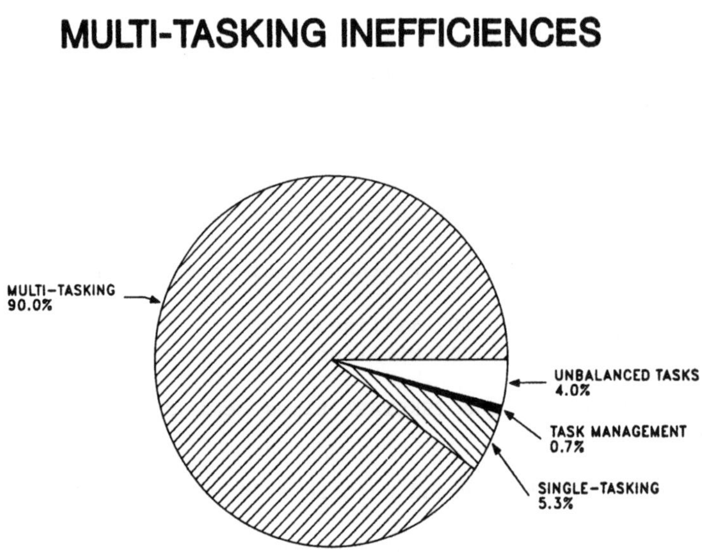

Figure 4

10. INEFFICIENCIES

By measuring the amount of time spent outside of the tasks, it can be seen how much of the code has been multi-tasked and therefore what additional improvements might be made in the future (see Fig. 4).

The TSKWAIT time reported in the previous section was the minimum observed i.e. for the case where the master task completed after the started task and was therefore not held up in the synchronising process. By examining average TSKWAIT times, it is possible to obtain estimates of how imbalanced the pairs of tasks are. Figure 4 shows that these imbalances account for about 4% of the overall model time. Most of the imbalance was observed in TASK1. TASK 2 and TASK 3 imbalances were smaller by a factor of 9.

There are at least 2 reasons for this imbalance. One concerns LOCKS and will be discussed below. The other concerns the nature of the computation in grid-point space (part of TASK 1). Although the amount of work done for each latitude line is exactly equal for the dynamics part of the code, this is not always true in parts of the physical parameterisation. Convection and condensation calculations are affected by synoptic conditions and will therefore vary in space and time. The maghitude of these variations in terms of computing expense has not yet been measured.

LOCKS are used to protect critical regions of code in some 20 places, mostly for statistic gathering purposes. These locks all occur in TASK 1 and are mostly insignificant in time. However some random I/O is carried out to a single dataset which is common to both tasks. In the current Cray software, a lock is applied whenever I/O is initiated to any dataset, so that the strategy of splitting this dataset into 2 will not be useful until this high level lock is moved to the level of the dataset. Indications are that this causes most of the imbalance observed in TASK 1.

11. FUTURE IMPROVEMENTS

Since the target time for a 10 day forecast is approximately 5 hours, there remains substantial optimising to be done before the spectral model is fast enough for operational use. However, significant improvements have already been made (see Fig. 5). By reducing the single-tasking time and by attacking the out of balance inefficiency, it should be possible to improve the multi-tasking performance. It may be possible to bypass the I/O lock and hence substantially reduce the imbalance due to locking. There is also some scope for optimising at the loop level (loop unrolling etc).

Alternative multi-tasking trategies can be tried in order to reduce the number of synchronising points and hence the imbalance time. Unfortunately there is little scope for this effort given the

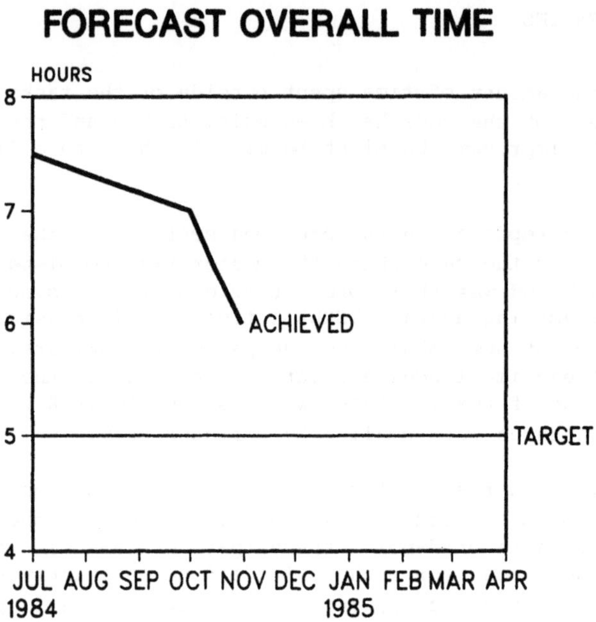

Figure 5

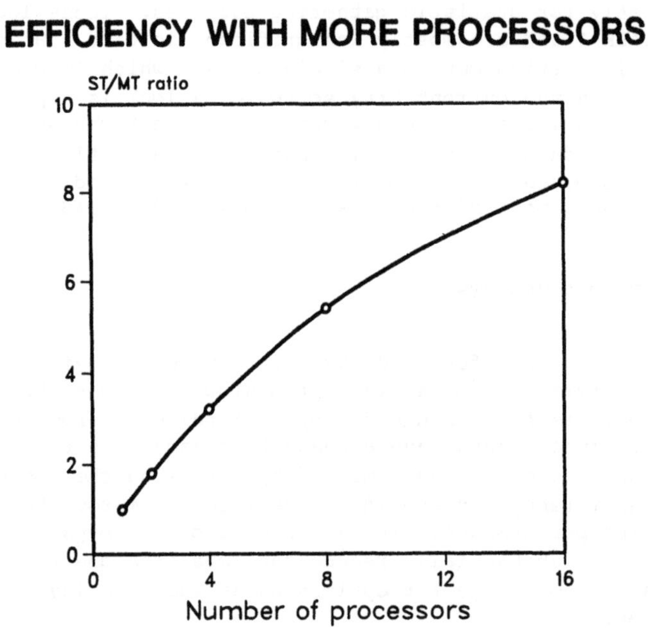

Figure 6

constraints of central memory and SSD space. EVENT synchronising is known to be more efficient than TSKSTART-TSKWAIT and this could be implimented easily in at least part of the application. However, since the task overhead is relatively small, this is unlikely to be useful.

It is interesting to speculate on the model's performance when executed on future hardware with additional processors. Code already exists in the model for N processors but it is largely untested to date. It is based on the current multi-tasking strategy and therefore performance estimates may be made based on the measurements reported earlier in this paper. Fig. 6 indicates that a multi-tasking efficiency of about 3.2 could be achieved with 4 processors, but with 16 processors a speedup of only 8.2 would be achieved over the single-tasked model.

12. SUMMARY

The ECMWF spectral model is a large application which has been successfully adapted to a multi-tasking environment. The task overheads are small compared to the task sizes. Improvements must be made in the elapsed time before the model can be used operationally, but these appear to be achievable in the time frame required.

The code provides the basis for execution on future machines but in order to make good use of increased number of processors some refinement of the multi-tasking strategy is necessary.

UK Meteorological Office's Plans for Using Multiprocessor Systems

P. W. White and R. L. Wiley

UK Meteorological Office, Bracknell, United Kingdom

1. COMPUTING REQUIREMENTS OF THE METEOROLOGICAL OFFICE

1.1 Demand for powerful Computers

Numerical modelling of the atmosphere is now an important tool in many areas of meteorology such as:

- Forecasting the weather on time scales of hours, days or weeks
- Investigations of general circulation for research into matters such as climatic change
- Gaining a deeper understanding of physical processes.

Leading organisations in the development and application of relevant models have always been able to use the most powerful machines that technology could offer at any time and see ways of making good use of even greater processing power. It is quite clear that this position is unchanged and that there is a growing demand for much more powerful computers than existing machines such as the CYBER 205 and CRAY X-MP. Past experience suggests that useful improvements in numerical models, for example to improve the forecast quality for a target time 48 hours after observations to the same level as previously achieved 24 hours after observation time, require an increase in effective computing power of well over an order of magnitude.

1.2 The next major Change

Financial constraints are such that the Meteorological Office can only replace its major computer system at intervals of ten years with only minor upgrades between replacements. A CYBER 205 was installed in 1981 and will probably be replaced in the early 1990s. At that time something like thirty times the power of the CYBER 205 will be required to handle the range of computing tasks encountered by the Meteorological Office. Peak processing rates may not give a good indication of whether this can be achieved because of the difficulties of mapping numerical models of the atmosphere on to the architecture of machines that may be available in the early 1990s. More main memory will be required to allow for larger models and to allow for a number of processes to run concurrently without paging. About 500 Mwords will be needed.

1.3 Problems in achieving the necessary performance

During the 1970s the most powerful machines that were commercially successful worked in a generally serial manner and something close to peak performance could be obtained in a properly configured system. This situation changed with the introduction of vector-oriented machines in the 1980s because only part of any model could be fully vectorised. The questions now are:

- How will technology provide substantial increases in power?
- What resulting problems will face those implementing numerical models?

2. THE IMPACT OF TECHNOLOGY

2.1 Fast Processors

Technological improvements will certainly allow the speed of existing designs to be improved. However, the obvious possibilities, such as reducing gate switching time and circuit lengths, do not offer very much hope of meeting the required targets by the early 1990s, at least in commercially available machines. In practice an improvement in speed by a factor of three is all that can be expected from technology alone. How then is a further factor of ten to be achieved?

2.2 Architecture

The most promising prospect is to use a number of processors simultaneously, that is multiprocessing. There are already examples of multiprocessors in general purpose and powerful scientific computers, e.g. IBM 308X and CRAY X-MP. These, however, are only available with up to four processors and are most easily used when each processor handles an independent stream of tasks. In cases where all the power has to be applied to a single problem it is the user's responsibility to split the process into sub-processes. The user must also ensure that the sub-processes are properly synchronised and that boundary problems between the sub-processes are handled. Given that eight or sixteen processors are likely to be required to provide the necessary power, these problems could take on daunting proportions.

2.3 Alternatives

It has been proposed that machines should be built to match meteorological models rather than vice versa. An example is the Distributed Array Processor from ICL in which it is proposed that there should be one processor per grid point. This architecture is very satisfactory for purely dynamical models but it is difficult to achieve good performance overall and the structure of the model is constrained for the life of the machine. There are also suggestions that arrays of Transputers might be developed into general purpose super computers. These ideas are at an early stage of development and may not be relevant on the timescale of 1990. The multiple super computer appears more likely to meet the needs of meteorology in the 1990s.

2.4 Technological Progress

Over the past 25 years there has generally been a factor of thirty improvement in the processing capability of the most powerful scientific computers each decade. The cost of the most powerful computers at any time has been more or less constant. If this trend continues, a machine of 30 to 100 GFLOPS should be available in the early 1990s for around S20M.

3. MAKING EFFECTIVE USE OF MULTIPROCESSORS

3.1 The Workload Profile

Through the 1980s and 1990s the Meteorological Office will be running models concerned with short period forecasting, daily forecasting and global circulation. There will be associated streams of analysis, assimilation and preparation of output. Some scope therefore exists to allocate independent processes to individual processors and achieve increased throughput. However, this will not be enough to achieve the overall performance target and major models will have to make use of multiple processors.

3.2 The Requirement

The hardware and software in a multiprocessor system should be capable of taking a program in a high level language, such as extended FORTRAN, and generating a set of linked sub-processes allowing for the following points:

- a variable number of processors (to allow for failures and upgrades)
- Execution independent of particular processors
- Synchronisation of sub-processes
- Communication between sub-processes
- Preallocation or dynamic allocation of sub-processes to processors
- Conflicts between sub-processes accessing shared memory
- Interactive optimisation for maximum performance.

It may be reasonable to expect suitable features to exist in hardware available at the end of this decade but it is very doubtful whether software to deal adequately with the above problems will be developed within ten years. The commercial success of large multiprocessor systems in the supercomputer arena may well depend on the availability of systems that permit the user to achieve a high utilisation factor, over a number of processors, from a straightforward program.

3.3 Necessary Hardware Features

Assuming that a lot of the housekeeping necessary to apply multiple processors to a single task will be the responsibility of the user, there is a need for appropriate facilities in the hardware. In addition, the configuration must be balanced to permit maximum utilisation of the processors. It is likely that the following features will be needed:

- a mixture of very fast memory dedicated to processors and also fast shared memory;

- a mechanism for passing data directly between the dedicated memorys associated with any processors;

- global synchronisation flags;

- means to use processors interchangeably;

- a control processor to optimise allocation of processes whose execution time depends on data.

3.4 Using Multiprocessors

The Meteorological Office's plans for making use of multiprocessors are at an early stage of development as the requirement to do so, although inevitable, is not immediate. Preliminary thoughts on developing independent, relatively small processes and splitting larger ones into sub-processes are presented in succeeding paragraphs.

4. PRESENT AND FUTURE MODELLING ACTIVITIES

4.1 Large Scale Forecasting Models

The current forecasting models consist of a global model with a resolution of $1\frac{1}{2}°$ x 1 7/8° x 15 levels and a regional model covering the North Atlantic and most of Europe with a grid length of 3/4° x 15/16° x 15 levels. Both are grid point models and use a split explicit integration scheme with 4th order Lax Wendroff advection. A time step of 15 minutes is used in the global model and $7\frac{1}{2}$ minutes in the regional model. On the Cyber 205 the global model takes 4 minutes for each forecast day and the regional model 6 minutes. Both models obtain their initial data from a repeated insertion data assimilation scheme.

The forecast suite of programs has to be run to a strict schedule and, as this is governed largely by customer requirements outside the Meteorological Office, it is not likely to change greatly in the future. Consequently improvements in forecast models must be accommodated within the same time slots.

There is evidence that greater detail and more accurate short range predictions could be obtained from models having finer resolutions than are used at present, however it must be borne in mind that halving the grid length and doubling the number of levels will result in 16 times more computation. In order to provide more detailed initial data for such models it will be necessary to rely increasingly on satellite observations and on data obtained from other automatic observing systems. As more of these become available, the calculations in the data assimilation scheme will increase substantially. More elaborate numerical techniques will probably be required to enable accurate calculations in the neighbourhood of internal discontinuities, such as fronts, to be made in the finer scale models.

4.2 Local Short Range Forecasting

A non-hydrostatic meso-scale model is currently being tested for short range local weather forecasting. The area of integration covers the British Isles, and the model has a 15 km grid length, 16 levels and a time step of 1 minute. It uses a semi-implicit leap frog finite difference scheme. If the model is introduced operationally, it is thought that 12 or 18 hour forecasts might be produced every 3 hours, with detailed meso-scale analyses made every hour.

In order to represent some of the finer scale, but important, detail of the topography of the British Isles a 5 km grid length should be used rather than a 15 km one. Additional levels are also necessary to enable the boundary layer processes to be resolved more accurately. The size of the area of integration needs to be increased so that the lateral boundaries are further away from the region of interest. Such improvements would require a computer 30-50 times faster than the CYBER 205.

4.3 Climate Studies

The 11-layer atmospheric general circulation model designed for climate studies normally uses a $2\frac{1}{2}°$ x 3 3/4° global grid, though resolutions of 2° x 3° and 5° x $7\frac{1}{2}°$ are also used for some experiments. The model uses a leap frog explicit finite difference scheme and takes about 15 CPU hours on the CYBER 205 for a one year run. Several multi-annual cycle runs have been made for periods of up to 8 years. Work is currently in progress to couple the atmospheric general circulation model to a global ocean/sea-ice model. The ocean model will initially have the same resolution as the atmospheric model but an appreciably finer resolution is planned for the future.

The Meteorological Office is playing a full part in the World Climate Research Programme and it can be anticipated that research in this aspect of meteorology will expand over the next 20 years or so. Models of various degrees of complexity will be applied to the three streams of climate research (stream 1 - 1 month to a season, stream 2 - 1 to 5 years, stream 3 - decades to centuries). Finer mesh models with more accurate methods of representing physical processes will be needed as well as simulations for longer periods. More tests will be conducted on the likely effects of man's activities on both the global climate and on regional climates.

4.4 Research Models

The Meteorological Office has a number of research projects that involve running larger computer models in conjunction with field experiments or laboratory studies. The research is aimed at obtaining a better understanding of certain meteorological phenomena such as comulonimbus convection, frontal dynamics, boundary layer turbulence and diffusion, airflow over orography, fog and stratocumulus, atmospheric chemistry, rotating fluid flow etc. Greater understanding of these phenomena can be expected to lead to improvements in the way they are represented in large scale numerical models of the atmosphere. It is anticipated that work in this area will need to be maintained in support of the more complex global and regional models being developed.

5. SOME SUGGESTIONS FOR IMPLEMENTING PARALLEL COMPUTATION

5.1 Vectorisation

An important factor in the design of efficient code for the Cyber 205 is the organisation of the computation in terms of long vectors. To assist this, the data need to be arranged either as illustrated in Fig. 1(b) or as in Fig. 1(c) rather than as in Fig. 1(a) which gives short vector lengths. The present climate model and mesoscaled model use the vertical slice arrangement (Fig. 1(b)) while the forecast model uses horizontal fields (Fig. 1(c)) (which permits the longest vector lengths). We assume that the start-up time for vector operations on future generations of computers will be shorter than at present (so that there will be less advantage in using horizontal fields rather than vertical slices) but that short vectors will still need to be avoided.

5.2 Task Sectioning

A simple way of splitting a forecast model up into parallel computation streams is to divide the region of integration up into sections as illustrated in Fig. 2. Each processor deals with all the computation for a different sub-region. If the architecture of the computer is such that each processor has a large local memory, the bulk of the data could remain in place with new values overwriting old ones. Some overlaps will be necessary to allow the calculation to proceed at the edges of the sections and these would need to be copied at the start of each time step from the results in neighbouring sections.

A disadvantage of the technique is that the amount of computation may vary from one processor to another because different amounts of sub-grid scale physics will be involved in each section of the

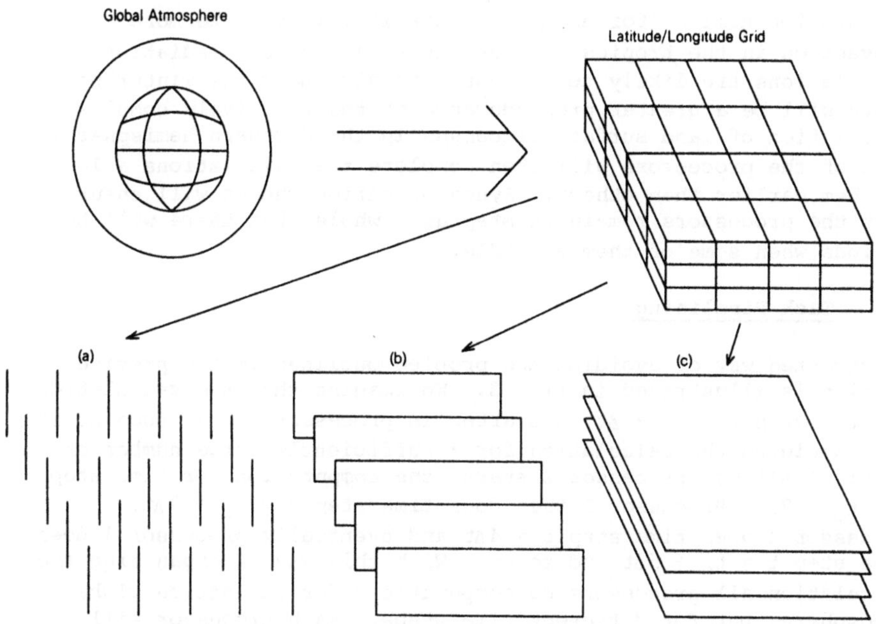

Figure 1 Different ways of organising data for vector computation (a) vertical columns, (b) vertical slabs, (c) horizontal fields

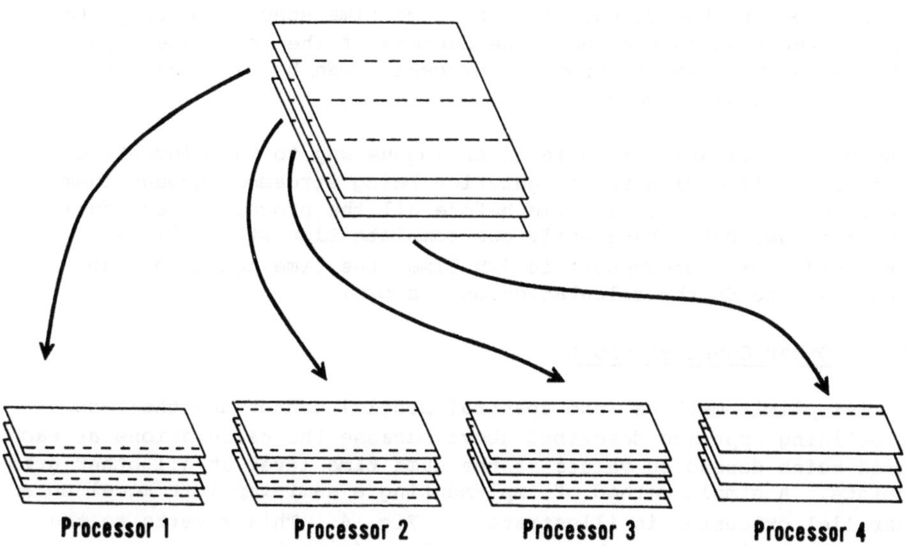

Figure 2 Task Sectioning

integration domain (for example, there is likely to be more convection in the tropics than at higher latitudes, radiation calculations are likely to be less extensive near the winter pole, there will be a greater preponderance of the relatively complex calculation of land surface processes in the Northern hemisphere). Some of the processors will then complete the calculations allocated to them earlier than others. Synchronisation checks will ensure that the processors remain in step as a whole, but there will be periods when some of them are idle.

5.3 Task Pipelining

A suggested way of avoiding the problem outlined in the previous section is illustrated in Fig. 3. We imagine that the calculation for the step $t = t^{o} + \Delta t$ is started in processor 1. As soon as it has completed the calculation for a sufficiently large number of vertical slices, processor 2 starts the computation for time step $t = t_o + 2\Delta t$. Processor 3 then does time step $t = t_o + 3\Delta t$, processor 4 does time step $t = 4\Delta t$ and eventually processor 1 does time step $t = t_o + 5\Delta t$ and so on. With this way of organising the calculation all processors do computations for the entire global atmosphere, but for different time steps. Each processor will therefore be evenly balanced with its neighbours. In principle the system can run without synchronisation steps, provided that checks are made to ensure that results required by each processor have been completed by its predecessor.

Only explicit (or split-explicit) finite difference schemes can be used with the method, since both semi-implicit schemes and spectral models require the calculation for each time step to be complete before the next is started. The success of the technique depends on the speed that results from one processor can be made available for input to another processor.

The processors are acting in an analogous way to the elements of a vector pipeline with the computation being streamed through them. There will be a start-up time before all the processors are fully functioning, but subsequently one complete time step will be produced from N processors in 1/N times the time taken for one processor to do the calculation on its own.

5.4 Other Considerations

Semi-implicit methods (and spectral models) cannot use the task pipelining approach described above because the calculations at each grid point depend on values at the same time level at all other grid points. A simple method of implementing a semi-implicit model on a parallel processor is illustrated in Fig. 4. This reverts to the task sectioning approach discussed above (with the same disadvantages) for the explicit part of the calculation while the second order partial differential equation that arises for the semi-implicit scheme is solved by decoupling the vertical modes and solving for them in separate processors.

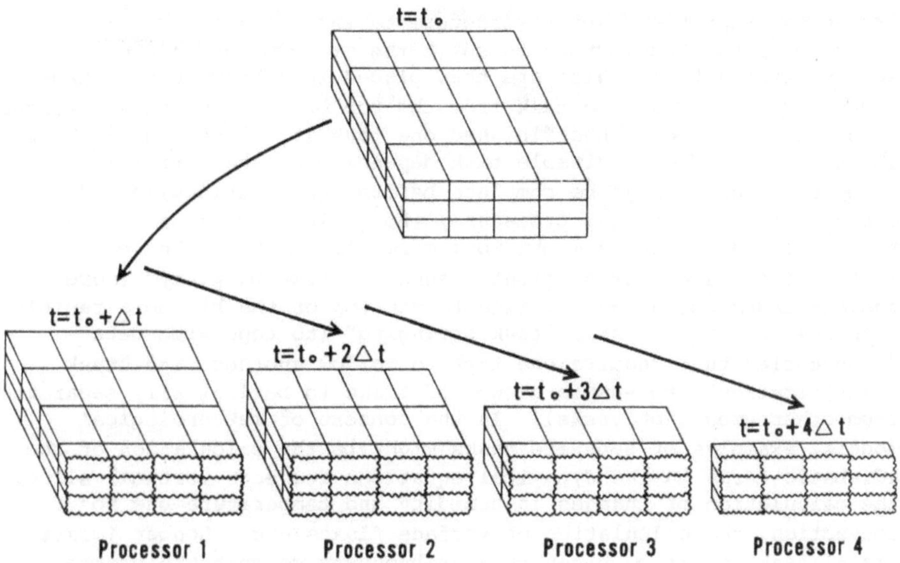

Figure 3 Task Pipelining

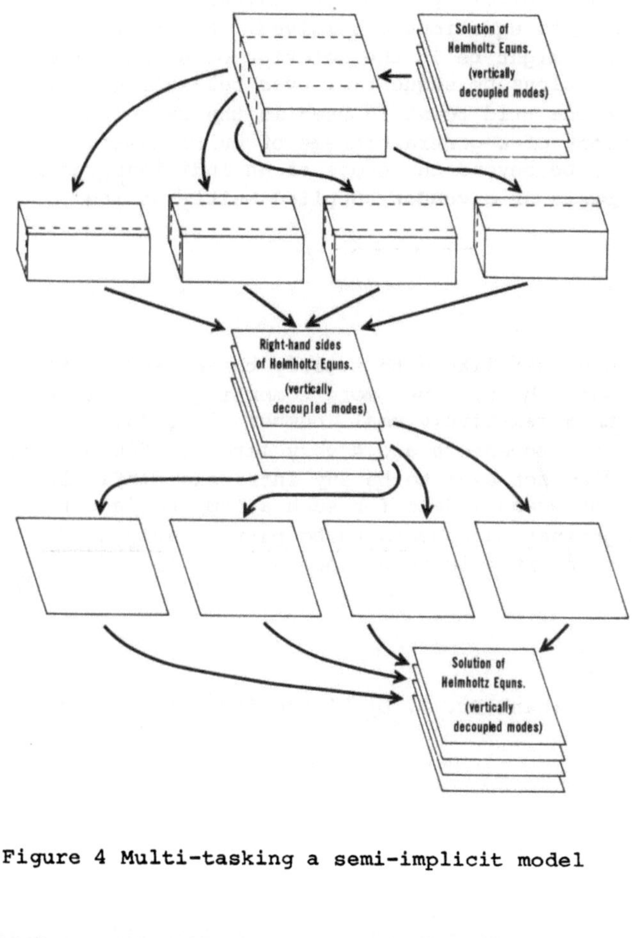

Figure 4 Multi-tasking a semi-implicit model

A more general approach to the problem of multi-tasking is to avoid trying to adapt the program structure to the specific architecture of the computer but instead to split the program into a number of self contained tasks which are then placed in a "task bin". Each processor then selects a task from the top of the bin and starts the calculation. When it has finished one task it selects another from the task bin. The inevitable task dependencies that arise (for example if task A must be complete before task B starts) could be achieved by a networking procedure, though in many cases a less rigid program structure might work more effectively. One could instead introduce task attributes such as "task buoyancy" (more buoyant tasks would tend to rise to the top of the bin more rapidly than less buoyant tasks), "task stringing" (to cope with data dependencies that require one task to follow another) and "task stratification" (to enable groups of tasks to be logically separated from other groups of tasks). In the context of meteorological models, examples of separate tasks might be the calculation of the adiabatic terms in the dynamical equations for each vertical slice, the calculation of changes in humidity and temperature due to convection, the calculation of surface fluxes etc. Longer (coarse grain) tasks might be given greater buoyancy so that the shorter (fine grain) tasks can fill in the gaps that might otherwise give rise to processor idle time at the end of each task stratum. Task stratification could be useful, for example, in the semi implicit scheme to ensure that the explicit part of the calculation is complete before the Helmoltz equations are solved. An example of the use of task stringing might be in the solution of second order partial differential equations by sequential over relaxation where the result obtained for one grid point is used in the calculation of the next. The development of a general scheme of multi-tasking of this sort would probably be beyond the scope of an individual user but it could form the basis of a vendor supplied multi-tasking operating system.

6. CONCLUSIONS

The computer architecture most likely to satisfy the Meteorological Office's need for at least thirty times more computing power in the early 1990's is one with a relatively small number (8 or 16) state-of-the-art parallel processors and a very large (500 Mwords) common memory. There does not seem to be any intrinsic difficulty in adapting atmosphere or ocean models for such a computer and the necessary program reorganisation appears to be particularly simple for explicit or split-explicit grid point models.

7. ACKNOWLEDGEMENT

The authors would like to thank Dr. A. Dickinson for providing many of the ideas in section 5.

Canadian Meteorological Services' Plans for Using Multi-Processor Systems

ANDREW STANIFORTH

Atmospheric Environment Service, Environment Canada, Dorval, Québec, Canada

1. INTRODUCTION

The Atmospheric Environment Service (AES) of Environment Canada has a mandate to provide comprehensive information to Canadians concerning past, present, future and possible conditions of the atmosphere. To fulfil this mandate AES has developed, and is continuing to develop, numerical models of the atmosphere for both research and real-time operational applications in the broad areas of

(1) weather prediction,

(2) climate, and

(3) air quality.

A powerful computer is evidently required for this kind of numerical modelling, and late in 1983 a Cray 1-S computer was installed at AES's facility in Dorval, Montreal, to replace a Control Data Cyber 176. An upgrade to a Cray XMP-22 (two-processors, two million words of memory) is scheduled for the Fall of 1986, and AES is currently examining how best to convert model codes from the 1-S and optimize them for the XMP.

Weather prediction codes that need to be run on the XMP include those for objective analysis and regional and global forecasting. The objective analysis code uses optimal interpolation techniques, the global codes use the spectral technique, the regional codes use finite-element and semi-Lagrangian techniques, and the global and regional forecast models have comprehensive parameterizations of sub-grid-scales physical processes.

The general circulation model code used for climate simulations is very similar to the global spectral weather prediction code. Model codes used for air quality studies and the long-range transport of pollutants include transport modules based on Eulerian and semi-Lagrangian techniques, as well as modules for the chemical interaction of many chemical species.

AES has an agreement with the National Sciences and Engineering Research Council of Canada to make 10% of the Cray computer's capacity available to researchers in many disciplines in Canadian universities. As a consequence many other codes are also run on AES's Cray in addition to those previously mentioned. They include, for example, diverse codes in oceanography, astrophysics, nuclear physics and engineering.

In this paper we mostly restrict our attention to coding considerations for global and regional forecast models, since these form the bulk of the computational load on AES's vector computer. Lacking any concrete experience of our own with a Cray XMP, an analysis (summarized in the following sections) was performed about a year ago with a view to anticipating (and avoiding) some of the difficulties likely to be encountered due to the architectural differences between the Cray XMP and the Cray 1-S. Now that the Cray XMP has been available to users at various sites for a year or so, we are very interested to hear their experiences (both good and bad), and the present workshop provides an ideal opportunity to do so. In this way we hope to be able to avoid some of the problems encountered by the first users. As can be seen from the content of the paper, an adequate amount of memory (both main and SSD) has been identified as a major concern.

2. MEMORY CHARACTERISTICS OF WEATHER PREDICTION MODELS

For a given level of accuracy spectral models require fewer degrees of freedom (i.e. less memory) than competing methods, but at the expense of more CPU time/degree of freedom, and their use of orthogonal basis functions permits a natural "slicing" of model computations such that I/O overhead is minimized. On the other hand, regional models trade-off accuracy at later time-frames for increased accuracy in the short time-frame (up to 2 days) by a redistribution of the degrees of freedom and computing effort. Such models use techniques that inherently require more memory than global or hemispheric spectral models, but this is generally compensated for by the fact they use less CPU time/degree of freedom.

In order to improve the accuracy of weather element forecasts at all time ranges, more sophisticated parameterizations of the unresolved physical processes are being incorporated into atmospheric models. To make the parameterizations more complete additional prognostic and diagnostic fields must be introduced. For example, a parameterization of turbulent fluxes presently being implemented requires two further 3-D prognostic variables, viz the turbulent kinetic energy and the turbulent mixing length, while a more accurate treatment of clouds requires prognostic cloud variables. All these extra variables improve the accuracy of the models, but they make it increasingly difficult to overlap I/O operations with CPU calculations for a given memory configuration. One of the current research thrusts in AES is to increase the efficiency of the timestepping algorithm, with the expectation of being able to reinvest the computational time saved in increased resolution. Thus it appears that pressure on the amount of available memory will further increase.

3. SOME MEMORY CONSIDERATIONS WHEN CONVERTING CODES FROM A CRAY 1-S TO A CRAY XMP

At any given instant in time a two-processor Cray XMP will hopefully be working on twice the number of operands as a single processor Cray 1-S. Furthermore, because they have multiple paths to memory (compared to the single path on the Cray 1-S) they complete their calculations in fewer clock cycles and are thus ready sooner to operate on the next set of operands. If these operands are unavailable then the CPU will be idle. This problem would never arise if the main memory were sufficiently large to store all the fields required for subsequent calculations. However in the real world, main memories are both limited in size and costly and we have to use a backing store, which on a well-configured Cray XMP is an SSD.

Some coding considerations that arise because of the architectural differences between the Cray 1-S and the Cray XMP are (and these are particularly important for the coding of regional models):

(1) the high overhead incurred to obtain the first word in a main memory/SSD transfer makes it advantageous to transfer data in large chunks (of the order of a hundred thousand words or more): the relative overhead is much higher than that of main memory/IOP transfers on a Cray 1-S because the first word still takes as many clock cycles to arrive on a Cray-XMP as on a Cray 1-S, even though subsequent words arrive 10 times as fast because of the increased channel speed;

(2) transferring in large chunks means we must have large chunks of main memory available to accept them while the CPU is busy operating on other large chunks;

(3) because there is a 6-fold increase in memory bandwidth (from memory to CPU) for a 2-processor Cray XMP compared to a Cray 1-S, the CPU's are ready for new operands much sooner and it is therefore more difficult to ensure all operands are available (after transfer from the SSD) when required by the CPU:

(4) performing calculations in large chunks increases vector lengths which further enhances CPU performance and increases the chances that the CPU will be waiting for main memory/SSD transfers to complete.

We conclude from the above that for a given code having I/O overlapped with CPU execution (as is generally the case with weather prediction codes) that substantially larger amounts of main memory will need to be made available to a given code, otherwise the code is likely to become I/O bound.

4. MULTI-PROGRAMMING CONSIDERATIONS

If a computer system permits multi-programming, there will in general be an increase in the total amount of useful work performed during a given time period when compared to the same machine executing the same jobs sequentially; it also permits improved turnaround for short jobs. This is achieved by an increased parallelism and a more sustained usage of the various computer components (CPU, channels, memory, etc.), and is of particular importance in an environment that must process a large number of jobs having significantly different memory and CPU requirements, which is the case in AES.

Most centers that use a super-computer for atmospheric modelling fall into one of two classes. The first is characterised by being a real-time operational environment serving a relatively limited number of users running a limited number of specialized programs (e.g. NMC, ECMWF), while the second is characterized by being research oriented with few, if any, real-time constraints (e.g. NCAR, GFDL). In either case it is acceptable (and current practice) to permit individual jobs to operate fairly close to the limits of memory. For the first class there are a limited number of users who tune their models to simultaneously use as many of the resources of the machine as possible and only a limited multi-programming capability is required. For the second class it is of paramount importance to get the result at the resolution required and, because it is not a real-time environment, such centers also have the luxury of permitting jobs to operate close to the limits of memory and accepting limited multi-programming. However, AES falls into neither of these two categories.

AES's Cray must support a vigorous research environment and a heavy operational one. Unlike ECMWF, the Canadian operational runs are spread fairly uniformly over 24 hours, including a heavy use of prime time, and account for approximately half of the total computer time; all other (research) jobs must be run either in parallel with the operational jobs or in the remaining time which is extremely fragmented. Furthermore the Cray computer configuration has to support university researchers as well as an ever-expanding number of AES users. There is clearly a need for multi-programming to increase machine throughput, in an environment that is both operationally and research oriented.

But multi-programming requires lots of memory, both main memory and backing-store memory (SSD). If we wish to simultaneously execute jobs in parallel, additional main memory is required otherwise we are unable to simultaneously (for example) perform I/O for one job while the CPU is busy with another.

Let us examine the situation where an operational model is in execution during prime time (consuming large amounts of both main and SSD memory) and we wish to run jobs which have large memory requirements but small execution times (of the order to seconds compared to an hour or so for an operational model). If there is a sufficiently large amount of SSD memory available, then a copy of the contents for main memory associated with the operational model can be "rolled out" to SSD memory and another similar-size (but short execution) job "rolled in"; it is imperative that these transfers be effected extremely fast to minimize overhead (CPU idle time) and this requires the use of SSD memory rather than buffer-memory. By proceeding in this fashion the R and D community can still continue to use the machine in prime time for "large-memory-but-short-execution" jobs while operational jobs are being processed (which is an almost continuous phenomenon during prime time, given that the operational run consumes approximately one half of the computer resources spread fairly uniformly throughout the day). Several-times-a-day turnaround during prime time for such "large-memory-but-short-execution" jobs is vital for a viable atmospheric R and D program. Each such job requires an image of main memory contents to be stored on the SSD as well as the usual amount of SSD memory associated with the job. Since several of these jobs need to be SSD resident it is clear that an adequate amount of SSD memory is necessary, otherwise the potential throughput of a Cray XMP in our environment will not be realized.

5. MULTI-TASKING CONSIDERATIONS

A multi-processor machine can potentially improve the total throughput of the machine for many users, or reduce the real-time execution of a job for a single user, or do both but to a lesser degree, and the optimum mix will depend on the goals of an organization. For the reasons mentioned in the previous section we anticipate that we will find ourselves comfortably (or perhaps uncomfortably) between the two extremes. We need adequate throughput to satisfy high demand, but at the same time operational results must not be unduly delayed, and these considerations have an impact on program design.

The fundamental question that needs to be addressed in designing programs for multi-processor execution is at what hierarchy level should the user(s) multi-task, that is partition the work-load into a set of more-or-less independent tasks for independent execution. Should it be at the level of

(1) the job (different jobs executing on different processors),

(2) the job-step (different programs of a single job executing on different processors),

(3) the program (different subroutines of a single program executing on different processors),

(4) the subroutine (different loops of a subroutine executing on different processors), or

(5) the loop (different parts of a loop executing on different processors)?

At one extreme (i.e. level (i) of the above hierarchy) the user relies on the operating system to optimize throughput and helps the system by minimizing resource requirements, such as memory: this extreme would perhaps be appropriate in an environment where turnaround is not an issue.

At the other extreme (i.e. level (5) of the above hierarchy) a single user tries to squeeze the maximum real-time juice out of the proverbial lemon.

The following factors are also important when adopting a multi-tasking strategy:

(1) the (scratch) memory overhead associated with splitting a job into a large number of small tasks rather than a proportionally smaller number of large tasks;

(2) the flexibility of the strategy (e.g. how easily can the program be adapted to other multi-processor machines having a larger number of processors);

(3) the balancing of the computational load across processors;

(4) the synchronization of tasks that depend on the completion of other tasks;

(5) the possible use of an alternative algorithm more suited to a multi-processor environment; and

(6) the programming effort required to achieve the desired result.

In our environment we tentatively conclude from the above considerations that it is appropriate to multi-task model codes at the highest possible levels (consistent with acceptable real time performance) using a relatively small number of tasks, and to rely on the system to optimize throughput across user jobs. The advantages of such a strategy appear to be 3:

(1) a small number of tasks will need less scratch memory (which is in short supply);

(2) the strategy is reasonably flexible inasmuch as it should be relatively straightforward to adapt the code to multi-processor machines having a larger number of processors, by either further dividing the computational load between processors at the same level of multi-tasking hierarchy (when possible), or by descending one level in the hierarchy for critical portions of the code (when necessary);

(3) the balancing of the computational load between processors for a given job need only be approximate (enough to give acceptable real-time performance) without needing to be optimal (since it is highly unlikely to be the only job in the machine and other jobs will interfere with it to some extent anyway); and

(4) the fewer the number of tasks, the easier it is to program and synchronize them and the smaller the multi-tasking overhead.

The only real disadvantages appear to be a degradation of real-time performance if run in isolation, and a reliance on system software to optimize machine throughput. This latter point brings us full circle back to the need to have adequate main and SSD memory. We can only hope the circle isn't vicious...

As regards putting the above theory into practice in the context of our weather prediction codes, a possible first attempt being considered for the dynamical calculations is to define the calculation of the right-hand side of the momentum, thermodynamic, continuity equations, etc. as tasks and send them to different processors. On the other hand, the majority of the physical parameterization calculations are horizontally independent and can be split up into sets of vertical columns, each set of which defines an independent task. It seem prudent to try to restrict the number of independent tasks between synchronization points to be less than the total number of processors, and thus avoid initiating too many tasks that generate scratch memory demands.

6. RESULTS

With respect to our actual experience with a "multi-processor", it is very limited. Two of our models (regional finite-element and spectral) have been run by Robert Wellck of Cray Research on a single-processor Cray XMP and execution times compared to that given by identical codes executing on a Cray 1-S; the results are summarized in Table 1. Although these results do not tell us what speed-ups we can expect to achieve with optimized codes on a multi (two) processor Cray XMP with respect to a Cray 1-S, they do at least provide lower bounds.

	Finite element		Spectral	
	CPU	wall-clock	CPU	wall-clock
Cray 1-S	1002	1745	172	427
Cray XMP (one processor)	558	736	94	149
Speed up, Cray 1-S/Cray XMP	1.8	2.4	1.8	2.9

Table 1 Execution times (in seconds)

7. SUMMARY

AES objectives are such as to require the modelling of a wide variety of meteorological scales. It is argued that regional ("small-scale") models achieve more accurate local forecasts (but valid for a more limited time period) than spectral ("large-scale") models, by using methods which inherently require more memory but use less CPU time/degree of freedom. The need to simultaneously support real-time operational execution of models as well as government and university R and D is also such as to require more memory (both main and SSD) than might otherwise be needed, because of multi-programming considerations. Furthermore there are also pressures on memory use due to multi-processor/multi-tasking considerations, all of which leads us to identify the amount of available memory as being a matter of concern, particularly for regional and mesoscale models.

In the context of our environment, it is further argued that it is appropriate to multi-task model codes at the highest possible levels (consistent with acceptable real time performance) using a relatively small number of tasks, and to let the operating system optimise machine throughput across user jobs. A tentative strategy is also given in the context of applying these principles to our weather prediction codes. In the longer term it is important to develop new algorithms (or adapt unused older ones) to take best advantage of the multi-processor architectures.

ACKNOWLEDGEMENTS

Discussions with Michel Valin and the expert typing of Maryse Ferland are gratefully acknowledged.

Multiprocessors: Requirements and Consequences Summary of the Discussion

PAM Prior

European Centre for Medium-Range Weather Forecasts, Shinfield Park, Reading, Berkshire RG2 9AX, United Kingdom

1. REQUIREMENTS FOR METEOROLOGICAL MODELS

The computer requirements sought by scientists for meteorological models in the late eighties, early nineties were discussed. There was a considerable difference between what was considered desirable for meteorological purposes and what was considered feasible within this timescale.

1.1 Computing Speed

It was considered that ideally a peak performance of 100 Gigaflops would be required in the early nineties, but realistically a CPU performance of 10-30 Gigaflops was sought.

1.2 Memory

100 Megawords was considered the absolute minimum necessity for readily accessible memory, that is, memory more readily accessible than disk. However, it was agreed that the possibility of running several models at once was desirable, and therefore 500 Megawords was considered a more realistic working target.

1.3 Algorithms

The possibility of splitting algorithms across processors was discussed. It was agreed that a fine balance would need to be struck between the advantage gained by the splitting of algorithms and the disadvantage of the supplementary coding and processing resulting from the additional I/O and control necessitated by such splitting. None of the group had any plans to create or investigate new algorithms.

1.4 Parallelism

The general opinion was that only a rather low level of parallelism would be available within this timescale, that is, total of 8-16 processors as a maximum. This was not seen as a cause for concern (see 2.1). It was also felt that algorithms used with hundreds rather than tens of processors would need intensive investigation.

1.5 Languages

It was generally agreed that it would be difficult to change from Fortran as the language of meteorological programming. However, it was strongly felt that new techniques could, and should, be grafted onto the original language. A good vector extension language would be a very useful addition, as would multitasking constructs, which, it was suggested, might be taken from ADA.

2. CONSEQUENCES OF REQUIREMENTS

2.1 Number of Processors

It became apparent during the discussions that the smaller the number of processors the better, the optimum remaining a uniprocessor running at the required speed, because of all the additional overheads caused by multiprocessing. The aim should therefore be to combine as few highpowered elements as possible to achieve the operating speed required. It was believed, however, that economic considerations will eventually force the change to large numbers of small processors, since they are likely to prove much cheaper to acquire. These small processors will become very specialised to the point where, eventually, the algorithm is built into the operating system. This level of specialisation would not, however, be appropriate to meteorological applications, since many of the algorithms are subject to constant development and change.

2.2 Memory Layout

It was noted that problems caused by the demands of multitasking upon memory resources had already been experienced: large areas of memory shared in read-only mode, memory bank conflicts, accidental memory overwrites etc. In general, it was felt that the more processors there were involved, the more problems were likely to arise.

2.3 Problem Decomposition

Currently used algorithms can all be decomposed fairly satisfactorily, however, memory management becomes a problem, without the introduction of hierarchies. There has not yet been enough development of software to protect one process from another and debugging poses even more complications. It was noted that the number of processes in problem decomposition was linked to the availability of memory resources: the more memory available, the higher the level of decomposition possible, which would in turn probably lead to a higher proportion of multitasking. However, once the number of processes exceeds the number of processors, the advantage begins to be lost and unbalanced tasks may become a problem.

2.4 Number of Processes

The optimum number of processes depends upon the memory layout of the machines available. It was impossible to come to any conclusions in this regard without having additional information from individual manufacturers.

3. THE TOOLS REQUIRED TO ASSIST IN THE PROGRAMMING OF MULTITASKING MACHINES

3.1 Languages

There are some good languages available at present. Meteorology, however, needs a language which is not only efficient but which can also be easily maintained by a number of individuals. Moreover, code produced should be easily transportable because of meteorology's international nature. The language should allow flexibility for the development of new models, yet at the same time be capable of efficiently running an operational model.

3.2 Memory Protection

One of the most difficult problems to be overcome with multitasking is the protection and efficient use of shared memory. It was considered essential that manufacturers devise a method of comprehensive built-in hardware protection of memory.

3.3 Debugging Facilities

Multitasking makes debugging extremely difficult and the greater the number of tasks running in parallel, the more complicated becomes the location of problems and errors. Users considered the development of sophisticated and comprehensive debugging facilities to be essential to the production of efficient multitasking applications.

4. FUTURE WORK

Finally, the group considered which of these areas needed further investigation. More needs to be learnt about problem decomposition: at what point exactly is the advantage gained by decomposition lost through synchronisation, I/O and other overheads? What is the best method of achieving task control? The tools required to aid multitasking programmers - Fortran extensions, debugging tools, etc., need closer investigation and users should attempt to make their requirements known to manufacturers. Finally, more needs to be known about the kind of memory structures which will be offered by manufacturers, so that this can be taken into account when considering degrees of parallelism.

Tools for Assessing Multiprocessing

DAVID F. SNELLING

ECMWF, Shinfield Park, Reading, Berkshire RG2 9AX, United Kingdom

INTRODUCTION

In the past few years, the computer industry has seen a dramatic increase in the number and variety of hardware architectures with parallel processing capabilities. These computer systems vary dramatically in terms of the size and number of processors, the type of the processor (scalar or vector), the memory structure, the processor/memory interconnect strategy, and the synchronization hardware. Parallel processors tend to fall into several different categories. Many systems fall into the **Parallel Von Neumann** category, where the system is a collection of processors acting on a single global shared memory. Larger numbers of processors can be connected, if the system is composed of a cluster of processors acting on a shared cluster memory and having connections to other **Clustered** processors. A further extension of this approach isolates each processor/memory pair in a completely **Distributed** system much like the Floating Point System's T-Series. Other categories, which will not be discussed here, are **Data Flow**, **Systolic Array**, and **Special Purpose** systems designed for specific applications.

Variation is even greater in terms of the software approach used by different manufacturers. The software approach is typically motivated by the given hardware, so that those facilities available at the hardware level are externalized by the software. As a result there is no accepted, design level model, like a flow chart, for describing parallel algorithms.

In current systems, parallel constructs are made available in the form of parallel languages such as OCCAM [1], in the form of macros where a facility is provided to pre-process Fortran [2, 3], or through the use of libraries such as the Cray and the ETA multitasking libraries [4, 5]. The aspect of parallel algorithm development most affected by the variety of software approaches is the control of data structures. In particular, global memory systems provide a facility for controlling access to critical sections, where global data may be updated. Other systems provide message passing facilities between processors or processes, and private memory for each process, thus preventing direct modification of another process's memory.

Below are examples of three different approaches to the externalizing of parallel processing hardware through user level software. The first is a partial list of the routines from the multitasking library provided by Cray Research.

```
CALL TSKSTART (...)
CALL LOCKON (...)
CALL EVPOST (...)
IF (LOCKTEST (...)) THEN
```

The second example illustrates the multiprocessing facilities provided by the Denelcor HEP system, in which extensions were made to FORTRAN. Asynchronous variables, added to the language, begin with a S sign and are used to control synchronization [6].

```
CREATE SUBR (...)
PURGE SI,SX
SI=SI+1
IF (FULL (SX)) THEN
```

The third example is a small parallel program to compute the fourth power of a variable in OCCAM.

```
PAR
   VAR x:
   SEQ
      CHIN ?  x
      CHTO !  x*x
   VAR y:
   SEQ
      CHTO ?  y
      CHOT !  y*y
```

In an effort to sort out some confusion caused by diversity in the multiprocessor industry, I have divided the problem into four sections which should be helpful in assessing multiprocessing:

1) Program design and development issues.

2) Metrics for understanding the performance characteristics of various multiprocessor systems and their software.

3) A brief discussion of some tools available for analyzing and extracting parallelism from programs.

4) A description of SPLIB, a portable parallel library used in the development of parallel benchmarks and other application codes.

Program design and development issues

The most important aspect of designing or developing a parallel program is data control. The data structures encountered in a parallel environment can be divided into five types. The five types of data are: **private data, invariant data, reduction data, result data,** and **message data.**

Private data is accessible only to a particular parallel process. Typically, data local to a subroutine is private to a process. On some machines data in COMMON is also treated as a private data. However, on machines with large global memories like the Cray X/MP, COMMON is a shared data structure.

The second type of data, invariant data, is available to all parallel processes, but does not actually change during the execution of a parallel section of code. Typical examples of invariant DATA are data statements or PARAMETER statements. Also, COMMON blocks that are changed outside the parallel section of a program and do not change during the parallel section are also considered invariant.

The third type of data is reduction data. This data, on a Cray X/MP for example, resides in COMMON, and may be changed by all parallel processes under the control of critical regions. Data of this type is involved in some form of reduction operation, such as summation, global maximum, or index extraction.

The fourth type is result data. Result data consists of those values created by parallel processes independently of each other. Data generated by one process is not referenced by other processes, although on a global shared memory system they may reside in the same data structure. A typical example of result data is an array in which different columns are generated by different processes.

The fifth type of data is message data. In many systems this simply amounts to a signal from one process to another that an event has taken place. However, in some cases, with machines that do not have an explicit shared memory, message data is used for all information exchange between parallel processes.

After all the data in a parallel application has been categorized into the above types, it is a fairly straight forward process to map these concepts onto a given hardware/software system. Each data type may be dealt with independently and strategies for manipulating them developed.

Once the question of data distribution and control has been answered, there are several other issues to be dealt with. The first is the distribution of work among processes. This involves first dividing the application into sections of computation which may be run in parallel and then selecting an optimal grain size. If the amount of work in the parallel section is too small, then the overhead of multiprocessing will constitute a significant portion of the calculation. If the grain size is too large, then inefficiencies may result because the parallel sections are out of balance. When designing a parallel algorithm it is important to isolate those places in the code, that will have to be execute serially. These can be major bottlenecks during execution and should be eliminated if at all possible.

After a parallel program has been designed, the development phase begins. In the development phase one of the most common techniques used for evaluating the correctness of the program, is to make serial test runs. Unfortunately, a serial test run does not test any of the multiprocessing aspects of an algorithm. In addition this can be very time consuming, and in cases where the program was designed for a large distributed system, it may actually be impossible to run the algorithm serially.

Programers must then resort to executing the program in parallel, and analyzing a postmortem dump when it fails. In many cases this process is aided by the use of full symbolic debuggers and parallel trace back utilities. These can give some indication of what has gone wrong, but in the end the programmer has a formidable task. This task can be made much easier if the problems typically encountered can be classified. Such a classification would help the programmer focus his thoughts on the constructs most likely to cause the given class of problem. I have divided these problems into five classes or effects. These classes are the **stampede effect**, the **accomplice effect**, the **deadlock effect**, the **irreproducibility effect** and the **Heisenberg effect**.

The stampede effect occurs when one process, for whatever reason, encounters an error. It is not possible to immediately stop all other processes, and as a result they may continue to execute and stampede over the evidence, leaving little information for subsequent debugging.

The accomplice effect occurs when one process corrupts another process's data space. In this case the corrupted process is the first candidate for debugging, when in fact it is actually another process, the accomplice process, that has caused the damage.

The deadlock effect is by far the simplest class of problem, in multiprocessing. When a synchronization mechanism, such as a critical section, has not been set up properly, all parallel processes come to a deadlock and stop executing. The result in this class of error is that all the parallel processes have actually stopped at the point where the error occurred. By examining the state of each process in turn the situation can usually be corrected easily.

The irreproducibility effect is probably the most frustrating of all. In many cases an algorithm can be coded that is essentially correct, but because of limited precision on computer hardware, the answers are not exactly the same each time the program runs. It can be argued that these differing results are all equally correct; but because they are not identical, detecting the presence of an error (if one is suspected), or even detecting the fact that it has been removed, may be impossible.

The Heisenberg effect takes place after the debugging process has begun, when additional diagnostics or instrumentation have been added to the code in an effort to track down clues. What typically happens is that the instrumentation perturbs the execution environment enough that the problem vanishes or manifests itself in a completely different form.

Since most all the problems encountered in developing parallel algorithms can be classified according to the above system, programmers may find it helpful to use this classification as a tool for analyzing problems. In any case this classification, should provide programmers with a language in which the problems encountered, when using multiprocessing, may be discussed.

Metrics

A variety of metrics are used for measuring the performance of computer systems. Some of these metrics are very primitive in nature and are used primarily for marketing purposes. These include MIPS, the instruction issue rate of a machine, usually defined as the inverse of the cycle time measured in microseconds; and Mops, which measures the peak operation rate of the machine. Often many Mops are produced as the result of a single instruction. However, the most frequently used primitive metric is Megaflops or MFLOPS. It is the peak floating point operation rate, and is usually equal to an even multiple of the MIP rate, where the number of pipes or the length of the longest functional unit chain provide the multiplier.

More important than these marketing metrics are some of the historical metrics used within the industry. A typical example is the Whetstone instruction rate. Like many performance measures, the Whetstone instruction rate is based on the performance of one particular benchmark. This benchmark was written originally for the Atlas, and each pass through the main loop required 1,000 instructions. This was then used as a basis for machine evaluation.

It was discovered in the early 1970s, with the advent of vector machines, that it was no longer useful to measure only the scalor performance of a computer system. Since then a variety of benchmarks targeted at vector processors have been developed. These tests almost always use MFLOPS as the metric. Examples include the Dongarra LINPACK tests, the Livermore Loops, and the Sandia Megaflops test program.

A more academically based metric for dealing with vector and parallel processing are r-infinity and n-1/2, developed by Dr. Roger Hockney for vector machines [7]. Later Hockney and Snelling developed S-1/2 for evaluating multiprocessing systems [8].

The metrics discussed here are used, by ECMWF, for the evaluation of parallel processing only and not the raw performance capabilities or I/O capabilities of a machine. These metrics are all based on the serial and parallel wall clock execution times of given benchmarks. No evaluation is made of any performance characteristics independent of a benchmark. There are essentially four basic measures used at ECMWF. These are the **speed-up**, the

efficiency, the **granularity** and the **relative speed-up**.
The relative speedup, which although related to the true speed-up, is a different concept which should not be confused with true speed-up.

Measured quantities

p	=	Number of processes
Ts	=	Serial run time for a given benchmark in wall clock seconds on stand-alone machine
Tp	=	Parallel run time for a given benchmark in wall clock seconds on stand-alone machine using p processes
GS	=	Grain size in milliseconds. A grain is the work done by each process between points of synchronization (barrier synchronization is used).

Derived quantities

Sp S'p	=	True and theoretical speed-up
Ep E'p	=	True and theoretical efficiency
Gp(x)	=	Granularity
RSp	=	Relative speed-up.
Sp	=	Ts/Tp
Ep	=	Ts/(Tp*p) or Sp/p
Gp(x)	=	GS such that Ep is greater than x
RSp	=	T_1/Tp

Theoretical speed-up and efficiency

ts	=	Time of serial code + Time of fixed multiprocessing overhead + Time of per process overhead run in parallel
tq	=	Time of parallel code run using q processes
to	=	Time of per process overhead run serially
S'p	=	Ts/ts + t2/2 + ... tp/p + p*to or Ts/ts + tp/p + p*to for balanced code
E'p	=	S'p/p

Analysis Tools for Parallel Programs

The first step in developing a parallel algorithm from an existing program, is to do an analysis of the serial code. This involves such utilities as cross references and flow traces, which give an indication of where the program is spending the greatest amount of time. More sophisticated tools include dependency analysis programs that will analyze the flow of data in a Fortran program and give information on what sections can be multiprocessed.

The second step involves the structuring of memory and data, probably the most important aspect of multiprocessing. There are essentially two extreme methods of managing data. In the first approach all data is treated as global and changes are made to shared data structures only inside critical sections. In the other extreme all data is private to the process and data sharing in accomplished through the exchange of data by means of some form of message passing construct.

Below is a list of some of the advantages and disadvantages of these two approaches.

Global shared data - advantages

Any data that the programmer may want access to are easily available.

Programming is easier with global data since the structures are more like those already familiar to FORTRAN programmers.

Global shared data - disadvantages

Programmers are susceptible to the stampede effect.

Programmers are susceptible to the the accomplice effect.

Programmers are susceptible to the the deadlock effect.

Programmers are susceptible to the the irreproducibility effect.

Programmers are susceptible to the the Heisenberg effect.

The user has no explicit knowledge of shared data, particularly after code modification has been made.

Private data - advantages

There is no stampede effect.

There is no accomplice effect.

There is a reduction in the Heisenberg effect.

Achieving reproducibility is easier since the user has exact knowledge of all the data exchange.

Private data - disadvantages

Programmers are susceptible to the deadlock effect.

In many cases excess data duplication may be necessary.

It tends to be easier to develop a code from scratch rather than modifying an existing program.

The third step is to select a multiprocessing strategy, a process simplified by the use of a conceptual library of multiprocessing templates. A template is a multiprocessing structure around which a parallel application can be built. Examples of templates include functional scheduling, pre-scheduling, self-scheduling, pipelined or large-grain data flow, conversational scheduling, divide and conquer approaches, and combinations which include any of the above.

Functional scheduling is used primarily when different aspects of the application can be separated from each other. An example would be the control system of an airplane which can be simulated independently of the engine system. Pre-scheduling is probably the most common of all approaches to multiprocessing. In a pre-scheduled algorithm different passes through a loop are allocated to different processes at compile time. In the self-scheduling approach passes through a loop are allocated at run time. A process, when it has finished one pass through the loop, will go and select the next pass not yet computed. As a result, each process may not compute the same number of passes through the loop. In many cases this allows for better computational balancing between processes. Pipelined and large-grain data flow parallel processing are useful when there are many stages to a calculation that are repeated for many sets of data. Such stages might be input data, first-phase calculation, second-phase calculation, graphics and output. By performing each of these stages on different sets of data in parallel, a some degree of parallelism may be achieved. In this approach the slowest stage in the process will typically form a computational bottle neck, thus limiting the parallelism achieved. Probably the most confusing approach to multiprocessing is the conversational approach, in which several parallel processes are communicating with each other at arbitrary points in the code as if in a conversation. This allows for a fairly flexible mode of programming but can be quite confusing. The divide and conquer approach is useful for recursive algorithms. For example a quick sort, in which the first level of the sort can be run serially and then the list split in half and one half given to a parallel process. If this is repeated high degrees of parallelism can be achieved. The combination of these approaches can be quite useful, particularly when a high degree of parallelism is required. For example, a program may be functionally divided at the outer level, and then each of these functional sections may be run as self-scheduled subroutines.

The fourth tool useful in evaluating and using multiprocessing systems is that of time stamping a parallel code. This can provide a detailed performance analysis and allow for the detection of bottlenecks and other bugs within the program. Most multiprocessing libraries currently on the market provide access to this information, although a detailed post-processing analysis is necessary to make meaningful use of this data. These post-processing tools are now provided by Cray Research, for example.

Portable parallel library (SPLIB)

Because of the complete lack of standardization in the use of language extensions, parallel libraries, and compiler directives, as means of providing access to parallel hardware, it was decided that a multiprocessing library would be developed that could provide access to the multitasking facilities on several different multiprocessor systems. To this end a portable parallel library named SPLIB (Standard Parallel LIBrary) was developed at ECMWF. By evaluating the various different multiprocessor systems available, a sub-set of parallel constructs was developed that could be supported on most machines. One major ramification of this attempt at standardization was the use of private memory rather than global memory for all data. Most systems provide some form of support for private memory, whereas only a few systems provide a global memory structure. One of the virtues of SPLIB is that it is simple and complete. SPLIB supports all the basic functions provided by other parallel libraries and some of the special facilities provided by parallel languages, such as OCCAM.

The remainder of this paper will discuss the basic philosophy of SPLIB, give a description of its syntax and usage, provide some examples, make comparisons with other multiprocessing libraries, and provide a brief discussion of some known weaknesses in the library.

There are four basic structures in the parallel library. The concepts of stream, barrier, channel and shared matrix will be discussed independently as well as a brief discussion of error handling and implementation philosophy.

The stream model

A stream is a parallel process, which is a sequence of independent operations, that may be created or terminated by the user. The user has the ability to name or identify these streams and they are not specifically assignable to a given processor. These streams may be started at library initialization time or they may be started subsequently in the course of program execution. The most important aspect of the stream is that all memory is private to the stream. In other words, no stream has access to any information used by another stream except through calls to the library.

The barrier model

A barrier is a global form of synchronization that allows a selected number of parallel streams to come to a particular point in the code, stop and wait for other streams to arrive at the same point in their code, and then all streams resume execution together. It is an all-in-all-out barrier, that is no streams may enter a barrier once streams have started to leave it. It is essentially equivalent to a fork/join construct although it tends to be much more efficient on most multiprocessor systems. It may be used by all the streams executing in a given program or it may be used by any sub-set of streams, thus it allows for control of nested parallelism as well as many other constructs. One useful variation in this implementation of the barrier allows the last stream into the barrier to execute a portion of code serially. When it has completed this serial segment of code it then signals all other streams to leave the barrier.

The channel model

The channel model provides for synchronous communication between streams. The two constructs are the ability to send or receive a block of data. Only one sender/receiver pair may communicate using a given channel at a time. In particular, the sender will wait for a receiver or a receiver will wait for a sender. Once the two have been paired, then a data exchange takes place. Only after the data has been transferred are the two streams allowed to continue executing. All streams within the program have access to all channels. There is no stream to stream connection defined for the channels. Therefore, except for the lack of point to point connections, SPLIB channels are functionally equivalent to OCCAM channels.

The shared matrix model

Unlike channels, the shared matrix provides for communication between processors and a large, logically, shared memory. This memory may be a large external memory, a portion of global memory, or distributed among private memories in a distributed memory system. This memory space is not accessible directly by the user but only through read and write calls in the library. Shared matrix data may be written or read by the library as elements, rows, columns, or as higher dimensional equivalents. A shared matrix is simply a four-dimensional FORTRAN-like array, which may be accessed through the library.

Each time a shared matrix is defined, its dimensions are specified and space allocated for its data. Library routines may then access any portion of the shared matrix. In addition to straight forward read/write access all, or part, of a shared matrix may be locked by a stream for exclusive access. One restriction in the shared matrix model is that in any request to a shared matrix, each dimension may be accessed in only two ways. Either a single index may be accessed or the entire dimension may be accessed, but not a portion of a dimension. In this way elements, entire rows, entire columns, or entire planes may be accessed.

Implementation notes

Streams, barriers, channels and shared matrices may be named for identification. All access to the library constructs are through FORTRAN syntactically correct subroutine calls. Each call has a fixed number of parameters and conforms to Fortran 77 syntax. This Fortran syntax allows for easy transportation of the library to different multiprocessor systems. It also provides the ability to use the multiprocessing library in a serial processing environment. As a result the portability characteristics of programs written using this library are much better than if vendor proprietary libraries, parallel languages, or FORTRAN extensions are used.

The primary consideration, after portability, in the development of SPLIB was performance. As a result error checking and reporting are implementation dependent and may be set as run time options.

The assumption for anyone implementing SPLIB is that correctly written programs will work correctly. No guarantees are made about the behavior of the library if it is used incorrectly, or if the program is not accurate. In the initialization call to the library a set of parameters is provided for the specification of error reporting and handling. All these parameters are implementation dependent. In addition, these parameters may be used to set up time stamping, tracing options and even memory allocation schemes for the shared matrix. This is particularly useful on large distributed memory systems, where memory allocation may be run time dependent. The only assumption about these parameters is that if they are omitted, correct programs will work.

Parallel Stream Creation

call crstrm (ID,subr)	- Create a stream
call wtstrm (ID)	- Wait for stream
ID=IDstrm()	- Return stream number

ID	is type integer and may be a Hollerith string not more than 1 word long.
subr	is an entry point declared EXTERNAL in the calling routine.
IDstrm	is integer valued, and will return 0 if called from the main program. Duplicate ID's are illegal, however error checking and handling are implementation dependent.

Barrier routines

call Barr (name,nstrms)	- Barrier synchronization
L = LBarr (name, nstrms)	- Logical barrier

name	is type integer, but may be a Hollerith string not more than 1 word long.
nstrms	is the number of streams which must call Barr, with the same name, before the barrier is released. The barrier is consistent in that all streams in a (just released) barrier must exit before any may re-enter. Exactly nstrms must call the barrier with Barr before the barrier will be released.
LBarr	is a logical valued function which returns False for all but the last stream. The last stream will not be held by the barrier and will be allowed to continue processing.
Barr	will release and reset the barrier when nstrms have called it.

Note: The following statements are equivalent:

```
call Barr ('name',nstrms)
IF (LBarr('name',nstrms)) call
Barr('name',nstrms)
```

where nstrms has the same value in all barrier calls.

Channel routines

Channels are used for communication between two parallel processes. Communication is always synchronous. That is, neither process continues until both have reached their respective calls, and the data has been transferred.

```
call Sendch (name, SRC, length)      - send data on channel
call Recvch (name, DES, length)      - receive data on channel
  L = LRcvh (name, DES, length)      - Logical receive data on
                                       channel
```

name — is type integer, but may be a Hollerith string not more than 1 word long.

Sendch — sends data stored sequentially in SRC out on the channel specified.

Recvch — works in the same way except that data from the specified channel is stored sequentially in DES.

length — It is an error if the lengths in the corresponding Sendch/Recvch calls do not match, but checking is an implementation issue.

LRcvch — Returns true if the transfer was performed and false if no corresponding send was ready (no data is transferred in the false case).

A Sendch (Recvch) call waits for a Recvch (Sendch) call on the same channel and then the data is transferred. A channel may be used only by one Sendch/Recvch pair at a time although there are no connections defined and therefore all streams may send and receive on any channel.

Shared Matrices

```
call opensm (name, DIM1, DIM2, DIM3, DIM4) - open (define) shared
                                             matrix
call clossm (name)                         - close shared matrix
call Writsm (name, SRC, I1, I2, I3, I4)    - write data to shared
                                             matrix
call Readsm (name, DES, I1, I2, I3, I4)    - read data from shared
                                             matrix
call Locksm (name, I1, I2, I3, I4)         - lock shared sub-matrix
call Unlksm (name, I1, I2, I3, I4)         - unlock shared
                                             sub-matrix
```

DIMi — is the size of the ith dimension within a shared matrix. A given shared matrix must be opened before it may be accessed.

Writsm — writes data stored sequentially in SRC to the shared sub-matrix specified. If Ii is positive only that index is used in specifying the sub-matrix. If Ii is '*' the whole dimension is used to specify the shared sub-matrix based on the corresponding DIMi used in the opensm call. In this way entire arrays or columns or rows of arrays may be transferred as well as single elements.

Readsm — works in the same way except that data from the specified sub-matrix is written sequentially into DES.

Note: The integer equivalent of '*' is not a valid dimension or index.

The user is expected to handle the synchronization. All transfers take place when the call is made. Waiting for shared matrix to be unlocked takes place only in Locksm and not in Readsm or Writsm calls. The sub-matrix that is to be locked cannot overlap with an already locked sub-matrix. Readsm/Writsm/Unlksm calls accessing locked memory proceed without error checking (although warnings may be issued in some implementations).

Initialization

call spinit (nstrms, mainsub, tunary, ntune) - initialize parallel library

nstrms — number of processes to create initially executing the main program mainsub.

tunary — an array of implementation dependent items to allow for tuning the library for optimum performance on specific systems.

ntune — is the number of items in tunary. Zero is a valid number in which case the library establishes default characteristics.

Examples

In the first example a matrix B is transposed and all its elements squared. This is accomplished using four parallel streams. The entire program is given below:

```
      EXTERNAL CONTROL
      CALL SPINIT (4,CONTROL,0,0)
      END

      SUBROUTINE CONTROL
      LOGICAL LBARR
      COMMON/DATA/B(100,100)
      IF (LBARR('START',4)) THEN
           CALL INITB (B)
           CALL OPENSM ('A',100,100,1,1)
           CALL OPENSM ('B',100,100,1,1)
           CALL WRITSM ('B',B,'*','*',1,1)
           CALL BARR ('START',4)
      ENDIF
      CALL MATQSQ
      RETURN
      END

      SUBROUTINE MATSQS
      COMMON/TEMPDAT/A(100),B(100)
      ISTART=IDSTRM()+1
      DO 20 IPART=ISTART,100,4
           CALL READSM('B',B,'*',IPART,1,1)
           DO 10 J=1,100
                A(J)=B(J)*B(J)
10         CONTINUE
           CALL WRITSM ('A',A,IPART,'*',1,1)
20    CONTINUE
      CALL BARR ('DONE',4)
      RETURN
      END
```

The second example shows how the library can be used where the multiprocessing strategy is run time dependent. In particular, the number of streams used is read in from I/O unit 5.

```
      EXTERNAL CONTROL
      CALL SPINIT (1,CONTROL,0,0)
      END

      SUBROUTINE CONTROL
      COMMON/DATA/NSTRMS,D(1000),X(10)
      EXTERNAL SUBR
      READ (5,*) NSTRMS
      DO 10 ID=1,NSTRMS
           CALL CRSTRM (ID,SUBR)
10    CONTINUE
      DO 20 ID=1,NSTRMS
           CALL SENDCH ('XMIT',NSTRMS,1011)
20    CONTINUE
      DO 30 ID=1,NSTRMS
           CALL WTSTRM (ID)
30    CONTINUE
      RETURN
      END
```

```
      SUBROUTINE SUBR
      COMMON/DATA/NSTRMS,D(1000),X(10)
      CALL RECVCH ('XMIT',NSTRMS,1011)
      ISTART=IDSTRM ()
      DO 10 IWORK=ISTART,ALLWORK,NSTRMS
         CALL WORK (IWORK)
10    CONTINUE
      RETURN
      END
```

Comparisons

Table 1 provides a summary of the various different libraries studied. In particular, the libraries are measured in terms of the number of routines present in the library and the number of parameters used by all routines combined. These figures provide a rough idea of the complexity of a library. It is the author's feeling that SPLIB with 15 routines and 58 parameters provides extensive capabilities without the magnitude of complexity present in the ETA and MTL libraries. (MTL is a multitasking library usable by CRAY and ELXSI).

LIBRARY	NUMBER OF ROUTINES	NUMBER OF PARAMETERS
SPLIB	15	48
HEP	11	15
CRAY	16	22
ETA	61	187
IBM	4	5
MTL	38	51

Table 1: A comparison of SPLIB and other parallel libraries

Known Weaknesses

As a result of using SPLIB in several different machine environments it has become apparent that there are some weaknesses. In particular, when the library itself was ported to an IBM machine some implementation problems arose. The IBM provides a precise implementation of the character data type, whereas the CRAY provides an extension to the Hollerith concept. Both are correct in the standard, but the process of identifying whether the "*" element is a character string or an integer, is difficult on the IBM. This does not mean that the facility is not usable, simply that the library is somewhat more difficult to implement. Secondly, because of the fact that the library is called in some cases with integers and with other cases with a character string means that the library does not conform to the execution standard for FORTRAN, although it is syntactically correct.

Probably the most serious weakness in the SPLIB library is the assumption that in both the channel model and the shared matrix model all variables are exactly one word long. This is not a problem on many machines, because most machines provide an environment in which integers and reals are the same size. However, it is frequently assumed that integers are 32 bits long and double precision arithmetic is performed with 64 bits words. This inconsistency produces some confusion when trying to develop codes for machines such as the IBM. As a result, one potential expansion to the library would be the facility of typing the shared matrices. This would allow an arbitrary specification of the size of elements. It is also believed that the increased complexity would not be severely hampering. One other aspect in connection with this is that channels also assume that a certain number of words are being transferred. This problem could easily be eliminated if channels transferred bytes instead of words. It would require the programmer to include in his parameter statements information on the size of real and integer variables. But it would increase the portability of the SPLIB model.

Conclusion

In conclusion the author believes that although parallel processing is a very difficult aspect of code development, it is not one for which tools or facilities are lacking. With the extensive capability that is provided by large cross references, time stamping, call tracing and data dependency analysis programs; the sorting out of which sections of code may run in parallel has become a less arduous task than it was when parallel processing was first introduced.

Also, SPLIB provides, at the very least, a benchmark tool for evaluating multiprocessor systems, but may be even a facility for producing code that could be used on several multiprocessor systems. This is particularly important at ECMWF, where information may need to be freely exchanged between the Centre and its Member States. At present, this is limited to information and knowledge and very little sharing of actual code is taking place. This is partially due to the very specific machine requirements imposed by multiprocessing.

Bibliography

[1] Hoare, C.A.R., ed. OCCAM Programming Manual, Series in Comp. Science, (Prentice Hall International, 1984).

[2] Arnold, C.N., "Proposal for Automatic Parallel Processing on the ETA 10", Internal discussion document, 1986.

[3] Lusk, E.L., and Overbeek, R.A., "Implementation of Monitors with Macros", Technical Memorandum ANL-83-97, Argonne National Lab., Mathematics and Computer Science Div., December 1983.

[4] Arnold, C.N., "Multitasking Library: Design Summary and Specification", ETA Internal document, July 1985.

[5] Cray Research Inc., Programming Library Reference Manual, SR-0113, (Cray Research Inc.; Mendota Heights, Minnesota), 1986, p. 14-1 to 14-31.

[6] Smith, B.J., "A Pipelined, Shared Resource MIMD Computer", Proceedings of the 1978 International Conference on Parallel Processing, p. 6-8.

[7] Hockney, R.W., and Jessope, C.R., Parallel-Computers: Architecture Programming and Algorithms (Adam Hilbers, Bristol, 1981).

[8] Snelling, D.F., and Hockney, R.W., "Characterizing MIMD Computers: e.g. the Denelcor HEP", Parallel Computing 83 (North Holland, Amsterdam, 1984), p. 521-6.

Dataflow Architectures and Implicit Parallel Programming

JOHN GURD

Department of Computer Science, University of Manchester, Oxford Road, Manchester M13 9PL, United Kingdom

SUMMARY

Dataflow is a technique for expressing and executing parallel computations that uses fine-grain message-passing to achieve asynchronous concurrent execution of graphical programs. Applications for dataflow machines are expressed in high-level, implicit, single-assignment programming languages. These concepts are introduced, and various specific dataflow machine/language systems are described.

1. INTRODUCTION

1.1 Parallel Computing Systems

We take the view that a parallel computing system comprises a **hardware architecture**, based on a model of computation with inherent parallelism, together with a **programming environment** that may or may not make the programmer aware of the hardware-level parallelism. Where the programmer must specify and control parallelism we call the resulting system **explicit**, otherwise it is **implicit**.

Dataflow is an implicit parallel computing scheme which uses a directed graph model of computation together with a single-assignment programming language. This paper introduces the implicit approach to parallelism, and describes the techniques used to implement an implicit system via the dataflow/single-assignment scheme.

1.2 Implicit Parallel Programming

It is important that applications programs are written in high-level programming languages that allow the programmer to abstract away from details of the machine structure. There is clear evidence that the programming task becomes significantly more complex and error prone when the programmer is required to take account of machine-level features, such as vectorisation, overlays, etc. This situation is exacerbated by the introduction of parallelism. It is therefore desirable that a parallel programming style be developed to retain the power of current programming languages and yet remove the necessity for explicit expression of parallelism, thus requiring no knowledge of the machine structure. This mirrors trends in conventional high-level programming languages, which hide details of complex architectural features (instruction pipelines, virtual memory, etc.) whilst reflecting the sequential nature of the machines that execute the compiled code.

In implicit parallel programming, it is intended that compilers should extract parallelism from programs automatically, and that the run-time environment should handle all workload distribution and resource management. Languages should be chosen so as to simplify implementation of these difficult tasks.

One option is to generate parallel code from a **conventional** high-level programming language. However, suitable analysis algorithms are highly complex because of the potential side-effects of program statements, due to explicit use of storage locations (accessed by the programmer as **variables**). Compile-time data dependence analysis is often impossible (e.g. when obscure array index expressions are used) unless assistance is available from the programmer (e.g. to indicate how the arrays will be accessed). However, the worst problem is that of **aliasing** via the use of unbounded arrays and arithmetic operations on pointers. No amount of compile-time analysis can help unravel devious or undisciplined use of such language features. The only method of control is to ban the facilities from the language.

Single-assignment languages (SALs) have no concept of sequential execution and no direct control statements such as the GOTO. To combat the ambiguities that might arise from reassigning values to variables, the languages allow each variable to be assigned just once in a program. Constructs that permit controlled reassignment in special cases (such as loops) are provided. SALs tend to use simple data structures, such as **arrays** and **streams**, that can be readily implemented. There are often strict type and scope rules. In particular, it is common to prohibit all forms of side-effect in re-entrant constructs. The net results are languages that provide ideal textual syntax for the description of dependence graphs.

SALs are similar to **functional** languages. These are based on the mathematics of functional algebra and have no concepts of storage state and assignment. They are sometimes referred to as **zero-assignment** languages. In fact, if assignment is restricted to occur only once for each variable in a program, the effect is the same as if there were no assignment at all, and assignment statements were treated as **definitions** of the variables. In this sense SALs and functional languages are identical, and it should come as no surprise to find that absence of GOTOs and side-effects are common to them both. However, functional algebra allows more powerful programming constructs than are used in SALs because it permits construction of higher order functions and abstract data structures. Consequently, the two groups are not directly equivalent. Nevertheless they have enough in common to make it attractive to implement full functional languages on parallel systems. There is similar interest in parallel implementations of **logic** programming languages.

1.3 Parallel Dataflow Computation

Dataflow notation expresses program parallelism in the form of **directed graphs** in which nodes describe operations to be performed and arcs describe paths along which data is routed from operation to operation. Although such graphs can be constructed at many different software levels, it is usual to consider nodes as machine **instructions**, rather than as **subroutines** or **procedures** some people use the term dataflow to cover all software levels, but this paper uses the term in the restricted sense of instruction-level dataflow. Execution of the directed graphs is **data-driven**, in that it proceeds according to the availability of data at the inputs to each node. Whenever a node has data available at all of its inputs, it executes, consuming the input data values from the input arcs, and subsequently producing its result, placing the output data value(s) on the output arc(s). Since the decision whether or not to execute is made locally at each node, and on the basis of information available only at that node, no communication other than that of data values is needed between nodes. The execution of nodes is thus entirely asynchronous, and it may be as parallel as desired within the constraints of the program.

Since 1980, active dataflow research has been concentrated in four major locations, at MIT in the USA, at the University of Manchester in the UK, and at the Electro-Technical Laboratory and the Nippon Telephone and Telegraph company in Japan. Two main themes can be discerned in recent work, the first concerned with implementation of high-performance **numerical** processors for scientific computation, the second concerned with parallel implementation of **functional** programming languages. This paper concentrates on the former area.

Amongst the active dataflow research groups there is some degree of consensus on the general approach to adopt. Three components are agreed to be involved: (i) a **fine-grain graphical model of computation**, normally based on **tagged message-passing**; (ii) a multiprocessor dataflow machine, conceptually comprising (a) a number of **processing elements**, (b) some form of specialised, distributed **data structure storage**, and (c) a **network** to interconnect these units; and (iii) a **high-level programming language**, normally based on the principle of **single-assignment** Styles of hardware architecture have been surveyed by Treleaven et al. (1982) and Watson (1984). Suitable languages have been surveyed by Ackerman (1982) and Glauert (1982).

The paper commences with a description of the single-assignment language SISAL, an example of a high-level implicit parallel programming language. The fundamental features of the dataflow model of computation are then presented. Various proposals for dataflow computer architectures are described. The efficiency of dataflow object-code generated from SISAL programs is discussed. Finally, the nature of parallelism found in dataflow/single-assignment systems is compared with that found in other styles of parallel computer.

2. THE SISAL PROGRAMMING LANGUAGE

2.1 Basic Concepts

SISAL is an expression-oriented, single-assignment language, designed by McGraw et al. (1983). It is weakly typed, i.e. types with the same structure are equal, and allows only explicit type conversion. A program written in SISAL is concerned with the definition and use of **values**. A value may be a **constant value** (such as '2.71828'), or a **value name** (such as **'time step'**) which is associated with the result of an **expression** evaluation. No value name may be redefined, and so all uses of a given value name in other expressions will always refer to the same value. If an expression uses a value name which has not yet been defined, evaluation of that expression cannot be completed until that value name is available for use. The **order** of execution of a SISAL program is thus determined solely by the availability of values for use in the computation of expressions, and will not affect the computed results.

An expression may yield more than one result. The number of results produced by an expression is referred to as its **arity**. In common with procedural expression oriented languages, the control constructs in SISAL are **structured expressions**. Any expression may be freely nested within any other expression, provided that its arity and the types of its results are correct for the context in which it is used.

Function calls may occur as part of any expression. A function definition simply encapsulates an expression within a parameter-passing mechanism. Functions, like the rest of the language, are completely free from side-effects, and have access only to their arguments and to other function calls.

2.2 Types, Values and Associated Operations

2.2.1 Scalar Types

SISAL provides the scalar types: **boolean, integer, real, double real and character.** Although SISAL provides for the declaration of value names, the typing system enables the type of any value name to be directly deducible from its defining expression, so declarations are only used when their inclusion improves program clarity.

The usual boolean operations and ( ), or (|) and not (~) are provided, together with tests for **equality** (=) and **inequality** (~=).

All the usual integer, real, and double real arithmetic and comparison operations are provided. In addition, a number of predefined functions are provided such as modulus (**mod**), magnitude (**abs**), maximum (**max**) and minimum **(min)**, and real and mixed exponentiation (**exp**). Explicit type conversion operations are provided. In each case the operator name is the name of the required type. These are necessary in SISAL since no implicit type coercions are performed.

The usual character comparison operations are provided, together with a type conversion operation to convert a character value to an integer. Character values consist of the 128 characters of the ASCII character set, and are mapped onto the integer subrange 0..127.

2.2.2 Arrays

An **array** has an implicit integer index type and components of an arbitrary but uniform type. Arrays have dynamic bounds, the size of the array being determined by evaluation of the expression defining the array value. Multidimensional arrays are represented as arrays of arrays. Some example array type definitions are:

```
type  a-real  = array [real]
type aa-real = array [a-real]
```

Creation operations create an array from component values. **Create empty** creates an empty array of a given type; **create by elements** creates an array by explicitly listing all its elements; and **create by fill** creates an array of a specified size with all elements equal to one given value:

```
A: = array aa_real [];                    % create  empty
B: = array a_real [0: 1.0, 10.0, 100.0]   % create by elements
C: = array_fill (1, 10, 0.0);             % create by fill
```

An empty array is represented by an upper bound one less than the lower bound; in particular, the create empty operation always sets the lower bound at 1 and the upper at 0. The create by elements operation takes a lower bound, followed by a list of the elements - the array type can be omitted, since it can be deduced from the element type. A special form of the create by elements operation is provided for creating an array of characters, using a character string enclosed in double quotes. For example:

```
D: = "string";
```
is equivalent to:
```
D: = array [1: 's','t','r','i','n','g'];
```

The create by fill operation takes the lower and upper bounds, followed by the value to be used in filling the array.

Access operations obtain some component part of an array value. **Array select** accesses an array element; **array_liml, array_limh** and **array_size** obtain the lower bound, upper bound and size of an array respectively:

```
A: = array [0: array a_real [], array [1: 1.0, 2.0, 3.0] ];
B: = A [1];                  % array [1: 1.0, 2.0, 3.0]
x: = A [1, 2];               % 2.0
i: = array_size (A);         % 2
j: = array_liml (B);         % 1
k: = array_limh (A[0]);      % 0
```

Transformation operations create a new array from an existing array value. The **replace** operation replaces a specified element in an array with a new value. For multi-dimensional arrays, replace may be used to change a complete element at the outermost level, or an element at any inner level. Several elements may be changed in one replace by specifying a list of subscript and value pairs:

```
S: = array [1: 2, 5, 7];     % 2 5 7
V: = S [2: 4; 3: 8 ];
```

The **concatenate** operation joins two arrays of the same type:

```
X: = array [1: 1 ];
Y: = X || V;                  % 1 2 4 8
```

Finally, the language provides an **adjust** operation that selects a subrange out of an array and **add** and **remove** operations that allow an array to be extended by one element or reduced by one element at either end of its range.

2.2.3 Streams

A **stream** is an ordered sequence or list of values of some arbitrary but uniform type. Unlike an array, components of a stream cannot be accessed randomly, but only in order, using operations such as **first** and **rest**. Stream type definitions are similar to those for arrays.

Creation operations create a stream from component values. **Create empty** creates an empty stream of the specified type; and **create by elements** creates a non-empty stream by listing all its elements:

```
type s_bool = stream [boolean]
type s_int = stream [integer]
S: = stream s_bool [];                % create empty
T: = stream s_int [2, 3, 5, 7, 11];   % create by elements
```

Access operations obtain some component part of a stream value. **Select first** accesses the first element of a stream; **stream size** gives the number of elements in a stream; and **test for empty** returns a boolean value indicating whether the stream is empty:

```
next := stream_first (T);             % 2
how_many := stream_size (T);          % 5
all_used := stream_empty (S);         % true
```

Transformation operations create a new stream from an existing stream value. **Select rest** discards the first element of a stream; **append** adds a new value to the end of a stream; and **concatenate** joins two streams of the same type:

```
U := stream_rest (T);                  % [3, 5, 7, 11]
W := stream_append (S, false);         % [false]
X := stream_append (W, true);          % [false, true]
Y := W || X;                           % [false, false, true]
```

2.2.4 Records

A **record** consists of one or more component fields of arbitrary types, each field being identified by a field name. There are three record operations: **create** a record value, **select** a field from a record, and **replace** a field in a record with a new value, giving a new record value. The type name for a record create is optional, since its type is deducible from the types of the fields and the language is weakly typed. More than one field can be changed by a single replace operation. For example:

```
type dest = record [Segment, Node : integer;
                    Input : character; MatchFn : array [character] ]
D        := record dest [ Segment : 2; Node : 2064;  % create
                          Input : 'L'; MatchFn : "EW" ];
in_point := D.Input;                                  % replace
F        := D replace [ Segment: 1; Node: 1405 ];     % replace
```

2.2.5 Unions

A **union** type consists of a set of alternative constituent types, each type being associated with a tag name. A value of a union type consists of a **tag value**, and a **constituent value**, of the type corresponding to that tag name.
For example:

```
type arith_op = union [ int_op:  integer;
                        real_op: real;
                        dble_op: double_real ]
```

A value of type arith_op will have a tag value of int_op, real_op or dble_op and a corresponding constituent value of type **integer**, **real** or **double_real**, respectively.

Unions may also be used to construct an enumerated scalar type, each tag name being used to represent one of the scalar values. A special constituent type **null** is provided for use with the tag:

```
type move = union [Lft: null, Rght: null, Up: null, Down: null]
```

A **null** type specification may be omitted, giving the following, more natural style for enumerated scalars:

```
type move = union [Lft, Rght, Up, Down]
```

In a **recursive** union type declaration, the structure will contain an element of the union type being defined. To illustrate this, consider the following type, defining a binary tree whose elements are of type data:

```
type TreeTp = union [Empty;
                     Full: record [ Element: data; lft, Rght: TreeTp ]
                    ]
```

A mechanism for accessing constituent values from union values is provided in the **tagcase** construct, which will be introduced in section 2.3.5. Apart from this, there are two union operations: **create** a union value, and **test tag,** which tests whether a union value was created with a specified tag:

```
opl := union arith_op [ int_op: 8 ]      % create
test_for_real := is real_op (opl);       % test tag
```

The type name for a union create is mandatory, since the same tag name may be used in different union types.

2.2.6 Error Values

Exceptions are handled in SISAL by generating **error values.** A special value **error** is a proper value of every type in SISAL. The production of an error value from an expression indicates that some exceptional condition prevented the normal evaluation of that expression. When an error value is generated for a component of a structured object, values for those components known to be correct are wherever possible preserved.

2.3 Control Structures

2.3.1 The **let** Block

The purpose of the **let block** is to define one or more value names, which are then used in the evaluation of an expression. The results of this expression are the results of the block. Value names defined in a block are not available for use outside that block. The following block calculates the two roots of a quadratic equation:

```
x_1, x_2 := let disc := b * b - 4.0 * a * c;
                root : = sqrt (disc);
                two_a := 2.0 * a
            in
                (-b + root) / two_a ,
                (-b - root) / two_a
            end let
```

2.3.2 The Conditional Construct

The **conditional construct** is used to select one of two **alternative** expressions for evaluation, depending on the result of a boolean test. The selected expression is evaluated, and its results are returned as the results of the whole construct.

The **else** branch must always be present to define the results of the conditional when the test yields **false**. Similarly, the expressions given in the two branches must **conform** in the **number** and **types** of results produced.

Finally, the conditional construct can be used to select one of a **number** of expressions for evaluation, without the use of nested conditionals, by using one or more **elseif** βranches:

```
result := if op_ch = '+'        then arg1 + arg2
             elseif op_ch = '-'  then arg1 - arg2
             elseif op_ch = '*'  then arg1 * arg2
             else                     arg1 / arg2
          end if
```

2.3.3 Two Iterative Constructs

Many programming languages provide an iterative construct for applying a certain operation to a number of elements. This construct often implies some form of timing: the operations are e.g. serialised. In SISAL there is no such concept of timing, the only restriction to parallelism is data dependence. In the **for all construct**, there are no data dependencies between cycles, and so it is possible to execute all cycles of the iteration concurrently. The **for init construct** is used to express data dependencies from one cycle to the next.

A **for all construct** comprises a generator which specifies the range of values for which the body will be executed, a **body**, and a **returns section** which specifies which value(s) will be **yielded. More than one result may be yielded by specifying a list of return values. The greatest** or **least** value over **all** cycles may be returned. Alternatively, the values from each cycle may be **reduced** into a single value using a **sum** or **product** operation. Finally, the returns section may **collect** all definitions of some value name and return them as an **array** or stream.

The **for all** construct is particularly useful for handling and constructing array and stream values. The generator may make the elements of an array available inside a loop without explicit subscripting. For example, assuming **Vec** to be of type **array[real]** the following **for all** calculates a new array **New_Vec** in which each element of **Vec** is reciprocated:

```
New_Vec := for elem in Vec
             recp := 1.0 / elem
           returns array of recp
           end for
```

Multiple ranges may be given, and combined using either **dot** or **cross** products. In a dot product, the ranges must have the same size and the **i-th** element of each range is used to drive the **i-th** body. The cross product is short hand for a nested **for all**. The next example performs matrix multiplication on two arrays **M1** and **M2** of type **array[array[real]]** of size a x b) and (c x b), respectively (the second matrix has been transposed before multiplication). The size of the array **M3** produced by the loop is determined by the cross product used in the generator, and in this case is (a x c).

```
M3 := for Row in M1 cross Col in M2
        M3_El := for R_El in Row dot C_El in Col
                    P := R_El * C_EL
                 returns value of sum P
                 end for
      returns array of M3_El
      end for
```

The **for init construct** is used in the following example to calculate the square root of a value x using the Newton-Raphson method:

```
sqrt_x := for initial est := 0.5 * x
          repeat est := 0.5 * (old est + x / old est);
                 diff := abs (est * est - x)
          until diff < tolerance
          returns value of est
          end for
```

The **initial section** defines initial values of **loop names** (**est** in the example) to start the iteration. Execution of this section is considered to be the first pass through the loop. The **loop body** is executed once for each subsequent pass, and defines new values for the loop names. The value of a loop name from the **immediately preceding** cycle may be used, and this is indicated by prefixing the use of the required loop name with the keyword **old.** The **termination test** is followed by the **returns** section which specifies the result(s) of the complete iteration; **value of** indicates that the value of the given value name when the iteration terminates is to be returned.

By placing the termination test **before** the body, an iteration that may execute the body **zero or more** times can be constructed. The keyword **until** in the test may be replaced by **while** to give the negated sense of the boolean expression. In addition to **value of,** the returns section provides the same possibilities as in the **for all** construct.

2.3.5 The Tagcase Construct

The **tagcase** construct is used to select one of a number of expressions for evaluation, depending on the tag of a union value. The following simple example updates a position in an x,y plane based on the direction of a given move

```
type move = union [Lft, Rght, Up, Down]
this_move := Get_Move (Player); % yields a value of type move
new_x, new_y := tagcase this_move
                  tag Lft:  x-1, y
                  tag Rght: x+1, y
                  tag Up:   x  , y+1
                  tag Down: x  , y-1
                end tagcase
```

The **tagcase** construct is also the mechanism for accessing constituent values of a union value. A new value name is introduced as part of the tagcase header, which may then be used within any arm of the tagcase to refer to the constituent value. This value name may only be used in the tag arms, and will have a type and value according to the tag value of the arm. The following program shows how a binary search tree can be built (function **Insert**) and searched (function **Retrieve**). In both functions the value name **Root** has type **null** (and consequently no value) when the branch tagged **Empty** is executed, and has type **Record[...]** when the branch tagged **Full** is executed.

```
Type TreeTyp = Union [ Empty;
               Full: Record [ ITKey, ITVal: Integer; L, R : TreeTyp ]
               ]
Function Insert( ITree: TreeTyp; Key, Val: Integer Returns TreeTyp)
 TagCase Root := ITree
Tag Empty    : Union TreeTyp [
               Full: Record [ ITKey : Val;
                              L : Union TreeTyp [ Empty ];
                              R : Union TreeTyp [ Empty ]
                            ]
               ]
Tag Full     : If Key  Root.ITKey
               Then Union TreeTyp [
                    Full: Root Replace [ L: Insert( Root.L, Key, Val ) ]
                    ]
               ElseIf Key  Root.ITKey
               Then Union TreeTyp [
                    Full: Root Replace [ R: Insert( Root.R, Key, Val ) ]
                    ]
               Else Union TreeTyp [
                    Full: Root Replace [ ITKey : Key; ITVal : Val ]
                    ]
               End If
 End TagCase
End Function

Function Retrieve( ITree: TreeTyp; Key: Integer Returns Integer )
 TagCase Root := ITree
    Tag Empty : Error [ Integer ]
    Tag Full  : If Key > Root.ITKey
                Then Retrieve( Root.L, Key )
                ElseIf Key < Root.ITKey
                Then Retrieve( Root.R, Key )
                Else Root.ITVal
                End If
 End TagCase
End Function
```

2.4 Modules and Programs

A SISAL program consists of a collection of type and function definitions. A program may be written as a number of distinct **modules**. Each module will contain a number of type and function definitions, together with a **defines list** of functions defined in the module (at the outermost level) which are to be made available to other modules, and a **globals list** of functions which are used in the module, but (possibly) not defined there. If a function in the globals list is not defined in the module, it is assumed that some other module will provide a definition of that function via a defines list.

In the SISAL implementation on the Manchester Dataflow Machine, there is a special **main** function, which can be called at run-time by supplying the appropriate values for its arguments as input to the program, and retrieving the results of the function call as the program output.

3. DATAFLOW MODELS OF COMPUTATION

A machine-level dataflow program is a directed graph in which **nodes** represent machine **instructions** and **arcs** represent paths along which data values travel from instruction to instruction at run-time. Simple program graphs, such as that shown in the figure below (which calculates the square root of $b^2 - 4ac$, given a, b and c, are **acyclic**. Such graphs are equivalent to the **expression evaluation** trees that are commonly constructed by compilers for high-level languages. Where a value is required in more than one part of the expression evaluation, the instruction marked **DUP** (for duplicate) in the figure is used to copy the input value to all the output arcs.

The essence of the dataflow computational model is that it defines a direct means for concurrent execution of these graph programs. Carriers for data values, known as **tokens**, are fired into the graph, and each instruction determines individually whether or not to execute on the basis of the availability of its required input data. Since this decision is made locally at each instruction, the overall pattern of execution is entirely asynchronous.

In order to measure the parallelism in a dataflow graph program, it is common to record the total number of instructions in the graph, known as **S1**, and the length of the critical path, known as **Sinf**. These values are derived from an abstract view of the execution of program graphs in which all instructions take the same period of time to execute and there is no delay between a **generating** instruction producing its result token and the corresponding **consuming** instruction receiving that token. Execution with p processors thus proceeds in **lockstep** fashion, with a maximum of p instructions being executed in each time-step, depending on the availability of input tokens. The number of time-steps required to execute the whole program is known as **Sp**. Hence, **S1** is equivalent to the number of time-steps required by a uniprocessor, and **Sinf** is the number of time-steps required by

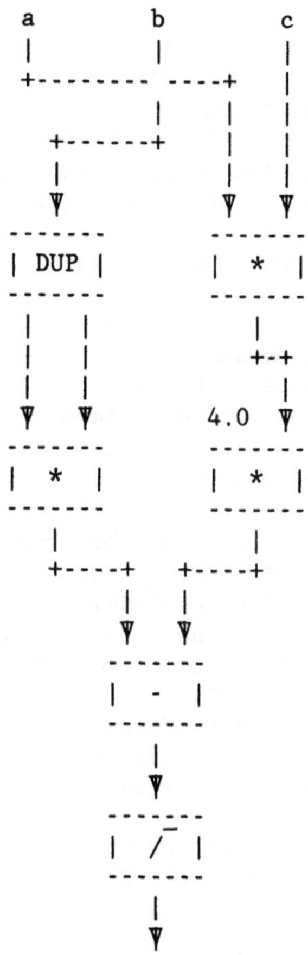

an **infinite** multiprocessor. The ratio **S1/Sinf** is a crude measure of the **average program parallelism** inherent in the graph. When the total number of floating point operations, F1, is also recorded, the value of **S1/F1** is known as the **MIPS/MFLOPS ratio** (MMR) which is a convenient measure of the overall efficiency of a numerical program (see section 5). In the above example, S1 = 6, Sinf = 4 and Fl = 5.

Graphs representing higher level constructs, such as **loops** or **functions**, contain **conditional** instructions and may be **cyclic**. For example, the program in the next figure calculates the area under the curve f(x) in the interval **x = low to x = high** (where **high = low + (n * h))** using a trapezoidal approximation formula on the n intervals of size h between **low** and **high**. A high-level textual version of this program would normally be expressed using an iterative construct, such as a **while** loop, with a loop counter variable to determine termination of the loop, and two further loop variables to determine the current values of x and the cumulative value of the integral to this point.

This low-level graph can be generated using straightforward data dependence analysis of the high-level program, such as is conventionally used in optimising compilers. The instructions marked **BRA** (for branch) are used to achieve conditional computation. A branch instruction accepts a boolean value on its right-hand input arc, and routes the incoming left-hand token to the left or right output arc according to the value of the boolean (left for **true**, right for **false**). The graph is somewhat inefficient since it calculates f(x) twice for each value of x.

The program illustrates two important characteristics of cyclic graphs that have led to the distinct **static** and **dynamic** approaches to dataflow computation. Firstly, note that the asynchronous nature of dataflow program execution can cause **race** conditions in cyclic graphs. For example, in the figure, the cycle defining the next value of the loop counter **i** can race ahead of the cycles calculating **x** and the integral, thus causing confusion between tokens belonging to successive cycles of the loop at the inputs to the left-hand **branch** instructions. Either this situation must be prevented from arising (the **static** approach, achieved by preventing an instruction from executing when any of its output arcs are occupied by tokens), or there must be some means for distinguishing between such tokens (the **dynamic** approach, presented below). Secondly, note that the two **calls** to function **f** can occur concurrently. Indeed, successive cycles of the loop can introduce additional concurrency by overlapping calls to function **f** in **pipelined** fashion. If such functions are implemented as the graphical equivalent of **subroutines**, with a single body of machine-code being shared amongst all calls to the function, then it must be feasible for several activations of the body to occur simultaneously.

The major dataflow projects now solve both these problems using a dynamic scheme known as **tagging**. Each token carries an extra field, called the **tag**, that is guaranteed to be unique in this area of the graph, and which therefore distinguishes between all possible activations of a particular piece of code. By separating tokens with different tags from one another, the asynchronous execution rules can be followed without confusion due to race conditions.

The tagging scheme is completely general, in that it allows all forms of parallel recursive functions to be implemented. An alternative scheme, with the same generality, is to copy the code representing the function body every time a function is called, thus expanding the program during run-time. The tagging scheme effectively simulates this expansion, and so both schemes have become known as **dynamic** dataflow. Some dataflow researchers feel that dynamic expansion of code is an unwarrantable expense and suggest that recursive functions are not essential in parallel machines. They are content to implement the static scheme, mentioned above, giving program graphs capable of implementing cyclic loops only. These issues are discussed more fully by Arvind and Gostelow (1978).

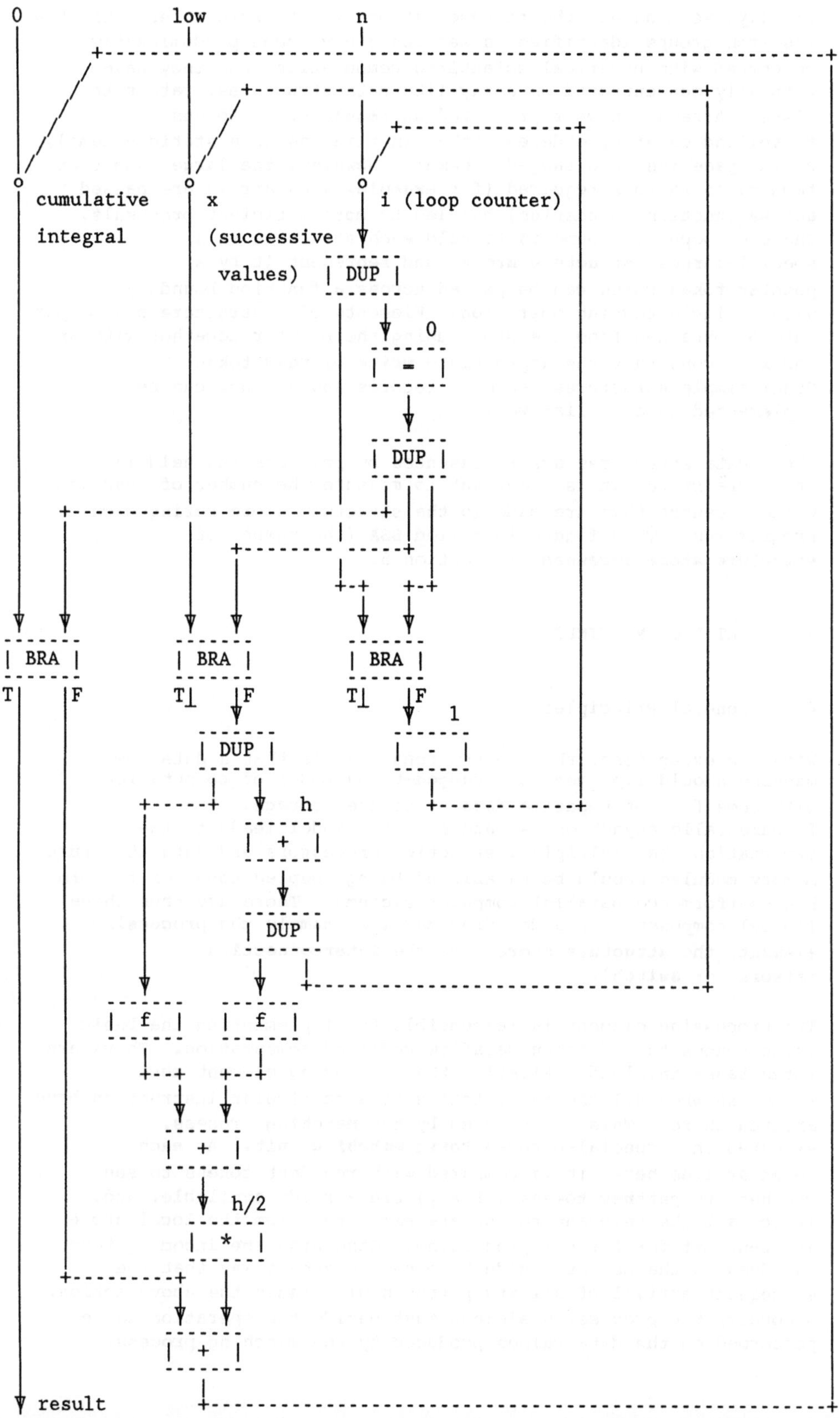

0
low
n
cumulative
integral
x
(successive
values)
i (loop counter)
DUP
0
=
DUP
BRA
T
F
BRA
T
F
BRA
T
F
DUP
1
-
h
+
DUP
f
f
+
h/2
*
+
result

Finally, we consider the representation of data structures. The key research groups identified in section 1 have been predominantly concerned with numerical scientific computation, and they have naturally concentrated on array-like data structures, rather than lists. Arrays can be represented as **sequences** of tokens travelling on arcs, ordered either in time (as in a static scheme), or in space (as in a tagged scheme). However, the large number of tagging operations required (for example, when arrays are passed across function boundaries) has led to more efficient proposals. The most popular scheme is to hold each structure in a special-purpose **structure store**, and represent it by a **pointer** token which can be passed across a function boundary with a single tagging operation. Elements of a structure can be put into or obtained from the store using the pointer together with an index to construct the appropriate **write** or **read** token. Other simple structures, such as records and unions, can be implemented in a similar way.

Where data structures are represented by pointers and held in a structure store, it is important to measure the number of read and write accesses that are made to the structure store during each program run. This figure is called **SSA** (the number of **structure store accesses**) in section 5.

4. DATAFLOW MACHINES

4.1 General Principles

With few exceptions, the present consensus is that a dataflow machine should implement the **tagged-token** model of computation with some form of separate data structure storage. The fundamentally asynchronous nature of the model leads to the expectation that multiple instruction processors and data structure memory modules should be capable of being coupled together to form high-performance parallel computer systems. There are thus three logical components in a dataflow machine, namely the **processing element**, the **structure store**, and the **interconnection network** (or **switch**).

The processing element is responsible for implementing the basic asynchronous tagged token dataflow model of computation. There are three tasks involved. Firstly, the processing element must recognise when all the input tokens to a particular instruction have arrived there. This is achieved by the **matching** process, executed in a special-purpose **token matching** unit. As each token arrives here, it is compared with resident tokens to see whether its **partner** tokens (if any) are already available, and, if so, all the relevant tokens are extracted from the local store and sent off for further processing. Otherwise the incoming token is added to the set of resident tokens in such a way that the subsequent arrival of all its partners will cause the above action. Secondly, the processing element must decide the operation to be performed on the data values produced by the matching process.

This is achieved by fetching an instruction from the **instruction store,** which holds a complete copy of the graphical program. A program is normally expressed as a set of **nodes,** each defining the **operation** to be performed and the **destination** node for each of the required output arcs. Finally, the processing element must be capable of performing the required operation on the data and generating the appropriate output token(s). This is achieved in a **processing unit** which contains the necessary functional capabilities to support the machine instruction-set.

The structure store is responsible for servicing access requests to data structures. The store is usually responsible for management functions, such as space allocation and garbage collection, as well as read and write accesses. All communication between structure store modules and processing elements is achieved by sending messages in the form of data tokens.

The interconnection network is responsible for sending token messages from one physical unit to another. These messages are always sent in **packet-switched** fashion. The network comprises a number of individual **mxm** switch elements, each of which contains buffers at its input to hold incoming data. Routing information within the data packet indicates which of the network outputs it is directed towards. The overall structure is effectively a set of parallel pipelines. An N-input, N-output network requires $\mathbf{N/m(log_m N)}$ switch elements, and any route through the network is a $\log_m N$ stage pipeline. It can be shown, assuming random contention for switching resources, that such a structure achieves a throughput approaching its maximum theoretical bandwidth. Nevertheless, its complexity in a practical implementation is manageable. Since the communication is uni-directional, the only price paid is the pipeline delay between processors. As long as the processors are kept active and the pipeline is kept full by parallelism in the program, no performance degradation should result.

A key factor in the design of a dataflow system is the ability to expand its power by adding extra processors via an extended communications switch. The desire to expand power by adding hardware is common to all dataflow systems designs. There is keen debate about the maximum size of switch that can be constructed (or that will be feasible in the forseeable future). There is an obvious relationship between the power of individual processors, the total power, and the size of the switch.

Three schemes are being studied for distributing the processing elements and structure store modules. The first scheme has the structure storage integrated within the processing elements. The advantages of this scheme are simplification of the hardware structure and, under the right conditions, reduction of the required network bandwidth. However, effective distribution of the workload is difficult. In order to minimise the required network bandwidth, the location of the processing workload must exactly match the location of the corresponding structure store workload. The second scheme has structure store modules and processing elements **clustered** together in relatively small numbers. This simplifies workload distribution at the expense of increasing hardware complexity and network bandwidth. The extent to which this scheme can be effective has yet to be determined, but the fundamental problem of matching processing and structure storage workload distribution remains and cannot be solved completely unless the structure store modules are entirely separated from the processing elements, as happens with the third scheme.

4.2 The MIT Static Dataflow Machine

A machine which uses the static dataflow model of computation has been implemented at MIT, as described by Dennis et al. (1980). It uses bit-sliced microengines connected via a general-purpose unidirectional routing network. Because the system uses conventional microprocessor software to emulate the dataflow model, it runs relatively slowly, and large switches will be needed for substantial applications (e.g. weather forecasting). It is envisaged in this design that switches of size 500x500 and more can be implemented using byte-wide 2x2 routers. Other researchers are less confident that switches of this size will be practicable and have concentrated on improving the execution rates of individual processors by designing their internal structure to be dataflow oriented. The single-assignment language VAL, described by Ackerman and Dennis (1979), is used for programming the system. The basic structure of the machine is illustrated below.

```
                +--------------+
+------------->|  PROCESSOR   |-------------+
|               +--------------+             |
|               +--------------+             |
|  +---------->|  PROCESSOR   |----------+  |
|  |            +--------------+          |  |
|  |                   :                  |  |
|  |                   :                  |  |
|  |            +--------------+          |  |
|  |        +-->|  PROCESSOR   |---+      |  |
|  |        |   +--------------+   |      |  |
|  |........|                      |.......|  |
|  |        |   +--------------+   |      |  |
|  |        +---|              |<--+      |  |
|  |        :   |              |   :      |  |
|  |        :   |   NETWORK    |   :      |  |
|  +-----------|              |<----------+  |
+--------------|              |<-------------+
                +--------------+
```

4.3 The MIT Dynamic Dataflow Machine

A second group at MIT is constructing a VLSI-based tagged-token dataflow processor, as described by Arvind and Kathail (1981). The main features of this system are that (i) data structure accesses are handled separately from ordinary token activities, and (ii) there is a two-tiered communication system. The processor structure is illustrated below.

The three major dataflow activities are handled by separate hardware. In particular, the token gathering operation is performed by a small associative store, known as the **waiting-matching section**. The structure storage is integrated within each processor, rather than being attached separately to the network. The two-tiered communication structure relieves the general communication switch of excess traffic as long as programs exhibit strong **locality** (i.e. processing activity is localised in subgraphs and processors, rather than communicating randomly with other subgraphs/processors). Locality also benefits the size of the unmatched token memory. The current plan is to implement a small (64-word) store.

Reduced traffic in the communications switch allows bit-wide data paths to be used. The proposed building block for this system is an 8x8 bit-wide module. Using program locality still further, large-size switches can be made rather less complex than the networks proposed for other systems. Because this design relies heavily on strong program locality, the programming language must be chosen carefully. The specially designed language **Id**, described by Arvind et al. (1978) has appropriate properties and is being used to program the system.

4.4 The Manchester Dataflow Machine

The dataflow research group at the University of Manchester has constructed a uniprocessor ring-structured dataflow machine, as described by Gurd et al. (1985). The three dataflow tasks (matching tokens together, finding the next instruction, and processing instructions) are implemented in three separate hardware modules. The individual actions in these modules are dependent solely on the module input data, so that successive actions may be overlapped by connecting the modules in a pipeline. One extra pipeline module is provided to queue excess tasks when highly parallel programs are running. The processor structure is thus a four-stage pipeline, as illustrated in the figure below.

A separate structure store unit is attached independently to the interprocessor communication switch. This unit comprises a microcoded processor with a large 40-bit wide data store, with facilities for deferring premature accesses (i.e. attempts to read structure elements that have not yet been written).

The critical part of the system is the token matching unit. The task of pairing tokens together is an act of association, so the unmatched token memory is (pseudo-)associative. It is implemented using a hardware-hashing mechanism. The average match time for the prototype version is 450 nanoseconds. This limits the instruction execution rate of the ringstructured processor to just over 2 million instructions per second (MIPS). The prototype **instruction processing element** is about twenty times slower than this, so a serially-activated parallel array of up to 20 such elements is provided in the processing unit.

As shown in the figure, the structure store consists of an **allocation unit**, a **clearance unit**, the **structure memory**, and a **deferred access queue**, all of which are connected to the communication switch. The processing ring exchanges messages with the structure store via the communication switch, in order to

```
                                +-------------+
                                | DEFERRED    |
                                | ACCESS      |
                                | QUEUE       |
                                +-------------+
                                      ||
                                +-------------+
     +--------------------------| DATA        |<--------------------+
     |                          | STRUCTURE   |                     |
     |                     +----| MEMORY      |<---+                |
     |                     |    +-------------+    |                |
     |                     |    +-------------+    |                |
     |                     +--->| CLEARANCE   |----+                |
     |                     |    | UNIT        |                     |
     |                          +-------------+                     |
     |                                ||                            |
     |                          +-------------+                     |
     |          +---------------| ALLOCATION  |<-------------+      |
     |          |               | UNIT        |              |      |
     |          |               +-------------+              |      |
     V          V                                            |      |
+-----------------------------------------------------------------------+
|                                                                       |
|                      COMMUNICATION  SWITCH                            |
|                                                                       |
+-----------------------------------------------------------------------+
     |                                                              ^
     |  +-------+  +-----------+  +-------------+  +-----------+    |
     |  | TOKEN |  | TOKEN     |  | INSTRUCTION |  | PROCESSING|    |
     +->| QUEUE |->| MATCHING  |->| FETCH       |->| UNIT      |----+
        |       |  | UNIT      |  | UNIT        |  |           |
        +-------+  +-----------+  +-------------+  +-----------+
                         ||             ||
                   +-----------+  +-------------+
                   | UNMATCHED |  | PROGRAM     |
                   | TOKEN     |  | GRAPH       |
                   | MEMORY    |  | MEMORY      |
                   +-----------+  +-------------+
```

allocate structures, to increment/decrement reference counts and to read/write data values. The **allocation unit** is responsible for managing storage space. The **clearance unit** performs garbage collection. The **structure memory** stores the structured data and allows asynchronous access to it in the normal data-driven fashion. When the required data has yet to be written, a read request is held in the **deferred access queue** until the data arrives. Deferred read requests for a store location will be released when the corresponding write request is processed. This scheme is similar to that proposed by Arvind and Thomas (1981) for the implementation of I-Structures. A structure store for a larger machine would contain several units of each type, configured in a more distributed manner.

The Manchester machine is programmed using SISAL (see section 2).

4.5 The ETL SIGMA-1 Dataflow Machine

Perhaps the most significant current dataflow project is the large-scale prototype, SIGMA-1, being constructed at the Electro-Technical Laboratory in Japan, as described by Shimada et al. (1986). This will be completed in 1987, with an estimated

```
+----------------------------------------------------------------+
|                                                                |
|                  GLOBAL COMMUNICATION NETWORK                  |
|                                                                |
+----------------------------------------------------------------+
                 | .............. x32 .............. |
+----------------------------+      +----------------------------+
|           10x10            | ...  |            10x10           |
| LOCAL COMMUNICATION NETWORK|      | LOCAL COMMUNICATION NETWORK|
+----------------------------+      +----------------------------+
   |  ...  |   |  ...  |              |  ...  |   |  ...  |
+---+    +---+ +---+    +---+      +---+    +---+ +---+    +---+
| PE| ...| PE| | SM| ...| SM| ...  | PE| ...| PE| | SM| ...| SM|
|  0|    |  3| |  0|    |  3|      |124|    |127| |124|    |127|
+---+    +---+ +---+    +---+      +---+    +---+ +---+    +---+
```

Each processor (PE) has the following internal structure:

```
                        +-------------+   +-------------+
                        |  MATCHING   |-->|  EXECUTION  |
                  +---->|   MEMORY    |   |    UNIT     |--+
        +--------+|     |    UNIT     | +>|             |  |
        | BUFFER ||     +-------------+ | +-------------+  |
from -->| UNIT   |----o                 |                  --> to
local   |        ||     +-------------+ | +-------------+  |  local
network +--------+|     | INSTRUCTION | | | DESTINATION |  |network
                  +---->|    FETCH    |--o->|   UNIT    |--+
                        |    UNIT     |   |             |
                        +-------------+   +-------------+
```

sustained performance of 50 to 100 MFLOPS for scientific applications. The system structure is half-way between the Manchester machine and Arvind's machine, since the processors (PEs) are **clustered** together with the data structure memories (SMs) in groups of four of each. The switch structure is hierarchical (tree-structured), it being intended that a total of 32 or 64 clusters will be constructed. The system architecture is illustrated below.

The SIGMA-1 is programmed using the locally developed single-assignment language DFC, which has C-like syntax.

4.6 Other Projects

Several other research projects are worthy of mention. A small uniprocessor machine was constructed at the University of Utah, as described by Davis (1978). A system was built at Texas Instruments using four microprocessors and a TI 990/10 host, connected together via a time-multiplexed communication ring and programmed using Fortran, as described by Johnson et al. (1980). These were both low performance systems which served to demonstrate some basic dataflow principles. A project team at CERT Toulouse in France produced a working machine based on their single-assignment language LAU, as described by Syre et al. (1977). A multiprocessor machine has been developed at NTT in Japan, as described by Amamiya et al. (1982). Their hardware has been operational since 1981, and the system is programmed in the functional language Valid. A second group at ETL in Japan has been constructing a dataflow machine, EM-3, to execute functional programs written in the language EM-Lisp, as described by Yamaguchi et al. (1983). Finally, a commercial dataflow product, the NEC PD7281 Image Pipeline Processor, has been manufactured.

4.7 Open Issues in Dataflow System Design

Two major areas remain to be thoroughly investigated, namely: **distribution of workload** amongst the hardware units; and **control of parallel activity** in the system.

The important features of an effective workload distribution scheme are to spread the processing/structure storage workload **evenly** across the available hardware modules and to minimise the resultant load on the interconnection network. The exact requirements for this naturally depend on the scheme adopted for distributing processing elements and structure storage. Natural distribution schemes for processing workload seem independent of natural schemes for structure storage workload, implying that the two should be treated separately. Study of appropriate schemes is at an early stage in all the machines discussed above.

It has been observed that highly parallel programs frequently use excessive amounts of data store within the processing elements. This occurs because most programs have, at least for part of their execution, **too much parallelism.** As a result, many partial results are created long before they are needed, and they end up waiting in the token matching unit. It is therefore necessary to (approximately) match the program parallelism to the parallelism available in the machine, a process known as **throttling.** This is another area in which experiments are still being conducted, and there is, as yet, no agreement on the optimum scheme to use.

5. DATAFLOW/SINGLE-ASSIGNMENT SYSTEM PERFORMANCE

5.1 Code Complexity

This section investigates the overall performance of a dataflow machine executing object-code generated from a single-assignment language. The Manchester Dataflow Machine and SISAL are studied since they are representative of this approach. A complete account of how SISAL code generation is accomplished on the Manchester Dataflow Machine is given by Böhm and Sargeant (1985). Their presentation includes a history of the code generator, showing how improvements and optimisations were made as the project progressed. This section concentrates on the performance results from that paper, using the measures S1, Sinf, F1 and SSA, introduced in section 3.

Results have been published by Gurd and Watson (1983) for a range of programs using the earliest, unoptimised, code. The table below shows how the efficiency of a fairly realistic program has improved over time. **Simple** is a 2-D hydrodynamics program containing about 1000 lines of SISAL code, originating from Lawrence Livermore National Laboratories. The problem size is the same (10) for each entry in the table. The **MIPS/MFLOPS** ratio (MMR) for a particular problem is roughly proportional to S1, but this relationship is not exact because some optimisations (such as common subexpression elimination) change the value of F1.

The code size has decreased steadily, except that **inlining** (replacing function calls by the appropriate function body)

Date	Code Size	S1	Sinf	MMR	SSA	Main Optimisation involved
Mid 84	22258	936371	2005	27.1		
Mid 84	16769	641798	1851	18.6	55210	Structure store used
Dec 84	11284	398571	1820	11.5	64546	Index-insensitive loops
Jan 85	10539	319214	1160	9.2	62353	Instruction set improved
Feb 85	8894	248153	1077	7.2	62393	TUPlicate
Jun 85	13035	203460	952	5.9	60924	Inlining
Jul 85	11697	165303	912	4.8	43031	Constant low array bound

increases it. In fact, the later figures are obtained using **total inlining** (all functions are combined into one). Much less inlining can be used without greatly increasing S1. Sinf goes down steadily, but more slowly than S1; spurious parallelism is removed. However, the program is still highly parallel for realistic data sizes. The comments in the last column are a gross over-simplification, since many other minor improvements have taken place in the meantime. The next table shows how other programs have improved.

Program	Original (MS) code				Optimised (SS) Code			
	Code Size	S1	Sinf	MMR	Code Size	S1	Sinf	MMR
LaplaceB	811	191984	915	37.0	295	40164	384	7.8
MatMult	657	100288	425	50.1	266	7292	133	3.6
Gauss	3201	215732	3620	70.0	1426	41241	1496	14.1
Splice	6957	5031909	165647	123.7	3569	819740	41738	20.1
RSIM/2	24314	1135912	62563	-	12606	468082	33753	-
IV/1	39091	126991	6571	-	18741	40649	3719	-

In some cases, the table underestimates the improvement. Because optimisations concentrate on innermost loops, the MMR of newer code tends to improve for larger data sizes. For instance, the MMR for the Gauss program drops to around 7. In addition, by rewriting programs according to some simple rules, MMRs of around 3 can be obtained for simple numeric programs such as Laplace and Gauss. The matrix multiply (MatMult) program drops to an asymptotic MMR of 1.5 quite rapidly. The most important rule is to avoid conditionals in innermost loops since these cause many branch nodes to be executed and make vectorisation less effective. These MMR figures should be compared with the corresponding figures for conventional machine/language systems. For example, values of around 4 are reported by Worlton (1985) as being typical of Fortran programs running on a CDC7600. Hence, relatively efficient code is being generated for numeric problems on the Manchester Dataflow Machine. It would be of interest to extend these results to a wider range of applications, but there is no widely accepted measure, such as the MMR, for quality of non-numerical code.

5.2 Other Characteristics

Viewed from a distance, the structure of the Manchester Dataflow Machine is similar to that of a multiprocessor vector-style supercomputer, with multiple function units plus their local workspace attached through high-bandwidth channels to multiple banks of store. Instruction storage is clearly separated from data storage, but this is the only major structural change compared to conventional practice. The token queue and token matching unit memories act in similar fashion to working registers, and the switch performs the function of multiport memory channels. The fundamental differences between the systems lie in the highly flexible, asynchronous packet-switched nature of the dataflow switch, and the close link between this hardware and the SISAL compiler/code-generator.

In a vector-style multiprocessor, the data memory interface is controlled by low-level microcode, and the modes of access to store are severely restricted (e.g. only constant-stride vectors can be accessed). It is the programmer's responsibility to situate data structures in appropriate locations in store and to ensure that they are accessed in a consistent fashion by the different processing elements. All these features are reflected in the explicit programming style required by languages such as Fortran (suitably extended to cope with parallel processes). Users have to be clever and precise if they are not to get into trouble with such systems.

In a dataflow multiprocessor, the structure store interface is controlled in a more flexible manner. The **system** is responsible for managing the data store, both at high-level (i.e. allocating and garbage collecting store areas) and at low-level (i.e. synchronising read accesses so that they take place after writing). These features are completely hidden from the programmer precisely because they are implicit in the statements of the high-level programming language. Distancing the user from these implementation concerns can only be beneficial, allowing him/her to concentrate on more important issues.

The achievement of dataflow research to date has been to show that these advantages can be obtained in uniprocessor machines without significant expense in terms of machine complexity and numbers of instructions executed per high-level operation. This contrasts with the achievement of vector-style multiprocessors, which has been to convince end-users that parallel programming is difficult, costly and error-prone. It is interesting that one of the principal partners in designing the SISAL language has been Lawrence Livermore National Laboratory. They report significant problems with nondeterminate behaviour of parallel programs running on Cray X-MP systems. Their motivation in studying SISAL is to develop **determinate** programs to run on whatever high-performance hardware is made available to them. Whilst not a direct motivation for turning to dataflow machines, this certainly indicates a desire to move away from explicit programming. It will be interesting to see how efficient the object-code generated from SISAL-like languages for Cray-like machines turns out to be.

Dataflow research has so far failed to demonstrate clearly that the uniprocessor results **scale** to large multiprocessor systems. However, this is a problem faced by all current multiprocessors. With vector-style machines, problems are presently being encountered with small numbers of processors, and these will take time to solve. The promise of dataflow is that the logical difficulties of scaling, once the initial efficiency issues have been settled, are minimal. The fine-grain approach offers simple solutions to the problems of workload distribution and throttling whilst giving no software problems at all.

As an example of the ease of workload distribution, curde initial experiments on a simulator for a multi-ring Manchester Dataflow Machine have already demonstrated that the maximum number of unmatched tokens in any processing ring during a program run can be made independent of the total number of processing rings.

The major difficulty for practical dataflow machines is likely to be construction of multistage switch networks with sufficiently high bandwidth to handle the expected quantity of fine-grain message traffic. With relative development costs ever favouring investment in hardware rather than software, it appears well worthwhile to tackle this problem in order to benefit from the simplicity afforded by implicit parallel programming.

6. CONCLUSIONS

Dataflow/single-assignment is a technique for expressing and executing parallel computation that shows great promise. It displays an impressive integration of high-level implicit programming concepts with low-level hardware implementation. Notwithstanding early criticisms of the dataflow approach, as expressed, for example, by Gajski et al. (1982), recent research, as surveyed by Gurd and Kirkham (1986), has demonstrated clearly that dataflow systems can be made as efficient as systems based on conventional explicit techniques. The inherent advantages of dataflow for multiprocessing will surely be demonstrated soon.

The immediate future for dataflow research is clear. The results for uniprocessor systems must be extended to encompass multiprocessor systems of significant size. This requires that workable schemes for workload distribution and throttling be developed and convincingly demonstrated. With such mechanisms in place, it will be possible to show high-performance dataflow multiprocessing in action on machines such as the 50-100 MFLOPS SIGMA-1 system which is expected to be operational in 1987.

ACKNOWLEDGEMENTS

The author acknowledges the contributions of his colleagues in the Dataflow Research Group at the University of Manchester. Particular thanks are due to Wim Böhm, Vicky Bush, John Foley, John Glauert, Chris Kirkham, Adrian Parker, John Sargeant, Bryan Saunders and Ian Watson. Contributions from researchers outside Manchester, especially the SISAL language design and implementation teams from Lawrence Livermore National Laboratory, Digital Equipment Corporation and Colorado State University are gratefully acknowledged. Construction of the prototype Manchester Dataflow Machine and implementation of the SISAL compiler were funded by the Science and Engineering Research Council of Great Britain. Applications studies have been supported by Digital Equipment Corporation and the UK Alvey Programme in advanced information technology.

REFERENCES

Ackerman W.B. (1982) Dataflow Languages. IEEE Computer 15:2, 1087-1095.

Ackerman W.B. Dennis J.B. (1979) VAL: A Value-Oriented Algorithmic Language - Reference Manual. Technical Report LCS-TR-218, MIT.

Amamiya M et al. (1982) A List-Processing-Oriented Data Flow Machine Architecture. AFIPS Proceedings 51:143-151.

Arvind et al. (1978) An Asynchronous Programming Language and Computing Machine. Technical Report ICS-TR-114a, University of California at Irvine.

Arvind, Gostelow K.P. (1978) Some Relationships between Asynchronous Interpreters of a Dataflow Language. In: Neuhold E.J. (ed.) Formal Description of Programming Concepts, North Holland, Amsterdam, 95-119.

Arvind, Kathail V. (1981) A Multiple Processor Dataflow Machine that Supports Generalised Procedures. Proceedings 8th Annual Symposium on Computer Architecture, ACM, 291-302.

Arvind, Thomas R.E. (1981) I-Structures - An Efficient Data Type for Functional Languages. Technical Memo LCS-TM-178, MIT.

Böhm A.P.W. Sargeant J. (1985) Efficient Dataflow Code Generation for SISAL. Technical Report UMCS-85-10-2, University of Manchester.

Davis A.L. (1978) The Architecture and System Methodology of DDM1: A Recursively Structured Data Driven Machine. Proceedings 5th Annual Symposium on Computer Architecture, ACM, 210-215.

Dennis J.B. et a. (1980) Building Blocks for Data Flow Prototypes. Proceedings 7th Annual Symposium on Computer Architecture, ACM, 1-8.

Gajski D.D. et al. (1982) A Second Opinion on Data Flow Machines and Languages. IEEE Computer, 15:2, 58-69.

Glauert J.R.W. (1982) High-Level Languages for Dataflow Computers. In: Wallis P. (ed.) State of the Art Report on Programming Technology, Pergamon-Infotech, Maidenhead, 173-193.

Gurd J.R. Watson I. (1983) Preliminary Evaluation of a Prototype Dataflow Computer. In: Mason, R.E.A. (ed.), Information Processing 83, North Holland, Amsterdam, 545-551.

Gurd J.R. et al. (1985) The Manchester Prototype Dataflow Computer. Communications of the ACM, 28:1, 34-52.

Gurd J.R., Kirkham C.C. (1986) Dataflow: Achievements and Prospects. In: Kugler H.J. (ed.) Information Processing 86, North Holland, Amsterdam, 61-68.

Johnson D. et al. (1980) Automatic Partitioning of Programs in Multiprocessor Systems. Proceedings IEEE Computer Conference, 175-178.

McGraw J.R. et al. (1983) SISAL - Streams and Iteration in a Single-Assignment Language. Language Reference Manual, Lawrence Livermore National Laboratory.

Shimada T. et al. (1986) Evaluation of a Prototype Data Flow Processor of the SIGMA-1 for Scientific Computations. Proceedings 13th Annual Symposium on Computer Architecture, ACM, 226-234.

Syre J.C. et al. (1977) LAU System - A Parallel Data-Driven Software/Hardware System based on Single-Assignment. In: Feilmeier M. (ed.), Parallel Computers - Parallel Mathematics, North Holland, Amsterdam.

Treleaven P.C. et al. (1982) Data-Driven and Demand-Driven Computer Achitecture. ACM Computing Surveys 14:1, 93-143.

Watson I. (1984) Dataflow Architectures and Performance. In: Chambers F.B. et al. (eds.) Distributed Computing, Academic Press, London 21-32.

Worlton J. (1985) Bottleneckology: Evaluating Supercomputers. Proceedings IEEE Computer Conference, 405-406.

Yamaguchi Y. et al. (1983) A Performance Evaluation of a LISP-Based Data-Driven Machine (EM-3). Proceeding s 10th Annual Symposium on Computer Architecture, ACM 363-369.

Transposing Large Arrays in Extended Memory

PAUL N. SWARZTRAUBER

National Centre for Atmospheric Research, Boulder, CO 80307, USA

Abstract

Three methods for transposing large arrays in cache, disk or solid state memory are examined. It is observed that an array of any size can usually be transposed in a single pass in which the array is read to high speed memory and then written back to extended memory in transposed order. Most meteorological models access a number of large arrays during each time step. The efficiency of the computation may depend on whether the array is stored by columns or rows and a transposition may be desired. The difficulty is that a transposition can itself require additional accesses which are time consuming. Using the methods discussed here, the arrays can usually be transposed without additional accesses. The methods are not limited to transpositions and apply to a large class of orderings called index-digit permutations which includes shuffles, digit reversal, and multi-dimensional rotations.

1. INTRODUCTION

The computations that are performed in a single time step of a meteorological model can be divided into several phases. Within each phase a set of arrays is accessed, updated, and then replaced. Whereas one phase may require an array to be accessed by (say) columns, another phase may require the same array to be accessed by rows. For example, a two-dimensional Fourier transform consists of several one-dimensional transforms, first on the columns and then on the rows. The FFT of a large one-dimensional array can also be posed on a two-dimensional array that must be transposed [2]. The transposition of a large array can be time consuming, particularly if the array is stored in cache, on disk, or in solid state memory. Multiple accesses of the data are usually required, either explicitly or implicitly by a virtual memory system.

Three methods for transposing large arrays are described in this paper. The first is straightforward but inefficient and is presented only as a benchmark for evaluating the remaining two methods which were developed by Fraser [1]. His methods apply to a general class of orderings called index-digit permutations which includes digit reversal, perfect shuffles and matrix transposition. The focus will be on matrix transposition with the hope that the extension to the other orderings will be evident. A surprising result is obtained from the third method; namely, that the transposition of an array of arbitrary size can usually be performed without accesses of the arrays over and above those that are ordinarily performed in large meteorological models.

We will consider the transposition of a large array that is stored in blocks in extended memory. It will be assumed that the blocks can be accessed randomly as they can on disk or solid state disk (SSD) but unlike tape which must be accessed sequentially. For clarity of exposition, the methods are described in terms of examples with the hope that the extension to the general case is evident. The goal is to present methods that reduce the number of times that the array must be accessed (passes) since this is the most time consuming part of the transposition. The first method requires 16 passes to transpose a matrix with 64 million elements and the third method requires a single pass.

2. ARRAY TRANSPOSITION, METHOD I

Without loss of generality, we will assume that the matrix X(i,j) is stored column-wise in extended memory and that we wish to convert it to row-wise storage. To this end, the first k columns are read into an internal memory buffer. The first l rows of this buffer are then moved into a second internal buffer that is large enough to hold l rows of the matrix. This process is repeated until the first l rows of the matrix reside in the second buffer which is then written to extended memory. Therefore the entire matrix must be read to construct the first l rows and it must be read again in order to construct the second l rows and so forth until all the rows have been written to extended memory. For a square matrix with 64 million elements, a total of 16 accesses of the matrix are needed to transpose the matrix if both internal memory buffers contain four million elements. The usual tradeoffs between multiple buffers, buffer lengths, and computer time apply.

3. ARRAY TRANSPOSITION, METHOD II

Assume for the moment that we wish to transpose a matrix that has only two columns. We begin by reading k elements of each column into two distinct memory buffers. The first k rows of the transposition are then constructed in a third buffer with 2k locations by alternately storing elements from the first two buffers. That is, the row buffer contains the perfect shuffle of the two column buffers. The first k rows can then be written to extended memory. This process is then repeated to construct the next k rows and so forth until all rows, each with length 2, are written to extended memory. The important aspect of this transposition, or perfect shuffle, is that it can be performed in a single pass for any column length.

Repeated use of the perfect shuffle can be used to transpose a matrix with more than two rows. To see this it is necessary to examine in detail the reordering of storage that is induced by the shuffle. Since memory is one-dimensional there must exist an implicit map of any multi-dimensional array onto a one-dimensional array. If the matrix has two columns and $i = i_{n-1}i_{n-2} \cdots i_0$(binary) then X(i,j) is located at position $ji_{n-1}i_{n-2} \cdots i_0$. Following the transposition this element is moved to position $i_{n-1}i_{n-2} \cdots i_0 j$. Therefore the effect of a perfect shuffle is a left cyclic shift of the digits in the binary representation of the element position.

Consider now the transposition of a matrix with eight rows and four columns. If $i = i_2i_1i_0$ and $j = j_1j_0$, then the element X(i,j) is located at position $j_1j_0i_2i_1i_0$. If two shuffles are applied sequentially to this array then the effects if two left cyclic shifts of the digits and hence X(i,j) is moved to position $i_2i_1i_0j_1j_0$ or to position X(j,i). Therefore the effect of two shuffles is to transpose the matrix. This process is described in Table 1 using index notation.

Table 1 Transposition of X(i,j) using perfect shuffles
$X(i_0, i_1, i_2, j_0, j_1)$
$X(j_1, i_0, i_1, i_2, j_0)$
$X(j_0, j_1, i_0, i_1, i_2)$

If the matrix has 2^l columns then it can be transposed with l shuffles. If there are fewer rows than columns then the shifts should be performed in the opposite direction using perfect unshuffles. For the example given in the discussion of method I, a matrix with 64 million elements would require 13 passes since the maximum-minimum dimension of a matrix with 64 million or 2^{26} elements is 2^{13}. Although this is not significantly fewer than the sixteen passes required by Method I, it can be easily reduced to seven by using a radix four perfect shuffle. The radix four shuffle requires four column buffers instead of two and, like the radix two shuffle, only a single pass is required for any column length.

Method II can be generalized to arrays whose dimensions are other than a power of 2. Let X(i,j) be an element of a two-dimensional array with M rows and N columns where MN has prime factorization MN $= P_0P_1 \cdots P_l$. Then X(i,j) can be mapped onto an l dimensional array $X(i_0, i_1, \ldots, i_l)$ that can be transposed using radix P_k perfect shuffles. The columns are read into P_k arrays and the shuffle is constructed in an additional array by a cyclic selection of elements from each column.

4. ARRAY TRANSPOSITION, METHOD III

We will now describe Fraser's method for matrix transposition which is a straight forward application of his general method for performing the index-digit permutation [1]. It will be described in the context of a particular example; namely, the transposition of a $2^5 \times 2^4$ matrix. If $i = i_4 i_3 i_2 i_1 i_0$ and $j = i_8 i_7 i_6 i_5$ then $X(i,j)$ is located at position $i_8 i_7 i_6 i_5 i_4 i_3 i_2 i_1 i_0$. The transposition is obtained by moving this element to position $i_4 i_3 i_2 i_1 i_0 i_8 i_7 i_6 i_5$. In index notation these orderings are designated by the first and last entries of Table 2.

Assume for the moment that there are two distinct processes that perform permutations of two proper subsets of the indices. Assume further that the union of these subsets is equal to all nine indices and the intersection is not empty. In our example, the first process will permute the last six indices and the second process will permute the first six indices. The intersection of the indices on which these processes operate is i_3, i_4 and i_5. The processes are combined in Table 2 to transpose a $2^5 \times 2^4$ matrix with 2^9 elements.

Table 2

Matrix transposition using Fraser's Method

$X(i_0, i_1, i_2, i_3, i_4, i_5, i_6, i_7, i_8)$
$X(i_0, i_1, i_2, i_5, i_6, i_7, i_3, i_4, i_8)$
$X(i_5, i_6, i_7, i_0, i_1, i_2, i_3, i_4, i_8)$
$X(i_5, i_6, i_7, i_8, i_0, i_1, i_2, i_3, i_4)$

In Table 2 the first process is applied to the original sequence to obtain the second entry. The second process is then applied to obtain the third entry and finally the first process is applied again to obtain the desired result or the fourth entry in Table 2. It remains now to describe the processes.

Assume that the 2^9 elements are stored externally as $2^6 = 64$ blocks, each with $2^3 = 8$ elements. The first process consists of reading the blocks in the permuted order $i_8 i_4 i_3 i_7 i_6 i_5$. After eight blocks have been read and the memory buffer is full then the second process consists of an internal sort in which the element at position $i_7 i_6 i_5 i_2 i_1 i_0$ is moved to position $i_2 i_1 i_0 i_7 i_6 i_5$. The block with internal address $i_8 i_4 i_3 i_2 i_1 i_0$ is then written to the block with external address $i_4 i_3 i_2 i_1 i_0 i_8$. The next eight blocks are then handled in the same manner and the process is repeated until the transposition is stored externally.

Since the blocks are written in an order that is different from the way that the were read, the transposition cannot be performed in place. However, if the correct order of the blocks is stored in a table then the blocks can be stored in any order. In particular, they can be stored in the order in which they were read or in place. They can also be subsequently read in any order by reordering the table.

For the example in Table 2, only one pass was required to transpose the matrix. In general, Fraser has shown that if the length of the sequence is $N = 2^n$, the block size si $B = 2^b$ and the memory buffer is $M = 2^m$ then the number of passes that are required to index-digit permute any sequence is:

$$n_p = \frac{n-m}{m-b} \quad \text{if } n \leqslant 2m$$

or

$$n_p = \frac{m}{m-b} \quad \text{if } n \geqslant 2m$$

In the example presented for the previous methods, the length was 64 million or $n = 26$. If a block size of a quarter million is chosen together with a memory buffer of four million, then $b = 18$ and $m = 22$. From the first of the equation above, it is seen that a transposition or any index-digit permutation can be performed in a single pass.

In most large models, in particular atmospheric models, the computations are performed out of memory. That is, a certain small number of passes is made corresponding to the various phases of the computations. Using Fraser's method, the data can be ordered so that the computations are as efficient as possible, usually without additional passes for the explicit purpose of reorganizing the data.

REFERENCES

[1] D. Fraser, Array permutation by index-digit permutation, J.Assoc.Comp. March., 22 (1976), pp. 298-308.

[2] P.N. Swarztrauber, Multiprocessor FFTs, Parallel Computing, to appear.

An Approach to Automatic Parallel Processing

BRUCE LEASURE

Kuck and Associates, Inc., 1808 Woodfield Drive, Savoy, IL 61801, USA

1. INTRODUCTION

This paper provides an overview of parallel processing from the view of a company deeply involved in compilation of numerical codes for high-performance computers.

Parallel processing has existed since the early days of computing in special purpose applications. The advent of cheap, high-performance processor chip sets and cheap, fast memory arrays has provided an economic push to exploit parallel processing techniques in a more general purpose, automatic, way.

2. A PARALLEL PROCESSING TAXONOMY

Parallel processing can be made visible to the programmer at any of these five different levels of parallelism:

1. between jobs
2. between subroutines
3. between blocks and loops
4. between iterations of a loop
5. between instructions in a basic block

As one proceeds from jobs, to subroutines, to blocks and loops, etc., the expected execution time of the work being parallel processed decreases, and the expected units of data sharing per unit of execution time increases. Similarly, the changes of needing synchronization to achieve the data sharing increases.

2.1 Parallelism Between Jobs

This level of parallel processing is a logical extension to multi-programming. It is quite common these days, especially since the advent of large memories, to always have several jobs ready for execution. The number of such jobs is called the multi-programming depth. In those environments where the multi-programming depth is greater than 2, it seems obvious to simply apply additional processors to the execution of jobs.

Parallelism between jobs places the least demands upon a parallel processing system. No user data sharing is needed, and the ordinarily large overhead of a processor dispatch is acceptable because such a cost would have been paid on a single processor system anyway.

Note that parallel processing between jobs does not change the execution time for an individual job. Throughput is increased because more processors are available. Average turn-around time is reduced because the amount of time a job spends waiting will be reduced (assuming that some bottleneck like I/O saturation does not increase this waiting time).

Parallelism between jobs is uninteresting from a compiler writer's or application programmer's point of view. It will not be considered further in this paper.

2.2 Parallelism Between Subroutines

This level of parallel processing is a logical extension of the procedure call. Many languages support the ability to start a task through a modified procedure call mechanism. Discrete event simulation of physical systems lend themselves to this type of parallelism because each physical entity is usually modelled in its own routine and because the state of the simulation is not updated until all of the physical entities have executed for this time step. Thus, the subroutine calls are independent of each other, and if enough resources were available, all could be executed simultaneously.

Parallelism between subroutines places one additional demand on a parallel processing system: data sharing and synchronization. Because the amount of data sharing performed is expected to be low, because the data sharing is usually well structured (at the beginning and the end of subroutine execution), and because the expected amount of computation is high, moderate to high overhead methods of data sharing and synchronization can be successful. Similarly, few additional demands are placed upon processor dispatch.

2.3 Parallelism Between Blocks and Loops

This level of parallelism is a logical extension of the type of parallelism exploited by look-ahead control units. Instead of instructions, collections of statements (basic blocks and/or loops) are scheduled across a parallel processor. In addition to data sharing occurring at the beginning and end of execution of the block/loop (as in an instruction schedule), data sharing can also occur at any time during the execution of the block/loop.

Parallelism between blocks and loops starts to place more demands upon the efficiency of processor dispatch, simply because the excepted execution time of the work being parallel processed has decreased. Also, because the blocks and loops come from the same routine, there is the additional problem of the "casual" sharing of variables that are local to the routine.

2.4 Parallelism Between Iterations of a Loop

This level of parallelism is a logical extension to the type of parallelism exploited by look-ahead control units. Instead of scheduling instructions to multiple functional units or a pipeline, code segments (blocks or loops) are scheduled across multiple processors. Data sharing and synchronization can occur not only at the beginning and the end of execution of each code segment, but also at any point in the middle of each code segment.

This level of parallelism can be thought of as a highly structured form of parallelism between blocks: each iteration of a loop becomes a block. From the structure comes a high degree of regularity in the patterns of synchronization and data sharing. This eases the task of deciding how many processes can be effectively utilized when executing this parallelism.

This level of parallel processing was first exploited by the vector supercomputer Illiac-IV in the early 1970's. To be sure, only the restricted case of a single assignment statement inside a loop was exploited, but it was a beginning.

Parallelism between iterations of a loop places one unique demand upon a parallel processing system: dividing the iterations of the loop up among the processors. The speed at which this task can be performed can determine the number of processors that can be effectively utilized. Also, because of the expected execution time of the work being parallel processed, low-overhead, efficient, data sharing, synchronization and processor dispatch must be implemented.

2.5 Parallelism Between Instructions in a Basic Block

This level of parallelism is a logical extension to the type of parallelism exploited by look-ahead control units. Instead of scheduling instructions to multiple functional units or a pipeline, statements, or small groups of instructions are scheduled across multiple processors. Data sharing and synchronization can occur not only at the beginning and the end of execution of each code segment, but also at any point in the middle of each code segment.

Some think that the lookahead control unit obviates the need for any compiler knowledge of the parallel processing performed, but the CDC Fortran compiler for the 7600 disproved this by demonstrating optimization techniques that improve utilization of the multiple functional units when running ordinary Fortran programs. With the advent of RISC architectures, these compile time optimization techniques have become even more important.

Parallelism between instructions places the most demands upon a parallel processing system. Very fast data sharing, and processor synchronization must be implemented, along with very low overhead processor dispatch.

3 A BRIEF LOOK AT PARALLEL PROCESSOR ARCHITECTURES

The methods employed in a parallel processor to share data has a significant impact on the level of parallel processing that can be employed with reasonable efficiency. While there are many other possibilities, these three cases span an interesting range: shared memory (e.g. Cray X/MP), network memory (e.g. N-Cube), and local/global memory (e.g. ETA-10).

The important differences in these architectures (for the purposes of this paper) are the overhead time required to perform a synchronization, the overhead involved in data sharing, and the number of processors that can be connected. Usually the shared memory architecture has significantly smaller number than the network memory architecture for all of these. The goal of the local/global memory systems is to achieve the low overhead of the shared memory architectures and the high number of processors of the network memory architectures.

3.1 Shared Memory

In this architecture, all of the processors execute out of the same physical memory. Data sharing involves little additional overhead since the processors can all address the same memory. Similarly, synchronization can be performed with an overhead of as little as a single instruction (not even that in the case of the Denelcor HEP).

This is the simplest architecture to parallel process simply because no explicit data motion is needed to achieve data sharing and because the overheads involved in data sharing and synchronization are so low.

This architecture is usually considered to be limited to small numbers of processors. There are two major reasons for this:

1. Any single memory cannot be made fast enough to support the simultaneous access of very many processors.

2. If large numbers of memories are used to support large numbers of processors, then the delay involved in the switch that connects the memories to the processors will be at least O(log(memories)).

3.2 Network Memory

In this architecture, each processor executes out of its own physical memory. Data sharing and synchronization is achieved by explicit transfers of information across a communication network.

This is usually the most difficult to utilize because of the usually high overhead involved in data sharing and synchronization. Processes must be discovered that have relatively large amounts of computation compared to data sharing and synchronization.

This architecture is usually the simplest kind to extend to large numbers of processors. The extensibility is determined by the communication network. There are several networks where architectures of thousands of processors seem reasonable.

3.3 Local/Global Memory

In this architecture, each processor executes out of its own local memory, but can also access a global memory that is shared among all of the processors. Data sharing and synchronization are performed much like on the shared memory architecture, though the overheads involved are usually larger.

This architecture is one of many points between the shared memory architecture and the network memory architecture. A global memory architecture can be more extendable than a purely shared memory architecture because part of the memory activity load is taken over by the local memories.

4. EXPECTED PERFORMANCE IMPROVEMENTS

Each of the five levels of parallel processing derives the exploitable parallelism from different regions of the programming environment. Each region has certain characteristics that are exploited to achieve parallel processing, and these characteristics determine the expected level of performance.

One important point to observe is that any of these levels of parallel processing can be recoded as another level (though the recoding may be difficult to do). Consider that parallelism between jobs can be rewritten as parallelism between routines by writing a very trivial "driver" to invoke each job as a subroutine. Similarly, parallelism between subroutines can be rewritten as parallelism between blocks and loops by macro expanding the subroutines in-line. Ordinary coding styles are assumed in this paper, not any possible coding technique.

Reduction in average job execution time (by parallel processing) will result in a reduction in average turn-around time, but will not result in increased throughput, since the amount of work performed is constant. In fact, it is possible to decrease average turn-around time and decrease throughput at the same time because of the effect of increased overhead.

4.1 Between Subroutines

There are two categories of parallelism within this level, each with different expected levels of parallelism. Limited empirical evidence suggests that the majority of numerical algorithms have a small degree (2 to 4) of parallelism available at this level. But, measurements are difficult to make and those that have been made have covered a very limited number of cases. An exception to this observation are discrete event simulations of large systems where often a high degree of parallelism is available at this level.

The degree of parallelism can be reduced by the effects of memory size, memory bandwidth, and I/O bandwidth.

In addition, the work being performed by each processor may not be balanced, so not all of the processors may be utilizable all of the time to improve performance on this program. This effect will be noticed most in programs that have a degree of parallelism that is about the same size, or less than the number of processors applied to the program. (Note that the unbalanced utilization of processors inside a program need not have a negative effect on the throughput of a computing system. If the system can utilize the idle processors to work on different jobs, then throughput will not suffer because of increased idle time).

If synchronization is needed to make the data sharing operate correctly, then the structure of the synchronization can easily limit the effective parallelism, regardless of the number of computational stages in the algorithm.

In any application there is a certain amount of "tissue" code that must be executed on a single processor. The amount of this tissue code is usually small at this level of parallelism, but it should not be ignored when estimating expected performance.

4.2 Between Blocks and Loops

The parallelism between blocks and loops is the first level of parallelism where coding style, rather than the structure of the problem or the machine effects the available parallelism. Measurements performed at the University of Illinois indicate that for shared memory parallel processors, the expected performance improvement is about 3. This study made several assumptions that are not exactly true in the real world: synchronization is as fast as arithmetic operations, data sharing takes no additional time, an unlimited number of processors are available, and it takes no time to dispatch a processor. Clearly, these assumptions will be violated in most real computers. This will reduce the effective degree of parallelism somewhat.

The observable degrees of parallelism can be reduced by system architecture features that are similar to those observed in the between subroutines level. Some of these features (time to dispatch a processor, and time to synchronize) have a more pronounced effect at this level because the expected execution time for a block or a loop is less than the expected execution time for a subroutine or job.

4.3 Between Iterations of a Loop

There are four classes of parallelism within this level, each with different expected degrees of parallelism.

The first class needs no data sharing or synchronization between iterations of the loop. This class of loop obviously has a maximum performance improvement equal to the number of iterations of the loop.

The second class requires synchronization to achieve the correct data sharing, but the synchronization would not be needed if all of the processors operated from the beginning of the index set. This class of loop obviously needs synchronization on most parallel processors (because they are asynchronous), but since this synchronization will usually be satisifed, very little delay will be introduced, yielding a maximum performance improvement roughly equal to the number of iterations of the loop.

The third class needs no sychronization except for those needed to support reductions such as summation, dot product, or maximum element. If the number of iterations of the loop is large when compared with the number of processors being applied, then the reductions involving associative operators (or in the case of addition and multiplication, nearly associative operators) can be partitioned and performed on each processor, then after all of the computation is performed, reduced the individual results from the individual processors. This yields a maximum performance improvement of somewhat less than the number of iterations of the loop. If the number of iterations of the loop is not large when compared to the number of processors being applied, then the time that it takes to perform the reduction will dominate the computation. This yields a maximum performance improvement of the Log (base 2) of the number of iterations of the loop. Additionally, if the reduction being performed is not an associative operator, or if it is on a nearly associative operator but nearly is not enough, then the operations involved in the reduction must be performed in the correct order. This yields a maximum performance improvement approximately equal to the amount of computation (the number of statements is one useful measure) being performed on one iteration of the loop.

The fourth class requires synchronization to achieve the correct data sharing, regardless of the restrictions placed on the processors or the index set assignment. This class of loop obviously has a maximum performance improvement equal to the amount of computation (the number of statements is one useful measure) being performed on one iteration of the loop. Any loop can be executed on a parallel processor with synchronization, but parallel processing this class of loops does not always help.

The success of vector machines in many different types of computation provides one indication of the number of loops that require no synchronization other than that needed to support reductions (class 1 and 4).

One important factor to consider when estimating the performance improvement from parallel processing the iterations of a loop is the overhead involved in passing out the iterations of a loop. If the iterations are passed out individually, then it is important to perform this operation very efficiently, as it quickly becomes the limiting factor in determining the degree of parallel processing.

4.4 Between Instruction in a Block

The parallelism between instructions in a block is the level of parallel processing that is the best understood. Measurements performed in the 1960's indicated that the expected performance improvement is a factor of three, and experience since then with lookahead control units, pipeline fill levels, and compilation studies all agree that three is still reasonable.

The observable degrees of parallelism can be reduced by system architecture features that are similar to those mentioned in the between iterations of a loop level. Some of these features (time to dispatch a processor, the time to share data, and the time to synchronize) have a more pronounced effect at this level because the expected execution time for an instruction is less than the expected execution time for an interaction of a loop.

5. APPLYING AUTOMATIC TECHNIQUES

There are three problems that must be solved in applying parallel processing at any level: discovery of the items of work to be parallel processed, allocation of memory resources, and insertion of data sharing and synchronization primitives.

All of the algorithms sketched below are based upon two graphs constructed from appropriate portions of the program being analyzed: the flow graph, and the dependence graph. Each of the graphs can be constructed with the nodes of the graph representing different sizes of code: an instruction, a statement, a block, a loop, a subroutine, etc. The descriptions given below is the terms that are the simplest to explain.

The flow graph is a directed graph where each node in the graph is a statement. An edge in the flow graph from node X to node Y indicates that when the statement corresponding to node X completes execution, the next statement to execute could be the statement corresponding to node Y. If multiple edges leave a node Z, then the statement corresponding to node Z must be an IF statement, or some other type of statement that makes a choice as to which statement will execute next.

The dependence graph is a directed graph where each nodes in the graph is a particular occurrence of a variable in an executable statement. An edge in the dependence graph from node X to node Y indicates that there may be some interaction between the corresponding occurrences of variables. An edge from node X to node Y can exist only if there is a path in the flow graph from the statement containing the occurrence of the variable corresponding to node X, to the statement containing the occurrence of the variable corresponding to node Y. In addition, an edge in the dependence graph can only connect two nodes that share at least one memory location. The difficult part is to build a dependence graph without too many extra edges. An exact solution for scalar variables has been presented in almost every book on compiler optimization (aho 1977). For insight into approximate solutions for arrays see Banerjee (1976, 1979).

There are three types of edges in the dependence graph.

```
A=      =A      A=
 =A    A=       A=

normal  anti-   output
```

A normal dependence edge from node X to node Y indicates that the value of the variable assigned by the occurrence of the variable corresponding to node X might be used by the occurrence of the variable corresponding to node Y. An anti-dependence edge from node X to node Y indicates that the value of variable used by the occurrence of the variable corresponding to node X might be assigned by the occurrence of the variable corresponding to node X might be reassigned by the occurrence of the variable corresponding to node Y.

Unlike many optimization techniques, it is not necessarily true that blind application of the parallel processing techniques in the following sections will not necessarily result in decreased execution time. It is necessary to generate an evaluation procedure so that the usefulness of a particular optimization can be compared against (at least) scalar code. In many cases, there are even multiple ways to parallel process a section of code.

5.1 Between Instructions in a Block

This is a well understood parallelism detection algorithm (Veidenbaum 1985). Usually, its use is restricted to those systems where a certain number of processors are known at compile time to be available for execution of this block. The author is not aware of any commercial compilers that implement this technique (on a multi-processor; the CDC 7600 compiler used this technique on a machine with multiple functional units), but believes that some compilers will be available in the next two years that do.

Consider first the basic block with no conditionally executed code, a single entry, and a single exit. A flow graph is built for the block of instructions (with an instruction as a node). Then build a dependence graph (with an instruction as a node).

In an abstract sense, the dependence graph is the schedule. The nodes of the graph are the work to perform, and the edges of the graph describe both a partial ordering for the work, and some additional work that must be performed if the nodes connected by an edge are not both executed on the same processors. One process could be assigned to each instruction, and the data communication (normal dependence edges) and synchronization (anti- and output dependence edges) and completely described by the dependence graph.

Scheduling this computation on a given number of processors can be done by any number of heuristics. A simple algorithm would assign a weight to each node corresponding to the instruction execution time and a weight to each edge corresponding to the data sharing/synchronization time, and then perform a job shop schedule for the given number of homogenous processors.

A more complicated algorithm might attempt to remove certain edges from the dependence graph by renaming (the creation of new temporary variables), before (or while) performing the same type of schedule. It is possible to remove both anti-dependence edges and output dependence edges from the graph under certain conditions. Additionally, some synchronization edges might not be necessary because satisfying one synchronization edge might insure that another edge is also satisfied.

To extend the algorithm to blocks of code that contain additionally executed code and multiple exits, but no cycles in the flow graph, add synchronization edges to the dependence graph from the node corresponding to the conditional branch to all nodes that are reachable (in the flow graph) from that node.

5.2 Between Iterations of a Loop

This is a well understood parallelism detection algorithm. It is not necessarily restricted to those systems where a certain number of processors are known at compile time to be available for execution of this block.

Consider first an iterative loop without additional exits. First a flow graph is built for the loop (with a statement as a node). Then build a dependence graph (with a statement as a node). The dependence graph must be a little more complicated now because it is necessary to distinguish between edges in that stay within a single iteration of the loop and those that go from one iteration to another. Since a single process will always execute at least one entire iteration, the partial execution ordering described by the edges that stay within a single iteration of the loop will always be satisfied. This means that only those edges that go from one iteration to another need to be considered.

The smallest process is a single iteration of the loop, and the data communication and synchronization need to insure correct execution is completely described by the dependence graph, at the statement level. There are additional techniques that can be utilized on shared memory architectures to reduce the amount of synchronization that is needed (Midkiff 1986).

There is one additional decision that must be made: how are iterations going to be assigned to processors? A method that works for every case is to have a piece of code (or hardware) that a processor can execute to get the next iteration. Using this method, one does not have to discover at compile time if each of the iterations is expected to take the same amount of time, or worry about synchronization patterns. There is one problem with this method, however. It is possible for the method used to assign iterations to be the limiting factor for the degree of parallelism, especially for loops that contain a small amount of computation.

Now, the additional complication of a conditional exit from the iterative loop can be considered. If a conditional exit might be taken on some iteration N, the next iteration (N+1) cannot begin until it is known that the Nth iteration did not exit. This can be assured by adding an appropriate edge to the dependence graph (from where it is known that the exit will not be taken, to the first statement in the loop on the next iteration).

Note that while loops present a difficult problem. Usually the condition that determines if the loop will be executed again must be the very last thing executed in an iteration. If this is true, then NO overlap is possible between iterations, and parallel processing does not help.

When there are multiple loops in a loop nest, it is possible to examine the loops from the outside in until a loop with an acceptable level of performance is encountered. Another possibility is to apply the technique of loop interchanging (Wolfe 78, Allen 84, Wolfe 86) in an attempt to move an acceptably parallel loop further out in the loop nest.

5.3 Between Blocks and Loops

This is a well understood parallelism detection algorithm (Veidenbaum 1985). The author does not know of any commercial compilers that contain this algorithm, but it is very similar to the algorithm described for parallelism between instructions in a block.

First a flow graph is built for the segment of code (with a statement as a node). Then blocks and loops are detected by examining the flow graph. A new flow graph (the block flow graph) is built by collapsing all of the nodes in the statement flow graph that are members of the same block or loop into a single node in the block flow graph. Duplicate edges are eliminated during this process.

A dependence graph is built (with a block/loop as a node). Arcs are added to this dependence graph to account for conditionally executed blocks/loops using a similar set of rules as in the between instructions in a block level of parallel processing.

The dependence graph is a schedule, and the only synchronization and data sharing that must be performed occurs at the beginning and end of the execution of a block or loop. At this point, the scheduling problem is identical to that described in the between instructions in a block level.

It is possible to perform synchronization and data sharing from anywhere in the middle of the execution of a block (the author is unsure about performing such synchronizations from inside a loop). In this case, a dependence graph using a statement as a node is used to determine synchronization instead of the dependence graph using a block/loop as a node. The author is unaware of any measurements on the effectiveness of this, more complex, algorithm.

5.4 Between Subroutines

While this level is often easiest for the programmer to consider, known automatic techniques are quickly overwhelmed with the details of the interaction between COMMON blocks, EQUIVALENCE statements, and parameter aliasing. Expect progress in this area, but ask those that claim something at this level for a demonstration.

6. A PROBLEM

At the moment, when a programmer constructs a parallel version of a Fortran program, that program can only be executed (in parallel) on one particular manufacturer's parallel processor. This slows the spread of existing parallel algorithms, and slows the coding of new parallel algorithms because of limited probability.

The most obvious cause of this portability problem is that each manufacturer has "private" extensions to Fortran for expressing parallel processing. KAI would like to propose a syntax capable of expressing each of the four levels of parallel processing. This proposal is provided as an appendix to this paper. The author would welcome comments, suggestions, and descriptions of experiences concerning this syntax.

7. REFERENCES

Aho, Alfred and Jeffrey Ullman, 1977: Principles of Compiler Design. Addison-Wesley Publishing, pp 429-441.

Allend, J.R., and K. Kennedy, 1984: Automatic Loop Interchange. Proceedings of the ACM SIGPLAN '84 Symposium on Compiler Construction, pp. 233-246.

Banerjee, Utpal, 1976: Data Dependence in Ordinary Programs. Rpt. No. UIUCDCS-R-76-837, University of Illinois.

Banerjee, Utpal, 1979: Speedup or Ordinary Programs. Rpt. No. UIUCDCS-R-79-989, University of Illinois.

Midkiff, Samuel and David Padua, 1986: Compiler Generated Synchronization for DO Loops. 1986 International Conference for Parallel Processing, pp 544-551.

Veidenbaum, Alexander, 1985: Compiler Organizations and Architecture Design Issues for Multiprocessors. CSRD Rpt. No. 520, University of Illinois.

Wolfe, Michael, 1978: Techniques for Improving the Inherent Parallelism in Programs. UILU-ENG 78 1722, University of Illinois.

Wolfe, Michael, 1986: Advanced Loop Interchanging. 1986 International Conference for Parallel Processing, pp 536-543.

8. APPENDIX: PARALLEL PROCESSING SYNTAX PROPOSAL

The purpose of this definition is to provide a common framework for implementation of parallel processing constructs upon the base of Fortran-77.

These are the goals of this parallel syntax proposal:

1. to be a logical extension to Fortran-77
2. to be compatible with Fortran-8x
3. to be able to exploit a variety of parallel architectures
4. to contain a subset that is easy for a naive compiler to translate to good code.
5. to be easy for programmers to use, especially if an optimizing compiler or preprocessor is available.

It is KAI's intention to provide a prepocessor that automatically detects parallel constructs from serial Fortran, and generates this extended syntax. This preprocessor will also be able to accept input in this syntax.

8.1 Additional Data Type

The EVENT statement is used to declare names to be of type Event.

```
EVENT <v> [,<v>]...
```

where <v> is a variable name, array name, or array declarator.

The Event constants are .WAIT. and .GO.. The only Event operator is assignment (=).

8.2 Additional Storage Classes

The PRIVATE statement is used to declare sections of memory that are private to a process. The SHARED statement is used to declare sections of memory that are shared by more than one process.

```
PRIVATE <v> [,<v>]...
SHARED  <v> [,<v>]...
```

where <v> is a variable name, array name, array declarator, or a common block name enclosed in //.

By default, variables, arrays, and common blocks are process PRIVATE. The PRIVATE declaration is provided for documentation purposes, and for those programmers concerned with creating portable code.

8.3 Synchronization

8.3.1 Statement Level Synchronization

```
WAIT SYNC( <event-name> [,<even-name>]... )
SEND SYNC( <event-name> )
```

where <event-name> is the name of an event variable or the name of an event array element.

SEND SYNC completes the event named by the indicated event variable or array element by setting the event variable to .GO.. This action also unblocks any process that is waiting on the completion of that event.

WAIT SYNC stops a process from executing any further until the indicated event(s) are completed. The process need not be suspended from execution if the indicated events are already completed when the WAIT SYNC is executed.

8.3.2 LOAD/STORE Level Synchronization

```
STORE( [<cause-event-name>], <ref> [,<wait-event-name>]... )
LOAD(  [<cause-event-name>], <ref> [,<wait-event-name>]... )
```

where <cause-event-name> is the name of an event variable or event array element that will be marked completed when the load or store is completed

<ref> is the item to be loaded/stored

<wait-event-name> is the name of an event variable or event array element that will be marked completed before the load or store is begun.

The LOAD function can be utilized around any operand that is coded in any place where the value of the variable at runtime will be used: on the right-hand-side of an assignment statement, in the subscript of an array element reference on the left-hand-side of an assignment, in a WRITE statement, etc.

The STORE function can be utilized anywhere that a variable reference is coded, and that variable is modified by the reference: on the left-hand-side of an assignment statement, as an actual parameter to a user routine, in the I/O list of a READ statement, etc.

8.3.3 Barrier Synchronization

```
[ ORDERED ] BARRIER ( <name> )
```

where <name> is a unique identifier inside this parallel construct

This construct can only be coded inside a PCASE or a DOACROSS.

Upon execution of a nonordered BARRIER, a process waits until all active processes (that are not destroyed) from the immediately enclosing PCASE or DOACROSS, are waiting at a barrier of this same <name>.

Upon execution of an ORDERED BARRIER inside a PCASE, a process waits until all active processes (that are not destroyed) with smaller logical process numbers have reached a barrier with this same <name>. Upon execution of an ORDERED BARRIER inside a DOACROSS, a process waits until all active processes (that are not destroyed) with earlier iteration numbers have reached a barrier with this same <name>).

8.3.4 Critical Sections

```
[ ORDERED ] CRITICAL SECTION [ ( <name> ) ]
<stmts>
END [ ORDERED ] CRITICAL SECTION [ ( <name> ) ]
```

where <name> is a unique identifier inside this parallel construct

In its simplest form, this construct assures that only a single process is allowed inside the bounded code segment.

If the optional <name> parameter is used, all of the code enclosed in CRITICAL SECTIONs of the same <name> are governed by an identical set of access restrictions. Hence, only one process will be allowed to execute inside any CRITICAL SECTION with the same <name> at any time.

If the ORDERED parameter is coded, then the CRITICAL SECTION must be coded inside a PCASE or DOACROSS.

Upon execution of an ORDERED CRITICAL SECTION inside a PCASE, a process waits until all active processes (that are not destroyed) with smaller logical process numbers have entered and left an ORDERED CRITICAL SECTION with this same <name>.

Upon execution of an ORDERED CRITICAL SECTION inside a DOACROSS, a process waits until all active processes (that are not destroyed) with earlier iteration numbers have entered and left an ORDERED CRITICAL SECTION with this same <name>.

8.4 Parallel-Processing Constructs

8.4.1 PCASE

```
PCASE [ (<process-id> = <number-processes>) ]
OF( <process-number> [,<process_number>]... )
    <stmts>
[ OF( <process-number> [,<process_number>]... )
    <stmts>
[ OTHER PROCESSES
    <stmts>
END PCASE
```

where <process-id> is the name of a scalar integer variable whose storage class is automatically PRIVATE. This number is in the range 1:<number-processes>, and is used to uniquely identify the process inside this construct.
<number-processes> is the number of processes that the programmer expects to be created at this statement.
<process-number> is used to select the processes that will execute the following statements. <process-number> is either an integer expression, or a relational expression involving <process-id>. For every <process-id> in the range, only a single OF clause may be selected.
<stmts> those statements to be executed for this process. <process-id> can be used as an integer variable in these statements.

When the PCASE statement is executed, it appears to the programmer as if the indicated number of processes are created, and each process executes the appropriate <stmts>, as determined by the OF clauses. The total number of processes to be created is determined from <number-processes> if it is coded, or form the maximum <process-number> coded in the OF clauses.

In each OF clause, the <process-number> must be coded as an integer constant if <number-processes> is not coded in the PCASE statement. For any given process number, no more than one OF clause may be selected.

If a process number does not cause selection of an OF clause, then the OTHER PROCESSES clause is selected if it is coded. It is an error to have a process number that does not cause selection of an OF clause if the OTHER PROCESSES clause is not coded. The OTHER PROCESSES clause cannot be coded unless <number-processes> is coded in the PCASE statement.

When all of the <stmts> for a single process are executed, that process is destroyed. Execution continues after the END PCASE when all processes created by the PCASE have been destroyed.

There are no restrictions placed on the statements in the OF clause or the OTHER PROCESSES clause. Note that appropriate support in the runtime environment will be needed to allow I/O statements, or intrinsic function invocations inside <stmts>.

All synchronization inside a PCASE construct must be supplied by the programmer using the LOAD and STORE functions and/or the WAIT SYNC and SEND SYNC statements.

Process PRIVATE variables may be defined by placing declarations for them inside the appropriate OF or OTHER PROCESSES clause. Any variables declared there are automatically PRIVATE.

8.4.2 DOACROSS

This construct is utilized to execute one or more perfectly nested DO loops on a parallel processor with each processor being given an iteration to execute.

```
DOACROSS ( <pi>=<ias> [,<pis>]...  )
   <stmts>
   END DOACROSS
```

where <pi> is an integer variable name that will hold the current value of the parallel index set. This name is a process PRIVATE name whose scope is the DOACROSS block.
<ias> is an integer arithmetic sequence that defines the parallel index set. It is coded as a simplified Fortran-8x array constructor (only one sequence):
[lowerbound:upperbound:increment] (:increment is optional -- defaults to 1).
<pis> is a <pi>=<ias> expression, but the parallel index name must be different from the names that occurred earlier in the DOACROSS.

When the DOACROSS statement is executed, it appears to the programmer as if one process is created for each element of the cross product of the parallel index sets described. Each process executes the <stmts>.

When all of the <stmts> for a single process are executed, that process is destroyed. Execution continues after the END DOACROSS when all processes created by the DOACROSS have been destroyed.

There are no restrictions placed on the statements inside a DOACROSS other than requiring that no branch out of <stmts> can be made except for subroutine calls or function invocations. Note that appropriate support in the runtime environment will be needed to allow I/O statements, or intrinsic function invocations inside <stmts>.

All synchronization inside a DOACROSS construct must be supplied by the programmer using the LOAD and STORE function and/or the WAIT SYNC and SEND SYNC statements.

Process PRIVATE variables may be defined by lacing declarations for them inside the DOACROSS. Any variables declared there are automatically PRIVATE.

8.4.3 CREATE

```
CREATE ( <event-name> ) <stmt>
```

where <event-name> is the name of the event variable that will be marked completed when the execution of <stmt> is completed.
<stmt> is any statement, but usually a CALL or assignment statement with a function reference.

The CREATE statement initializes the event variable named by <even-name> to .WAIT., and starts a new process executing <stmt>. When <stmt> is completely executed, including completion of any user and system routines invoked by the statement, the indicated event is marked completed.

8.5 The Subset

The subset is designed to provide sufficient functionality to describe useful work, without going to a lot of trouble to make it look pretty. These constructs should be considered a minimum for implementing explicit parallel-processing as an extension to Fortran-77.

1. EVENT data type, operations, and constants

2. SHARED statement and storage class

3. WAIT SYNC and SEND SYNC

4. PCASE construct

The Floating Point Systems T Series

D. A. TANQUERAY

Floating Point Systems (UK) Ltd., Bracknell, Berkshire, United Kingdom

Abstract

The Floating Point Systems T-Series is a massively parallel supercomputer for scientific and engineering applications. Each processing node combines the performance of fully pipelined vector oriented floating point arithmetic with the connectivity and parallel processing capability of the INMOS Transputer. Because each processor has its own memory rather than a switched path to a global memory, very large homogeneous configurations are possible. This paper describes the new architecture and gives an example of distributed processing with associated performance.

1. INTRODUCTION

The quest for increased computational power in scientific computing far exceeds the ability of single processor systems to satisfy it within the limits imposed by physical electronic devices. These limitations can be overcome by operating a number of processors in parallel.

The FPS T Series offers the promise of performance two orders of magnitude greater than can be achieved with conventional systems by the use of two central ideas.

The first is the provision of the highest performance vector processor which can be readily and cost effectively replicated. This is achieved through the use of very large scale integration (VLSI) technology.

The second uses the idea of a distributed memory based on the limited form of interconnection provided by a binary n-cube or hypercube. This avoids the expense and performance degradation exhibited by shared memory systems when scaled to large dimensions and consequently very large configurations containing up to thousands of processors become possible.

It has been shown by Fox (1984) and others that the topologies required to implement most important scientific algorithms on multi-processor systems are all simple subsets of such a binary n-cube.

2. T-SERIES ARCHITECTURE

The PFS T Series (Hawkinson, 1986) is a homogeneous multi-processor computer based on the binary n-cube interconnection scheme. The individual nodes are 64-bit floating-point computers that combine vector arithmetic, dual-ported memory, and fast communications links between nodes. The peak sustained vector performance of these nodes is 12 MFLOPS. The FPS T Series is built from modules containing eight of these nodes connected to each other and to a system support ring. These modules, with an aggregate performance of 96 MFLOPS, may be combined to form larger systems.

2.1 The Processor Node

Each individual processor or node in a T Series system contains a control processor, floating-point arithmetic, dualported memory, and communication links to other nodes (see Figure 1). The FPS design provides all of these functions on a single printed-circuit board. Each of the major elements of the node has been implemented with cost-effective VLSI technology.

2.1.1 Control

The control processor interprets and executes and user application code. It is also responsible for setting up operands and initiating vector operations in the vector arithmetic unit. The control processor can execute integer arithmetic and gather/scatter operations in parallel with the vector unit, and supports simultaneous inter-node communications.

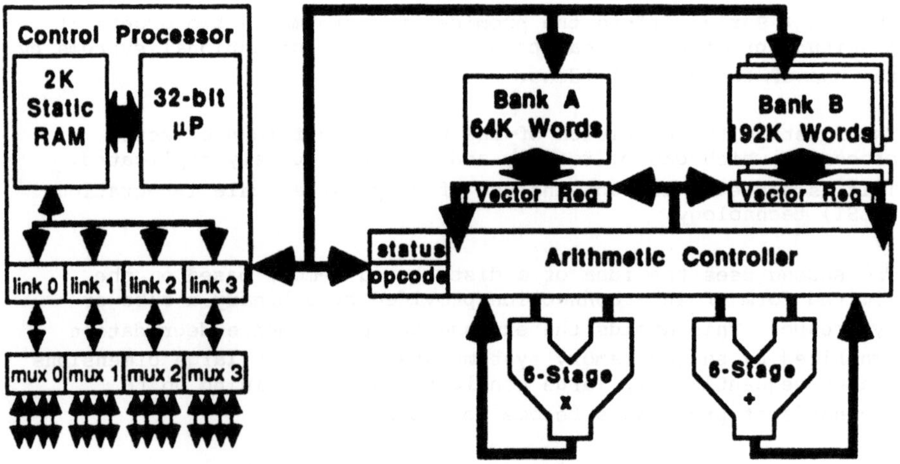

Figure 1. The FPS T Series Processor

The control unit in the FPS T Series is a 32-bit CMOS microprocessor (the Inmos Transputer) which has been specifically designed to support parallel processing in large networks of interconnected processors. The basic functional features are:

A 32-bit 7.5 MIPS microprocessor
Highly compact stack oriented instruction set
2048 bytes of on-chip RAM (single processor cycle)
Memory port capable of addressing up to 4 Gbytes
Four bi-directional serial communications links
Two-level process priority and interrupt services.

All features of the microprocessor are directly accessed through the Occam language, which provides for the execution of parallel, communicating processes. Processes can be created to execute sequentially or in parallel. Channel commands make direct synchronous data transfers between concurrent processes on the same node or different nodes. This combination of program structure and integrated communication allows Occam to describe the control and data flow for algorithms operating in a multi-processor environment.

The Transputer control processor includes instructions and hardware features specifically optimised to implement the parallel process and channel constructs of Occam. In this way the overheads normally associated with multi-process creation and switching, and handling communication links are avoided.

2.1.2 Memory

An essential feature of a computer's architecture is its central memory, which supplies both instructions and operands to the processing units. The main memory of each FPS T Series node comprises 1 MByte of dual-ported RAM.

The control processor and communications links read and write 32-bit words through a conventional random-access port, while the vector unit makes use of a collection of vector registers closely coupled with the main memory. A vector register can be loaded with an entire 1024-byte row of memory, in parallel (see Figure 1), in the same time that it would have taken to read or write a single 32-bit word. This leaves the memory free to be accessed by the control processor while the vector register data is being sequenced through the arithmetic unit.

The vector registers each hold 128 64-bit numbers or 256 32-bit numbers.

The control processor views the memory as a single contiguous address space of 256K 32-bit words. The vector arithmetic unit views the memory as two banks of vectors with 256 vectors in one bank and 768 vectors in the other. The division of memory into two banks permits two inputs in parallel to the arithmetic unit on each cycle (125 ns). The output of the arithmetic unit shifts results into either or both banks. Hence, operations such as SAXPY, Vector Add and Vector Multiply proceed at the full speed of the arithmetic components without being limited by available memory bandwidth.

The principal data paths and associated bandwidths are shown in Figure 2.

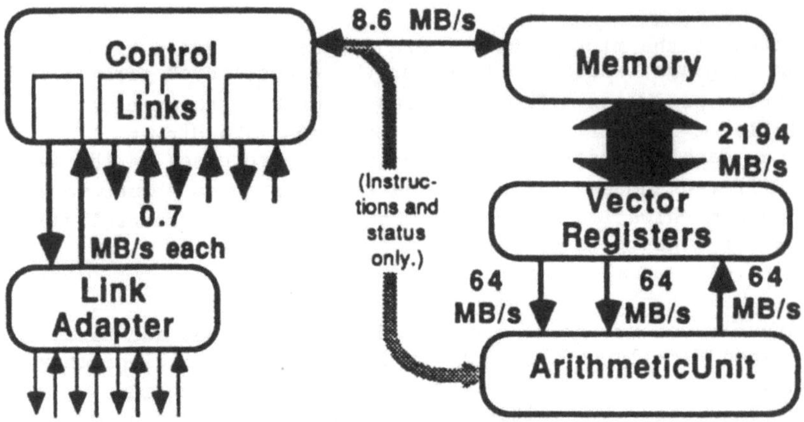

Figure 2. Processor Bandwidths

2.1.3 Arithmetic

The ability to perform high-speed arithmetic is essential in scientific computing. The arithmetic hardware in the FPS T Series consists of a floating-point adder, floating point multiplier, interconnection hardware, and a micro-sequencer. The adder and multiplier each can produce a 32- or 64-bit result every 125 ns, yielding a nominal performance of 16 MFLOPS per node. Overhead inherent in the architecture reduces the effective sustained peak to 12 MFLOPS.

The arithmetic units operate in pipelined mode. The adder has a six-stage pipeline, while the multiplier has a four-stage pipeline in 32-bit mode and a six-stage pipeline in 64-bit mode.

Scalars can be held in the input registers on each functional unit, and outputs from the functional units can be fed directly back as inputs to perform operations such as dot products and sums. The whole is controlled by a pre-programmed microsequencer which provides a wide range of vector forms without memory reference limitations.

The arithmetic unit operates in parallel with the node control processor, and only interrupts the controller when a vector operation completes.

2.1.4 Communications

In a distributed computer system, communications channels are required for passing data between processors participating in a common computational process.

The control processors of the T Series contain independent direct memory access (DMA) controllers for each direction of four serial, bidirectional links. The unidirectional transfer rate of each link is 0.69 MBytes/s. With all four links simultaneously active in both directions the total bandwidth is about 5 MB/s. With all links operating, the control processor performance is only slightly degraded. The startup time for a message sent between two processors over the links is about 9 microseconds on each end of the transmission.

Each links is switched four ways to provide a total of 16 bidirectional sub-links per node. Two sublinks are reserved for system communication, and two will often be utilised for mass storage and/or external I/O. This will typically leave 12 sublinks available for connection to other compute nodes, of which four can be simultaneously active.

2.2 System Description

2.2.1 Hypercube Interconnection

The members of the T Series consist of a number of node processors connected as a binary n-cube or hypercube. Such hypercubes may be constructed by a recursive doubling procedure. A cube of dimension N is made from two cubes of dimension N-1 with (only) their corresponding nodes connected and has the following characteristics. There are then 2^N processor nodes, each with N connections (links). Numbering the processors from 0 to 2^N-1, each processor is directly connected to all others whose numbers differ in only one binary digit. Since the maximum number of connections to traverse between any two processors is N, long-range communication costs grow only as $O(\log_2 N)$.

The binary n-cube can be mapped onto many important applications topologies, including meshes (up to dimension N), rings, cylinders, toroids, and even FFT butterfly connections of radix 2.

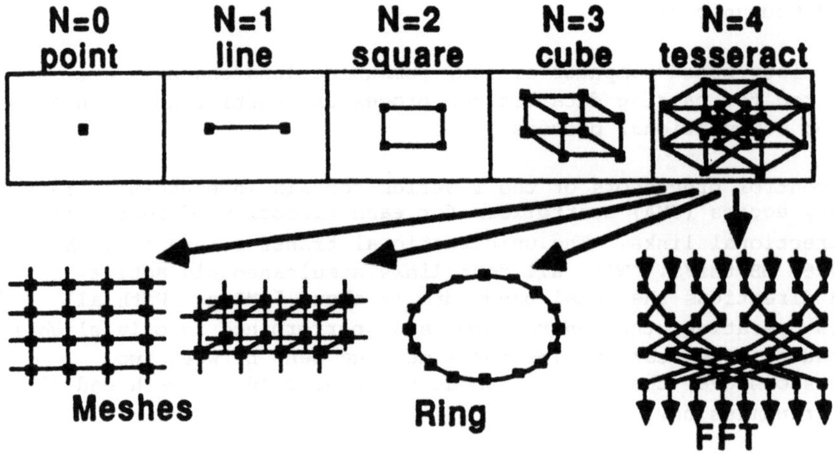

Figure 3. Binary n-Cube Mappings

2.2.2 Modules and System Ring

In the FPS T Series a processor node is contained on a single circuit board. Eight nodes are combined with disk storage and a system board to form a Module. Such a module has 96 MFLOPS peak vector floating-point performance, and 8 MB of user RAM. The module is the smallest unit (a model T 10) in the T Series range.

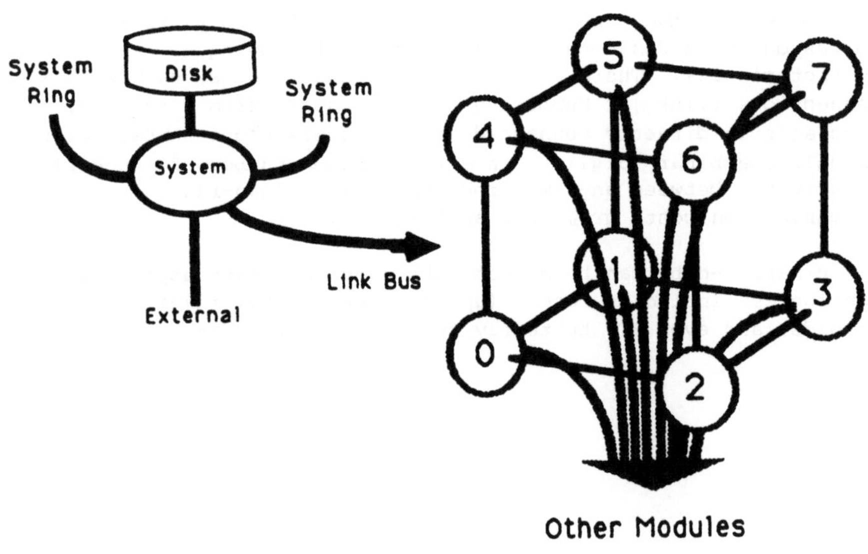

Figure 4. Eight Processor Module

The system board provides input/output and management functions. It is connected directly to the vector nodes by a bus accessed by one of the node sublinks. The system board contains a control processor, memory, disk controller and I/O interface. The four links are not mulitplexed. The system boards of multiple modules are themselves connected by links to from a System Ring that is independent of the binary n-cube network. The system board can support 0.7 MB/s to an external connection.

The primary functions of the system disk are to store user temporary files, record memory "snapshots" which checkpoint computations for error recovery, and to back up snapshots from other modules. Since each module has its own system board and disk, it takes the same time to checkpoint the largest configuration as it does for a single module.

Two modules (16 nodes) form a cabinet, or 4-cube (a tesseract). A cabinet is modular and self-contained with power supplies, disks, and air cooling fans, and requires no special facilities beyond normal air conditioning. Larger systems are simply assembled from these units by interconnecting cables. Connections up to 40 feet can be made without special consideration.

2.2.3 Larger Configurations

A four-cabinet (model T 100) system contains 64 nodes and has an aggregate peak vector speed of 0.768 GFLOPS and total user memory of 64 MBytes. This system contains 8 system disks and provides 8 links for external front-end interfaces.

There are enough links per node to permit a 14-cube to be constructed as the largest T Series configuration. Such a system would contain 16384 nodes. Using two links per node for external I/O and mass storage systems, a maximum-sized 12-cube consists of 4096 nodes in 256 cabinets. Such a system has over 49 GFLOPS vector processing capability and 4 GBytes of primary RAM.

By devoting a whole dimension of the hypercube, i.e., one sublink per node, to interfacing external I/O or mass storage systems, very large homogeneous disk systems could be built to provide the capacity and bandwidth to support large configurations. A similar approach can be taken for interfacing other external devices such as graphics devices, mass memories and other computer systems.

Because the system is homogeneous, with identical modules containing identical connections to other modules, programming is greatly simplified. This homogeneity also ensures that the balance between computing speed, main storage, mass storage, and external I/O can be preserved as configurations become large.

3. EXAMPLE: MATRIX MULTIPLY

The matrix multiply is one of the most widely used kernels in scientific programs. It also involves accessing data in a relatively global manner and as such is a useful algorithm for examining the performance of a distributed memory system such as the T Series. In this example a large matrix multiply is assumed to be distributed across all the nodes in the system which are then used to compute parts of the result in parallel.

3.1 Implementation

In this example two real square input matrices A and B are multiplied to produce a single square product matrix C, such that the (i,j)'th element of the result matrix C is equal to the sum of the products of the elements of the i'th row of the first input matrix A and the j'th column of the second input matrix B as follows.

$$c(i,j) = \sum_{k=0}^{n} (a(i,k) * b(k,j)) \text{ for all values of } i \text{ and } j \quad (1)$$

Since each element of the result matrix can be computed independently, the computation of the result matrix C can be divided into a number of parallel operations. Specifically the result matrix is divided into a number of sub-matrices C[m,n] which are each computed on a different processor. These are then computed by accumulating the sum of a series of sub-matrix multiples as follows.

$$C[m,n] = \sum_{p=0}^{n} (A[m,p] * B[p,]) \text{ for all values of } m \text{ and } n \quad (2)$$

The processors are all organised as a 2-dimensional mesh connected as a 2-D torus, i.e., the processor at the beginning of each row/column is also directly connected to the processor at the end of that row/column. The processor at mesh point M[m,n] then computes the sub-matrix C[m,n]. Input matrices A and B are also distributed across the processors in the same way.

Before processing beings the input matrices must be redistributed among the processors as follows if not already in the correct position.

> Sub-matrices of the A input matrix are shifted left toroidally by the number of processors equal to the processor row number (from 0).
>
> Sub-matrices of the B input matrix are shifted up toroidally by the number of processors equal to the processor column number (from 0).

The processor at mesh point M[m,n] then has sub-matrices A[m,n+m] and B[m+n,n] resident and is responsible for computing result C[m,n].

Each node in the processor array then executes the identical procedure.

Following initialisation of the result matrix the computation proceeds as a sequence of small matrix multiplies with accumulation of partial matrix products into the resident result sub-matrix. In each pass of the algorithm the contents of the resident sub-matrices A and B are passed toriodally to the East and South respectively, inputs for the next pass are read from the West and North, and a partial product accumulation is performed, all in parallel.

Within the small matrix multiply operation on each processor, the VSMVA (vector scalar multiply vector add) operation is used to accumulate the results into the output array C[M,n].

3.1.1 Notes on the Code

Figure 5 shows the essential parts of the algorithm expressed as an Occam procedure.

The indentation of the code lines is a required part of the Occam syntax which is used to indicate the structured nesting level of the code.

(a) Main Program

```
PROC matmul
  ...  Definitions
  SEQ
    torus.2d(processor, width, height, row, col, N, S, E, W)
    MatFileRead(Afile, processor, Abuffer, rows)
    MatFileRead(Bfile, processor, Bbuffer, rows)
    ...  Initialise [C] = 0
    ...  Redistribute Input Data
    SEQ i = [ 0 FOR meshsize ]
      PAR
        ...  Multiply Submatrix
        ...  Send 'Old' A East
        ...  Send 'Old' B South
        ...  Read 'New' A from West
        ...  Read 'New' B from north
    MatFileWrite(Cfile, processor, Cbuffer, rows)
```

(b) Redistribute Input Data

```
PAR
  SEQ i = [ 0 FOR row ]
    PAR
      ...  Read 'New' A from East
      ...  Send 'Old' A West
   SEQ j = [ 0 FOR col ]
     PAR
       ...  Read 'New' B from South
       ...  Send 'Old' B North
```

Figure 5. Matrix Multiply Program

The "..." construct is a feature of the Occam folding editor which offers a convenient way of examining and developing structured programs without having to see all the detail. Such a "fold" may contain any number of lines of code.

The PAR statement specifies that all processes contained within that nesting level are to be executed in parallel. In the main program five such processes are created on the control processor. One performs the arithmetic in the vector unit under control of the microprocessor, while the other four initiate parallel channel communication using the DMA link hardware. The Transputer waits for all parallel processes to complete before executing the next statement in the outer SEQ (sequential) construct.

The procedure "torus.2d" is a standard library routine which takes the processor number and the dimensions of the desired 2-dimensional processor mesh as parameters. It returns the position of the processor in that mesh and the numbers of the links to use to communicate with adjacent nodes in each of the four directions. This routine ensures that there is no conflict on the use of the multiplexed links. For example, on a 64 node T 100 the processors are allocated as shown in Figure 6.

row \ col	0	1	2	3	4	5	6	7
0	0	4	5	1	17	21	20	16
1	8	12	13	9	25	29	28	24
2	10	14	15	11	27	31	30	26
3	2	6	7	3	19	23	22	18
4	34	38	39	35	51	55	54	50
5	42	46	47	43	59	63	62	58
6	40	44	45	41	57	61	60	56
7	32	36	37	33	49	53	52	48

Figure 6. 2-D Mesh Processor Allocation for T 100

The MatFileRead and MatFileWrite procedures use the processor number to identify the relevant part of sub-file of the source files to read or write.

They could also be used to load the sub-matrices directly into the correct processors and avoid the initial data redistribution.

While the detail of the matrix multiply code has been described here, the user would normally make a single call to a Multi-Node Math Library routine to perform such a distributed function.

3.2 Results

The following table shows the results of running the distributed matrix multiplication on a number of different models of the T-Series.

Model	Nodes	Problem Size	Time (secs)	MFLOPS	MFLOPS /node
"T 1"	1	$(128)^2$	.405	10.4	10.4
T 20	16	$(512)^2$	1.755	149	9.3
T 100	64	$(1024)^2$	3.55	596	9.3
T 200	128	$2*(1024)^{2+}$		1229	9.6

+In this case two 1K matrix multipliers were computed and their results summed, (i.e. R = A*B + C*D).

4. CONCLUSIONS

The FPS T Series combines the use of high speed vector processing and large scale parallelism to achieve the promise of performance orders of magnitude greater than can be achieved with conventional systems.

With nodes organised into eight-processor modules, each with system disk and I/O services, the new architecture can be viewed as a homogeneous system, which can be expanded easily in a modular fashion.

An example has been shown of a distributed matrix multiply in which the data movement is fully overlapped with the computation such that performance close to peak speeds can be achieved with near linear speedup.

References

Floating Point Systems Inc., 1987: "Programming the FPS T Series: Program Development Guide." FPS Publication No. 860-0001-003.

Fox, G.C., and Otto, S.W., 1984: "Algorithms for Concurrent Processors." Physics Today, 37, 5, pp. 50-59.

Hawkinson, S., 1986: "The FPS T Series: A Parallel Vector Supercomputer." Floating Point Systems, Beaverton, Oregon.

Gustafson, J., 1986: "Programming the FPS T Series." Checkpoint, FPS Technical Journs, 4, 6, pp. 2-8.

Inmis Ltd., 1984: "OCCAM Programming Manual." Prentice/Hall International.

Parallel Processing on an IBM 3090 with Vector Facility

DAVID B. SOLL

IBM Corporation, Data Systems Division, Kingston, New York, USA

ABSTRACT

The IBM 3090 processor supports tightly-coupled multi-processor vector processing resulting from adding the Vector Facility as a feature to the base multi-processor configurations of the IBM 3090 models 200 and 400. The IBM VS FORTRAN Multitasking Facility permits a single VS FORTRAN application program to simultaneously using multiple processors in a multi-processor system.

This paper presents an overview of the IBM 3090 processor with optional Vector Facility, the VS FORTRAN Multitasking Facility, and performance results for several selected FORTRAN applications executing in combined vector/parallel mode.

PARALLEL PROCESSING ON THE IBM 3090 WITH VECTOR FACILITY

THE IBM 3090 WITH VECTOR FACILITY

The IBM 3090 family of processors presents a new opportunity for computation in the Scientific and Engineering arena. The functional design of the 3090 represents an evolutionary step in performance over the IBM 308x family. As shown in figure 1, the 3090 models 200 and 400 are tightly coupled multiprocessor configurations, with all processors sharing the central storage. The newly announced 3090E upgrades also provide 3 processor and 6 processor configurations in the models 300 and 600. Optional Expanded Storage provides electronic paging memory for virtual storage activity.

IBM 3090 Processors

	Model 200	Model 400
Central Processors	2	4
Vector Facilities	0, 1, 2	0, 1, 2, 3, 4
Central Storage	64 Mb	128 Mb
Expanded Storage	0, 64, 128 Mb	0, 128, 256 Mb
Channels	32, 40, 48	64, 80, 96
- Up to 3 Mb/Sec	All	All
- Byte 95-275 Kb/Sec	0-4	0-8
Consoles		
- 3180	2-5	3-5
- System	1	1
- Service	1	1

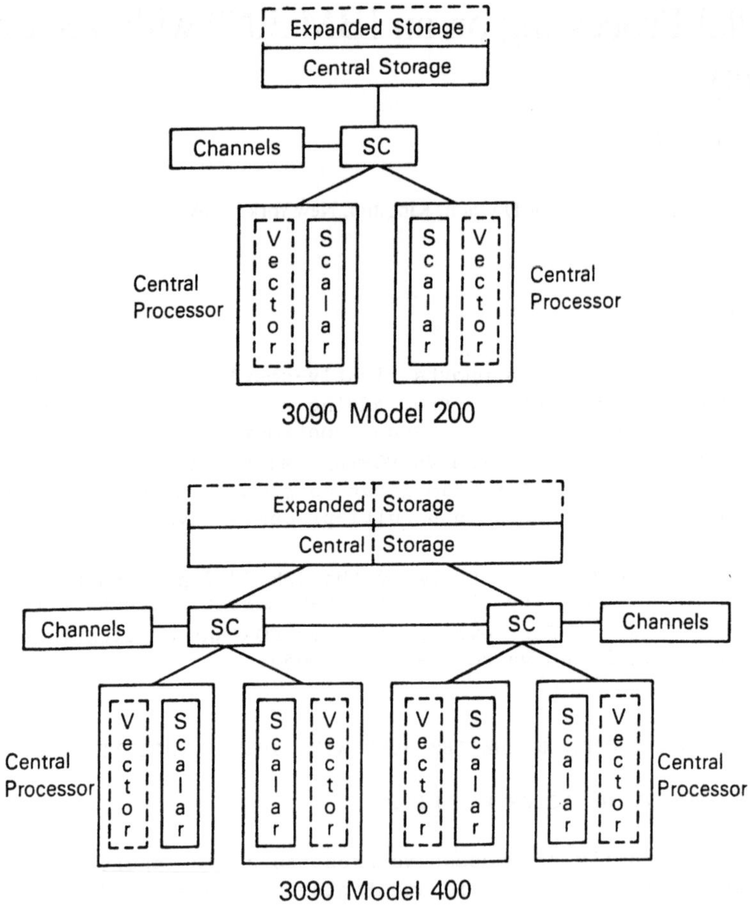

FIGURE 1

The optional Vector Facility integrates directly into each central processor. Table 1 lists some of the features of the 3090 models 200 and 400. A detail of the scalar design characteristics of a central processor is shown in figure 2. When the Vector Facility is installed in the central processor, as shown in figure 3, vector instructions are decoded from the same instruction stream as scalar instructions by the instruction element, and sent for execution to the vector execution element. The operating system provides CPU affinity by rescheduling the execution of vector instructions decoded by a central processor without an available Vector Facility, to a processor in which it is available. Each vector facility is functionally illustrated in figure 4. The 8Kb of vector register storage is configured as 16 vector registers, each of which contains 128 32-bit data elements. These 16 vector registers can be coupled to 8 vector registers each containing 128 64-bit data elements.

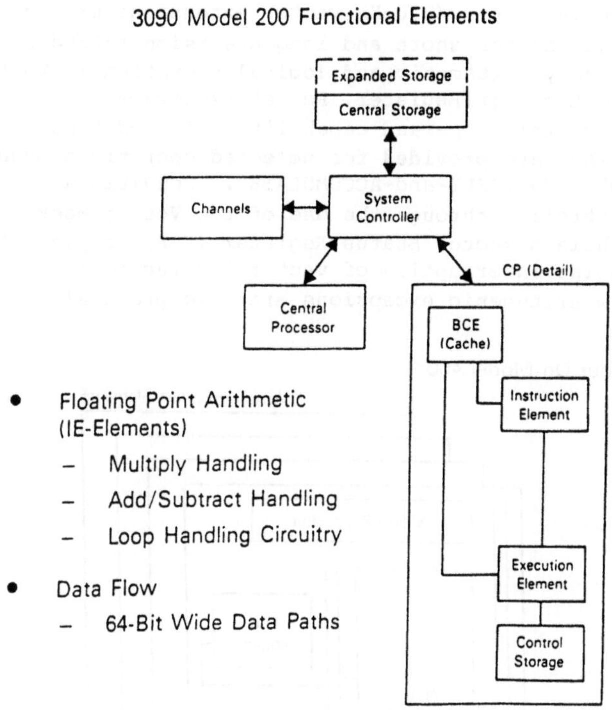

FIGURE 2

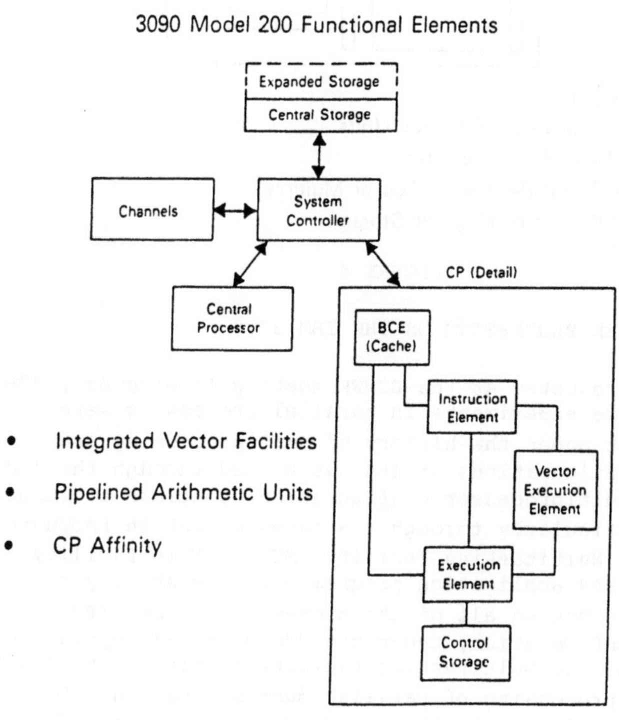

FIGURE 3

The IBM System/370 and System/370-XA vector extensions provide 171 new vector instructions for short and long precision (32/64 bit) floating point, binary arithmetic and logical operations. Vector operations provide Register/Register, Register/Storage, Storage/Register and three operand capability. In addition, compound instructions are provided for selected operations such as "MULTIPLY-and-ADD", "MULTIPLY-and-ACCUMULATE". Conditional operations are controlled through the use of the Vector Mask Register (VMR), while a Vector Status Register (VSR) provides the capability of precise interruption of vector instructions. Overflow/Underflow arithmetic exceptions are also precisely detected.

Up to Four On Model 400

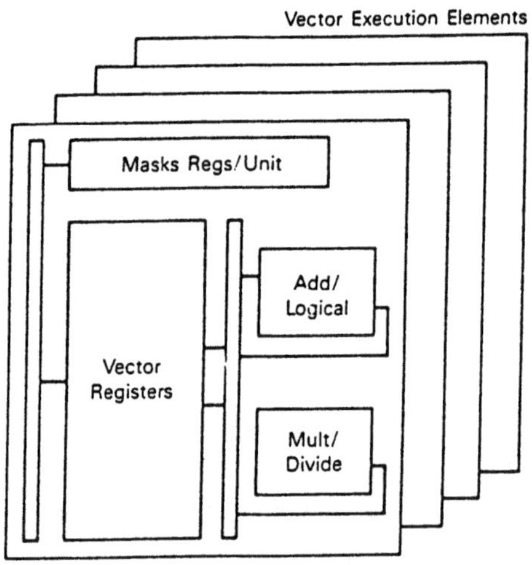

- Per Facility
  - 2 Pipelined Arithmetic Units
  - Vector Mask Register
  - 1 Result Per Cycle (Add or Multiply)
  - 8 Kb Vector Register Storage

FIGURE 4

VECTOR AND PARALLEL PROCESSING ON THE IBM 3090

In reference 1, presented at the ECMWF meeting in December, 1984, the results of some experiments in parallel processing were reported. In that paper the history of multiprocessing and multiprocessor configurations at IBM was traced through the IBM 3084. The 3090 multiprocessor configurations continue to support multitasking, particularly through the announcement in February 1985 of the VS FORTRAN Multitasking Facility (MTF). This facility provides the FORTRAN application program with the ability to simultaneously execute on all of the processors of any IBM multiprocessor configuration, under the MVS or MVS-XA operating systems. Although the multitasking facility is not limited to the IBM 3090, vector processing of parallel subtasks may only be performed on 3090 models with the Vector Facility installed.

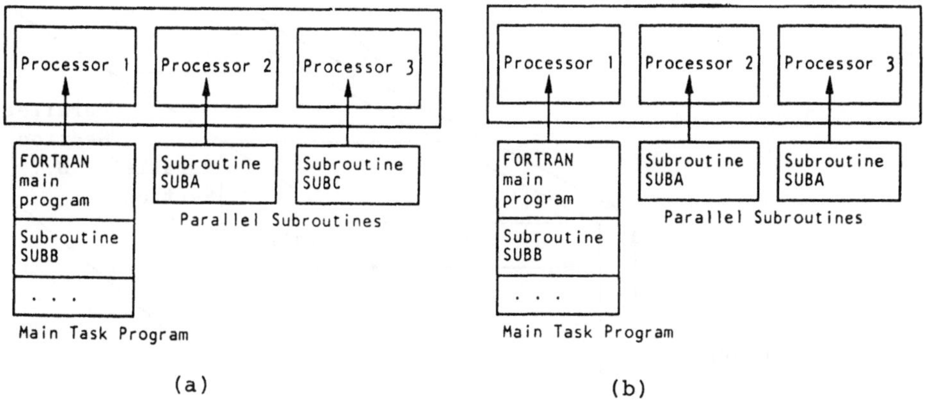

FIGURE 5

Several styles of multitasking applications are shown in figure 5. The multitasking facility permits the FORTRAN application program to schedule the execution of one or more FORTRAN subroutines which are (logically) independent (figure 5a). Alternately, multiple instances of the same FORTRAN subroutine may be scheduled, operating on independent sections of the data, as shown in figure 5b. Figure 6 shows the interpretation of a FORTRAN "DO" loop structure, say for a 3-dimensional data array, as consisting of a vector direction and a parallel or multitasked direction, depending on, and according to the independence characteristics of the data over the operations involved.

The Multitasking Facility provides three FORTRAN interfaces which are integrated into the VS FORTRAN Version 1 and 2 Libraries. These are, "DSPTCH", used to schedule multitasked subroutines, "SYNCRO", used to synchronize the execution of those tasks, and "NTASKS", which returns the number of subtasks set by the Multitasking selection keyword, "AUTOTASK", in JCL.

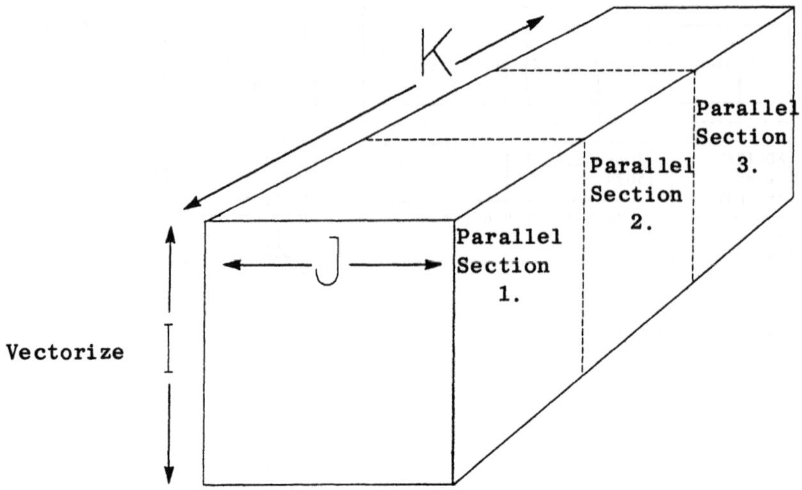

FIGURE 6

PERFORMANCE EXAMPLES

As discussed in reference 1, three applications had been selected for parallel execution. These are: 1) TBLADE, and application program that simulates the compressible air flow between two blades of a rotating turbine; 2) VA3D, which solves the aerodynamic flow field around a simple geometry, such as a projectile; and 3) BOAST, a black oil reservoir modelling simulation application. Some of the pertinent characteristics related to the modification for parallel execution are given in table 2. The effort to effect these changes, as measured by the lines of code (LOC) and number of parallel FORTRAN subroutines created, is shown to be application dependent. In addition, these data show that it is possible to perform such modifications without changing the entire program.

APPLICATION CODE MODIFIED FOR PARALLELISM				
Appl.	Size (LOC)	Parallel (LOC)	%Parallel (LOC)	%Parallel Routines
TBLADE	3500	677	19	13
VA3D	3800	1350	36	15
BOAST	4200	1021	24	5

TABLE 2

MEASURED APPLICATION PERFORMANCE 3090 With Vector Facility SERIAL				
Appl.	Scalar CPU Time Min.	Vector CPU Time Min.	Speed-up Ratio Scal/Vec	Speed-Up Ratio 30S1 Scal/ 3090 Vec
BOAST	19.2	8.4	2.3	4.9
VA3D	24.5	11.6	2.1	5.6
TBLADE	30.1	10.6	2.8	7.1

TABLE 3

Table 3 shows the performance improvement for these three application resulting from vector execution over scalar execution on the 3090/200 and compares that performance to an IBM 3081. The vector performance speed-up is measured by the CPU time rations between scalar and vector execution. The 3081 is a dyadic processor which bears the same relationship to the IBM 3084 (a 4 processor configuration) that the 3090/200 has to the 3090/400. The vector speed-ups of 2.1 to 2.8 are within the expected performance of the vector facility, and depend on the application.

MEASURED APPLICATION PERFORMANCE 3090 Model 200 2-Way Parallel				
Appl.	Serial Elapsed Time Min.	Parallel Elapsed Time Min.	Speed-Up Ratio Ser/Par	Percent Parallel
		SCALAR		
BOAST	19.2	10.4	1.8	92
VA3D	24.5	13.4	1.8	91
TBLADE	30.2	17.7	1.7	82
		VECTOR		
BOAST	8.4	5.3	1.6	75
VA3D	11.6	6.9	1.7	82
TBLADE	10.7	6.3	1.7	82

TABLE 4

The parallel (multitasked) performance on a 3090/200 (2-way) is shown in table 4 for both scalar and vector execution. The parallel speed-up ratio is measured by the elapsed time performance improvement due to parallel execution. The data indicates that the parallel performance ratio for vector execution may decrease relative to parallel-scalar execution. This results from diverting some of the data independence in the application program from parallel to vector mode. In fact, the performance is affected by a combination of several factors: 1) The amount of data independence available for vector and/or parallel execution; 2) the style adopted to distribute this data independence between vector and parallel modes; and 3) the granularity of the tasks, relative to the cost of scheduling tasks, and the vector-scalar performance ratio. These three factors are different for each of the three selected applications, and can be used to explain the performance behaviour presented here. The absolute performance is, of course, significantly better when the multitasked FORTRAN subroutines are executed in vector mode. Thus the parallel performance increase by itself is not a complete measure of the improvements when executing in vector/parallel mode.

Similar performance data for the same three applications on the 3090/400 (executed in 4-way parallel mode) are shown in table 5. Comparing the 2-way and 4-way parallel performance indicates that the performance increases, but not linearly. Again, the parallel portion of the performance increase is lower for vector mode execution, but the absolute performance continues to increase.

MEASURED APPLICATION PERFORMANCE 3090 Model 400 Vector/Parallel 4-Way Parallel				
Appl.	Serial Elapsed Time Min.	Parallel Elapsed Time Min.	Speed-Up Ratio Ser/Par	Percent Parallel
		SCALAR		
BOAST	19.2	6.2	3.1	90
VA3D	24.5	7.4	3.3	95
TBLADE	30.2	11.2	2.7	84
		VECTOR		
BOAST	8.4	3.4	2.5	80
VA3D	11.6	4.1	2.8	86
TBLADE	10.7	4.1	2.6	82

TABLE 5

MEASURED APPLICATION PERFORMANCE

(A) 3090 Model 400 Vector/Parallel 4-Way Parallel

Appl.	Serial Scalar Elapsed Time Min.	Parallel Vector Elapsed Time Min.	Speed-Up Ratio Ser. Scal/ Par. Vec
BOAST	19.2	3.4	5.6
VA3D	24.5	4.1	6.0
TBLADE	30.2	4.1	7.4

(B) 3090 4-WAY VECTOR/PARALLEL VS 3084

	3084 Ser.	3090 Vec	Speed-Up Ratio
BOAST	50.2	3.4	14.8
VA3D	70.2	4.1	17.1
TBLADE	80.4	4.1	19.6

TABLE 6

Finally, bottom line comparisons for the three applications are shown in tables 6a, b. Vector/parallel performance for 3090/400 4-way vector execution is compared to 1-way, serial, scalar performance on the 3090 and 3084 mainframes. A significant turn-around time performance advantage is realised compared to purely serial-scalar performance -on the 3090 itself. In addition, the performance improvements due to the 3090 relative to the 3084 are also achieved.

[1] Lim, A. and D. Soll (1984) Using an IBM Multiprocessing System. In this publication.

Multi-Tasking with a Memory Hierarchy

MASAYUKI IKEDA

FUJITSU Ltd., Kawasaki, Kanagawa, Japan

Abstract

Multi-tasking on multi vector-processors connected to GSU (Global Storage Unit) is discussed as one of the solutions to the market requirement for large scientific calculation.

The key points to obtaining high efficiency and performance from such multi-tasking systems are (1) vectorization, (2) granularity, (3) memory localization.

Vectorization is a technique to vectorize each task in order to obtain as much efficiency as possible from each vector processor.

Granularity is the length of time for executing a task which is independent of the tasks that are executable concurrently. Granularity depends on division of tasks that use global (common) variables. Large granularity reduces overhead for synchronization.

Memory localization depends on the division of tasks, assignment of tasks to processors, and assignment of memory to each local memory and global memory. The data movement overhead for starting a task from other processor's local memory or from global memory lessens the efficiency of multi-tasking.

There are already many techniques for vectorization to make efficient use of vector processors. So the techniques for the other two key points are important.

This paper assumes a typical hardware model. First, the behaviour of the bench mark code "SHALLOW" on this model is analyzed from three key points of view. It is shown that a code that is multi-tasked on ten vector-processors is from two or three times as fast as that on a single vector-processor. Then various techniques for optimizing the code for multi-tasking system are discussed. Finally, it is shown that for this bench mark code the optimization is very effective, and that the modified code that is multi-tasked on ten vector-processors is about seven times as fast as original code on one vector-processor.

1. INTRODUCTION

Improved architecture and progress in technology have let us use high performance computers that enable such high speed calculation and have such large capacity as we could not imagine.

In particular the idea of the vector processor on architecture has made it possible for us to have increased performance by a factor of 100 at once. But the market requirement for performance in large scale scientific calculation, including the meteorological field, is very high and it will be still higher in the future.

In this paper multi-tasking on multi vector-processors connected to a GSU (Global Storage Unit) is discussed as one of the solutions to the market requirement for large scientific calculation.

In Section 2, the multi-tasking system on multi vector-processors is presented as the solution of the problem, after the gap between market requirements and the computer system that is feasible by simple extension of today's technology is recognized.

In Section 3, key points for obtaining high performance on such systems are discussed.

In Section 4, the speed of benchmark code "SHALLOW" on VP-400 is measured in order to estimate the performance on a single vector processor.

In Section 5, one hardware model of a multi-tasking system is formed and some parameters for estimation are assumed.

In Section 6, the behaviour of the code "SHALLOW" on the model of Section 5 is assumed and the performance is estimated. The performance of the modified code after techniques for getting high performance are embodied is estimated.

In Section 7, it is reiterated that tuning of code and memory localization are important for multi-tasking.

2. MARKET REQUIREMENTS AND GSU

It is said that in the 1990's the requirement for computing power in the largest scale calculation in each scientific field will be more than 10 GFLOPS (giga floating operations per second). Against the requirement, if the improvements in architecture are not so great and the progress in performance must be due only to technology, the performance of a single vector-processor will be a few GFLOPS. In order to break the bounds, it will be necessary to connect multiple vector-processors running concurrently, which is called multi-tasking. In this case consideration of the fitness or adaptation to general application programs and of the overhead on multi-processing leads use to a system which is constructed with a small number (less than about ten processors) of high performance vector-processors.

On the other hand, the bigger the calculation size, the bigger the requirement for memory capacity. In the 1990's the requirement for memory capacity is said to be more than several GB (giga bytes). However, a reasonable memory capacity with feasible volume is less than the requirement, for the increasing memory capacity makes the volume large, which increases memory access time and lessens the performance of the processor. When program size is bigger than main memory, one must use the area in disk memory for "work file". Usually this lowers performance for I/O bounds. There are two solutions for this problem. One is to add a memory hierarchy between main storage and disk memory. In the view of memory hierarchy this main memory is called LSU (Local Storage Unit), and upper hierarchy memory is called GSU (Global Storage Unit). Access speed to GSU is slower than that to LSU. But it must be much faster than access to disk memory. And the capacity of GSU must be bigger than that of LSU.

The other solution is to tightly connect several processors each of which has as large memory as possible. For application programs that have remarkable access locality, most operations are executed using the data in the LSUs. If the communication overhead is negligibly small, this system looks like a system which has a larger main memory, by a factor of the number of processors. Sometimes this is more effective than that improvement in execution speed due to parallel processing.

The solutions for these two problems, which are performance limit and memory capacity limit, is multi-tasking on multiple vector-processors connected to GSU. One example of such system configuration is shown in Figure 1. In this figure GSU must satisfy following requirements:

(1) it is shared by multi vector-processors,
(2) it can be accessed by each vector-processor much faster than disk memory,
(3) it is larger than each local storage.

In the next section the key points to obtaining high performance from such multi-tasking systems are discussed.

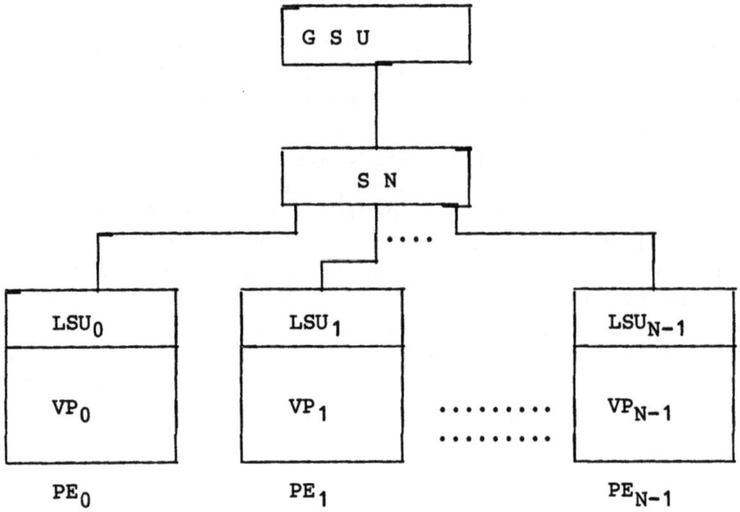

GSU: Global Storage Unit
SN: Switching Network
LSU: Local Storage Unit
VP: Vector Processor
PE: Processing Element

Figure 1 System Configuration

3. KEY POINTS FOR MULTI-TASKING

In order to get high performance from the systems assumed in the previous section, it is necessary to execute each divided task with higher speed and to lessen the overhead for communication among tasks and synchronization among tasks. In other words, the key points are (1) vectorization, (2) granularity, (3) memory localization.

Vectorization is important to execute each task with higher speed.

Granularity and memory localization are important to execute tasks concurrently.

Vectorization is a technique to vectorize each task in order to obtain as much efficiency as possible from each vector processor.

Granularity is the length of time for executing a task that is independent of the other tasks that are executable concurrently. Granularity depends on division of tasks that use the variables in GSU. Large granularity reduces overhead for synchronization.

Memory localization depends on the division of tasks, assignment of tasks to processors, and assignment of memory to each LSU and GSU.

The overhead for data movement for starting a task from another processor's LSU or from GSU lessens the efficiency of multi-tasking.

There are already many techniques for vectorization to make efficient use of vector processors. So the techniques for the other two key points must be discussed.

4. SINGLE-TASKING EVALUATION [SHALLOW ON VP-400]

To confirm that the benchmark code "SHALLOW" can be vectorized easily, the speed on a VP-400 is measured. The results are shown in Table 1. Since peak performance of the VP-400 is 1.14 GFLOPS (giga floating operations per second), about a third of peak performance can be obtained for a 64x64 grid, and about half of peak performance can be obtained for a 256x256 grid.

Table 1 Performance of SHALLOW on VP-400

	Operations per element				64x64 grid model		256x256 grid model	
	±	*	div	total	VP400 cpu time (ms)	Performance (M FLOPS)	VP400 cpu time (ms)	Performance (M FLOPS)
LOOP100	12	11	1	24	0.239	411	2.41	653
LOOP200	18	8	0	26	0.248	429	2.80	603
LOOP300	9	6	0	15	0.215	286	2.28	431
TOTAL	39	25	1	65	0.702	379	7.49	569

5. HARDWARE MODEL OF MULTI-TASKING SYSTEM

Since a multi-tasking system was not at hand, the performance of multi-tasking was estimated manually. System configuration is shown in Figure 1.

On this model the following parameters were assumed.

(1) Number of vector processors: N(from 2 to 10)
(2) Peak performance of each vector processor: 1G FLOPS
(3) Throughput between each PE and SN: 1G bytes/s
(4) Throughput between SN and GSU: MG bytes/s
M=2,4,8
(5) Set up time for data movement between PE and SN: 1 microsecond
(6) Overhead for synchronization: 6 microseconds.

These parameters are not based on any fixed technology background but are assumed only for discussion.

6. MULTI-TASKING EVALUATION

Behaviour of the bench mark code "SHALLOW" (64x64 grid model) on the multi-tasking system assumed in the preceding section is estimated. Parameter N is the number of PEs and M is throughput between SN and GSU.

6.1 Task division and task assignment to PE

There are various ways to divide into several tasks. In this paper, task division is decided by using a loop slice method with J (suffix of outer loop) direction, and task assignment to processor is static as shown in Figure 2. LOOP100, LOOP200, and LOOP300 are similar.

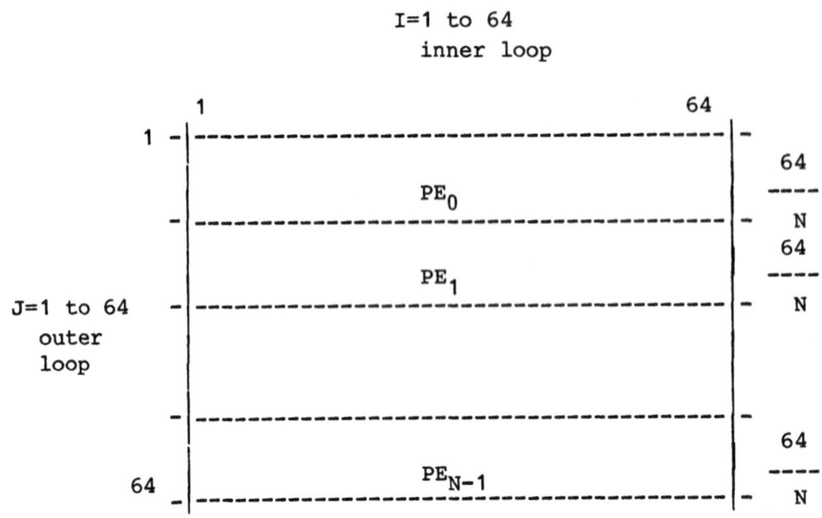

Fig. 2 Task division by using outer loop slice method (static task assignment to each PE)

As shown in Section 4, for a 64x64 grid the total of the three loops, the code is executed in 702 microseconds per one time step on a VP-400. Then it is executed on the single PE which has peak performance of 1 GFLOPS in

702x16/14=802 microseconds.

In the case the performance is

(65x64x64)/802-332 MFLOPS

This is performance of single tasking on a PE with a peak performance of 1 GFLOPS.

6.2 Memory assignment and performance

This paragraph shows how much the performance depends on the assignment to memory.

6.2.1 All data in GSU (E-1)

In this method all matrices are assigned to GSU. When a PE executes some operation, it requests the operands that are used for the operation. After finishing the operation the results are moved to GSU. LSU is used as buffer storage between GSU and PE. This method is very simple, but the amount of data movement between GSU-LSU is very large. When all tasks are executed in parallel on all N processors, the execution time is 802/N microseconds per PE per one time step. (For simplicity, fractional overhead is ignored).

Since synchronization before LOOP100, LOOP200, and LOOP300 is necessary, the time for synchronization is 6x3=18 microseconds per one time step. As one matrix consists of 65x65=4225 elements, the time for the movement between GSU-LSU with the throughput of 1GB/sec is

$$(4225 \times 8)/1000 = 33.8 \text{ microseconds.}$$

This does not include set-up time.

Matrices that are read for execution of LOOP100 are P, U, and V. Matrices that are written as results are CU, CV, Z and H. Similarly in LOOP200 seven matrices are read and three matrices are written, and in LOOP300 nine matrices are read and six matrices are written. So in one time step thirty-two matrices are moved.

Rough estimation of time for data-movement is as follows.

If SN-GSU throughput is bound, the time is
$(33.8 \times 32)/M$ microseconds

If PE-SN throughput is bound, the time is
$(33.8/N+1) \times 32$ microseconds

(For movement between PE and SN, set-up time, which is assumed to be 1 microsecond, is necessary).

Then the total time per one time step per PE is

$$t_1 = 802/N + 18 + \max[(33.8 \times 32)/M, (33.8/N+1) \times 32].$$

Performance is $P_1 = (65 \times 64 \times 64)/t_1$.

The relation between N(number of PEs) and performance P_1, plotted for M (throughput between SN and GSU) as a parameter is shown in Figure 3. Since data movement between PE and GSU is very large, the performance depends on M sensitively for large N. When N=10

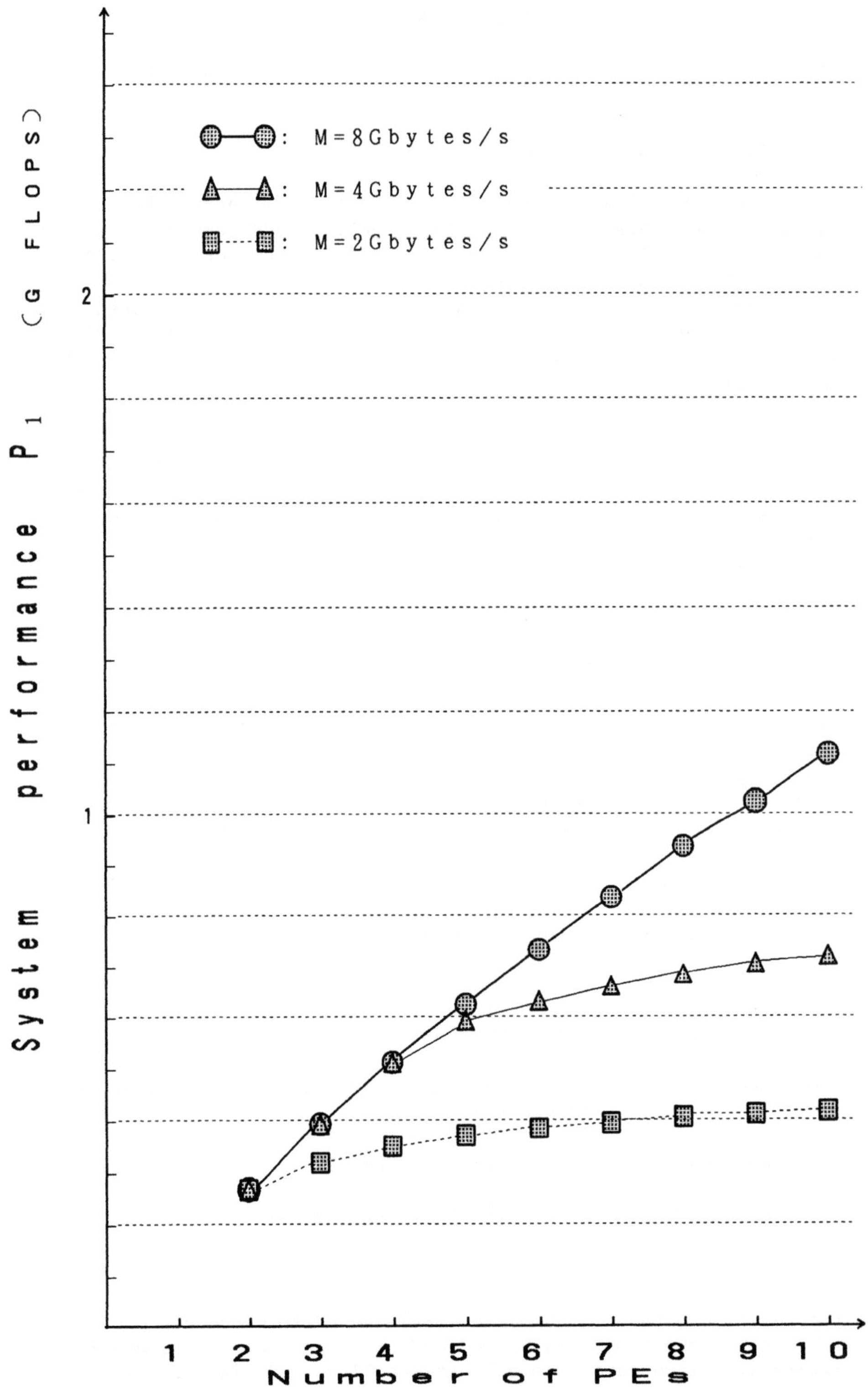

Fig. 3 Performance of multi-tasking (all matrices in GSU (E-1))

and M=8, the multi-tasking system is about three times as fast as a single vector processor. But when N=10 and M=2, it is only about 1.2 times as fast.

6.2.2 Load/store in each time step (E-2)

In this method, the matrices, P, U and V are assigned in GSU and the others are partitioned and assigned to each LSU. The matrices in each LSU are buffered in each LSU. All results of P, U, and V are written after execution of LOOP300 and read before execution of LOOP100 in each time step. When some PE doesn't have the elements of matrices that are necessary for the PE to execute operation, the elements are sent via GSU. In order to estimate the amount of data movement it is necessary to analyze the relation between definition and reference of each element. In LOOP100 the element CU(I+1,J) is calculated by P(I+1,J), P(I,J), U(I+1,J). As tasks are partitioned along "J" direction, the elements P(I,J) for all "I" are in the same LSU for fixed "J". For the data movement, the first suffix "I" can be ignored. The relation between definition and reference is denoted as follows.

```
CU(J)::P(J),U(J).
```

This shows all elements in column CU(J) are calculated by some elements of column P(J) and U(J). Using this notation the relation between definition and reference in LOOP100 is as follows.

```
CU(J)  ::P(J),U(J).
CV(J+1)::P(J+1),P(J),V(J+1).
Z(J+1) ::V(J+1),U(J+1),U(J).
H(J)   ::P(J),U(J),V(J),V(J+1).
```

Similarly the relation in LOOP200 is as follows

```
UN(J)  ::UO(J),CV(J),CV(J+1),Z(J),Z(J+1),H(J)
VN(J+1)::VO(J+1),CU(J),CU(J+1),Z(J+1),Z(J+1),H(J+1).
PN(J)  ::PO(J),CU(J),CV(J),CV(J+1).
```

(UN, UO, VN, VO, PN and PO indicate UNEW, OULD, VNEW, VOLD, PNEW, and POLD in code respectively.

The relation in LOOP300 is as follows

```
UO(J)::U(J),UN(J),UO(J)
VO(J)::V(J),VN(J),VO(J)
PO(J)::P(J),PN(J),PO(J)
U(J) ::UN(J)
V(J) ::VN(J)
P(J) ::PN(J)
```

If each band, which is a group of columns of a matrix, shown in Figure 4 is assigned to one LSU, the elements loaded in LOOP100 are all data of P(J), U(J), V(J) (J is from a to b) and elements stored in LOOP100 are columns CU(a), CV(b+1), Z(b+1), H(a), which are boundary data of CU(J), CV(J+1), Z(J+1), H(J) (J=a, a+1, a+2, a+3, ..., b).

```
P(J),U(J),V(J): for buffer
CU(J),CV(J+1),Z(J+1),H(J)
UN(J),VN(J+1),PN(J)
UO(J),VO(J),PO(J)
(J : from a to b)
```

Fig. 4 Memory assignment (type A) to LSU

Similarly the data loaded in LOOP200 are CU(b+1), CV(a), Z(a), H(b+1), VO(b+1) and the column stored in LOOP200 is VN(b+1). The column loaded in LOOP300 is VN(a), and stored data are all elements of P(J), U(J), V(J) (J is from a to b) and column VO(a). Thus there are 6 matrix movements which consist of 4225/N elements and 12 boundary data movements which consist of 65 elements per one time step per PE. The time for movement of 65 elements between GSU and PE is (65x8)/1000=0.52 microseconds excluding set-up time. Rough estimation of time for data movement is as follows. If SN-GSU throughput is bound, the time is (33.8x6+0.52x12xN)/M. If PE-SN throughput is bound, the time is (33.8/N+1)x6+1.52x12.

Then the total time per one time step per PE is

$$t_2 = \frac{802}{N} + 18 + \max\left[\frac{33.8x6 + 0.52x12xN}{M},\ \left(\frac{33.8}{N}+1\right) x6 + 1.52x12\right].$$

Performance is P_2=(65x64x64)/T_2. A graph of performance is shown in Figure 5.

6.2.3 All data in LSU (E-3)

In this method all matrices are partitioned and assigned to LSU. If one PE doesn't have the required elements, they are moved from other PEs via GSU. Matrices partition and assignment to LSU is similar to (E-1) in Figure 4. Some example helps to embody the method.

Example in case of N=8 is shown.

PE0 is assigned indices from 1 to 8 for suffix J of each outer loop,
PE1 is assigned indices from 9 to 16 for suffix J of each outer loop,
PE2 is assigned indices from 17 to 24 for suffix J of each outer loop,

:

:

and PE7 is assigned indices from 57 to 64 for suffix J of each outer loop. For PE0, the value of "a" is 1 and the value of "b" is 8. For PE1, the value of "a" is 9 and the value of "b" is 16. The columns moved via GSU are P(1), P(9),.....,P(57),U(1),(U9),...., U(57),V(1),V(9),.....V(57),CU(1),CU(9),.....,CU(57),CV(1),CV(9), ...,CV(57), and so on.

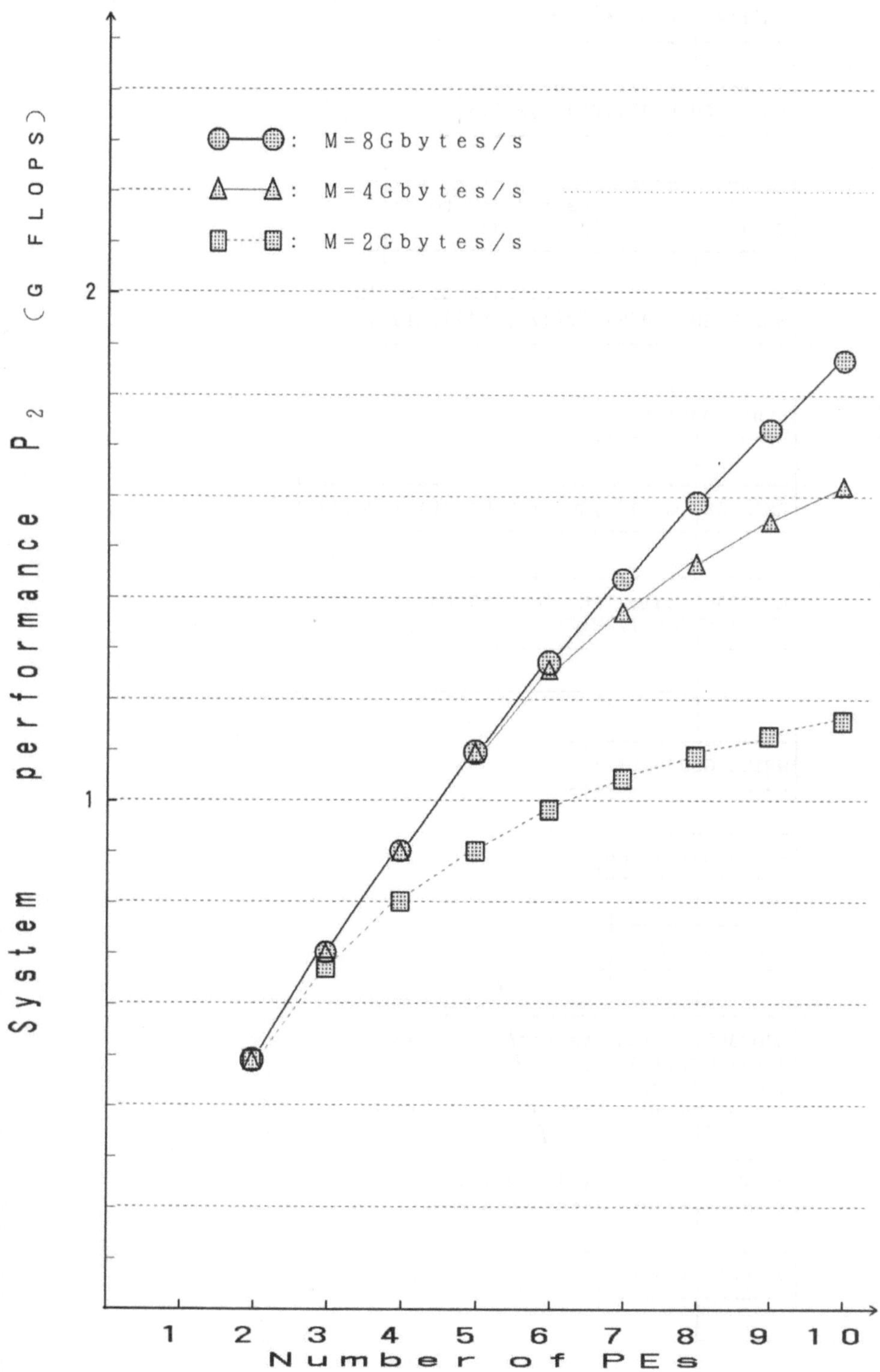

Fig. 5 Performance of multi-tasking
(load/store in each time step to GSU (E-2))

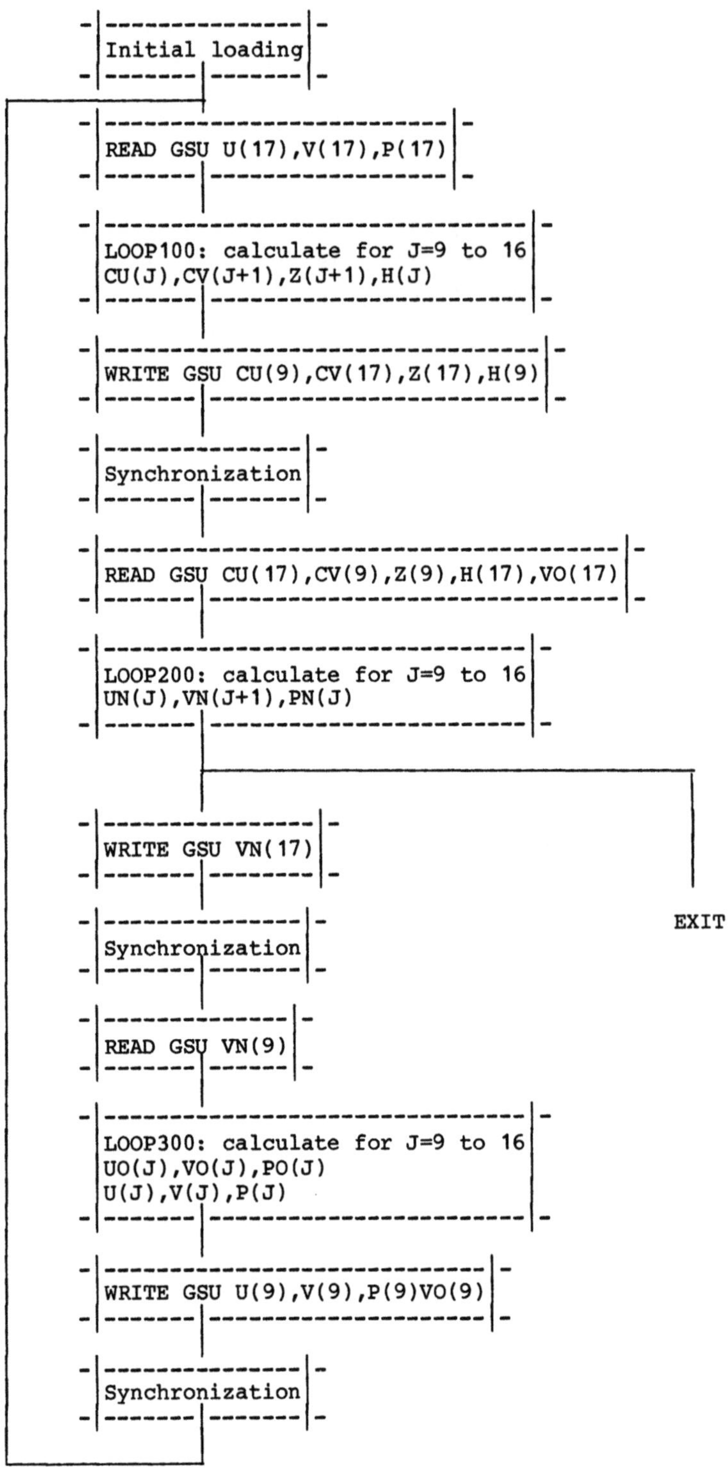

Fig. 6 Flow chart on PE1 for 8 PEs multi-tasking

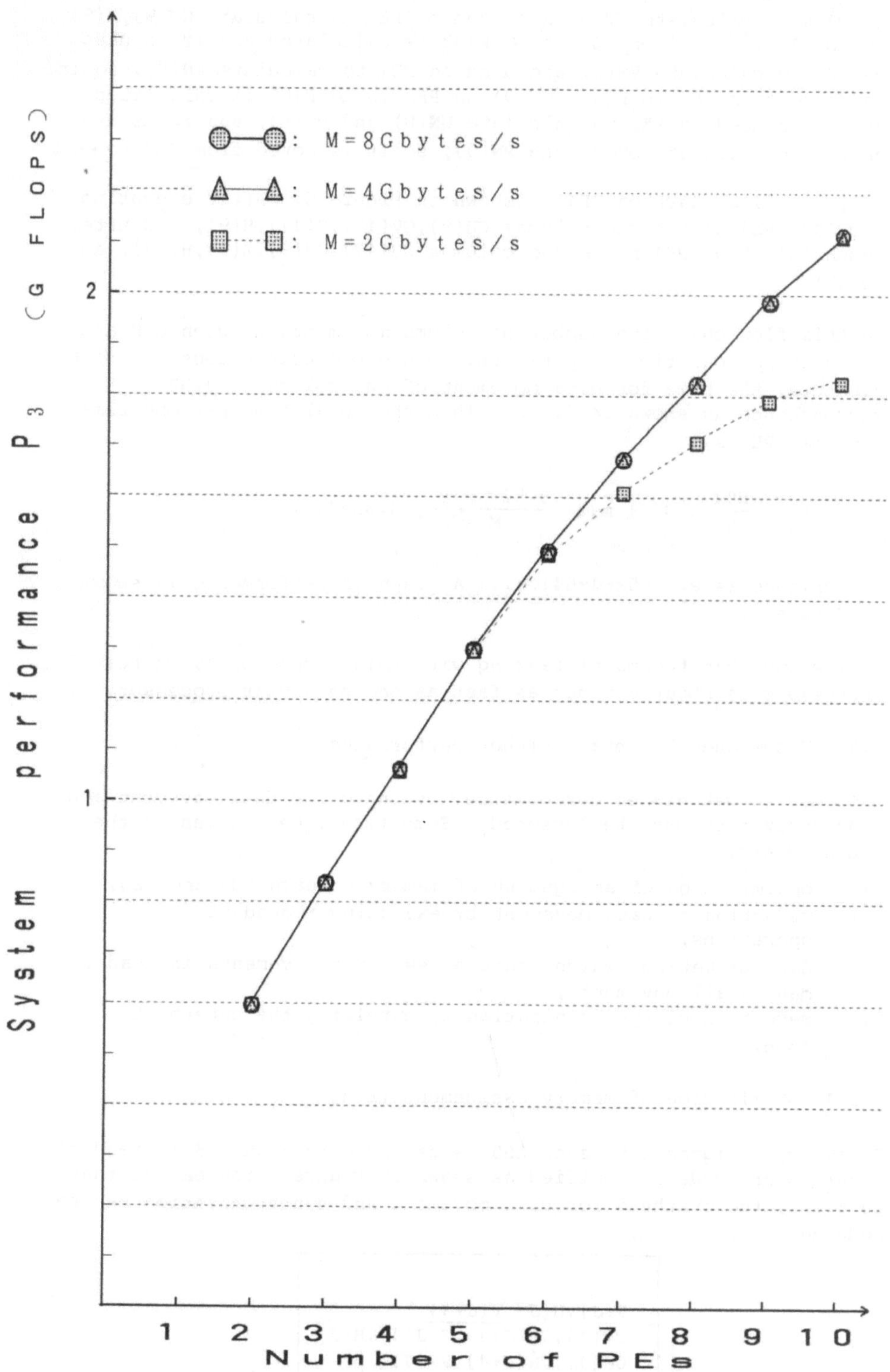

Fig. 7 Performance of multi-tasking (all matrices in LSU (E-3))

P(9) on PE1 in LOOP300 is calculated and it is used on PE0 in LOOP100 to calculate CV(9) and used on PE1 to calculate CU(9),H(9) in LOOP100. CU(9) on PE1 in LOOP100 is calculated and it is used on PE0 to calculate VN(9) and used on PE1 to calculate PN(9), so it is moved from PE1 to PE0. CV(9) on PE0 in LOOP100 is calculated and it is used on PE0 to calculate UN(8) and PN(8), and it is used on PE1 to calculate UN(9) and PN(9), so it is moved from PE0 to PE1.

A simple flow chart on PE1 is shown in Figure 6. After executing LOOP100, PE1 writes the columns CU(9),CV(17),Z(17),H(9), and after synchronization PE1 reads the columns CU(17),CV(9),Z(9),H(17), and VO(17).

In this flow chart the number of column movements between GSU and PE is 18 per one time step per PE. Since one column consists of 65 elements, the time for data movement of one column is 0.52 microseconds as shown in 6.2.2. Then the total time per one time step per PE is

$$t_3 = \frac{802}{N} + 18 + \max\left[\frac{0.52 x N x 18}{M} , 1.52 x 18 \right].$$

Performance is $P_3=(65x64x64)/t_3$. A graph of performance is shown in Figure 7.

This shows that the multi-tasking with this method on 10 vector processors is about 6 times as fast as one on vector processor.

6.3 Techniques to obtain higher performance

In order to get higher performance, overhead for data movement and synchronization must be lessened. Some techniques to lessen the overhead are,

(1) optimization of assignment of memory to each LSU and GSU,
(2) reduction of data movement by executing redundant operations,
(3) data packeting, which means a few large movements instead of many small movements,
(4) reduction of synchronization by unrolling the outermost loop.

6.3.1 Optimization of memory assignment (E-4)

If memory assignment to each LSU is as shown in Figure 8 instead of Figure 4 and code is modified as shown in Figure 9 instead of that in 6.2.2, the overhead for data movement and synchronization can be reduced.

```
P(J),U(J),V(J+1)
CU(J),CV(J+1),Z(J+1),H(J)
UN(J),VN(J+1),PN(J)
UO(J),VO(J+1),PO(J)
 (J ; from a to b)
```

Fig. 8 Memory assignment (type B) to LSU

```
LOOP300     UO(J)  ::U(J),UN(J),UO(J)
            VO(J+1)::V(J+1),VN(J+1),VO(J+1)
            PO(J)  ::P(J),PN(J),PO(J)
            U(J)   ::UN(J)
            V(J+1) ::VN(J+1)
            P(J)   ::PN(J)
```

Fig. 9 Code modification in LOOP300

Then the flow chart on PE1 in case of eight PEs is changed to Figure 10 and synchronization between LOOP200 and LOOP300 can be deleted. Since the number of data movements is 14 and the number of synchronizations is 2, the total time per time step per PE is

$$t_4 = \frac{802}{N} + 12 + \max[\ \frac{0.52xNx14}{M}\ ,\ 1.52x14\ ].$$

Performance is P_4=(65x64x64)/t_4. A graph of performance is shown in Figure 11.

6.3.2 Reduction of data movement

Modification of code can sometimes reduce the overhead for synchronization or data movement.

If the code is modified as shown in Figure 12, the synchronization between LOOP100 and LOOP200 is not necessary, for all the data needed in LOOP200 are calculated in LOOP100 on the same PE. It must be noted that the number of operations in LOOP100 is increased by one column which is indicated by iteration count of suffix J. Since this added operation is also executed on another PE, it is redundant. But it can reduce the above overhead. This method is effective when the reduction of the overhead for synchronization and data movement is larger than the increase in execution time.

In this modified code, the number of data movements is 12 and the number of synchronizations is 1 per time step. Increase in execution time is only in LOOP100 and the ratio of LOOP100 in time step 239/702, from Table 1.

Since the number of columns that are executed in one PE in original code is 64/N, the number of columns in modified code is 64/N+1. Then the total time per one time step per PE is

$$t_5 = \frac{802}{N} x \left(1 + \frac{N}{64} x \frac{239}{702} \right) + 6 + \max[\ \frac{0.52xNx12}{M}\ ,\ 1.52x12\ ].$$

Performance is P_5=(65x64x64)/t_5. A graph of performance is shown in Figure 13. When M=4 and N=10, multi-tasking of the modified code(E-5) is about 7 times as fast on a single vector processor. When M=4 and N=10, the modified code (E-5) is 3.4 times as fast as original code (E-1).

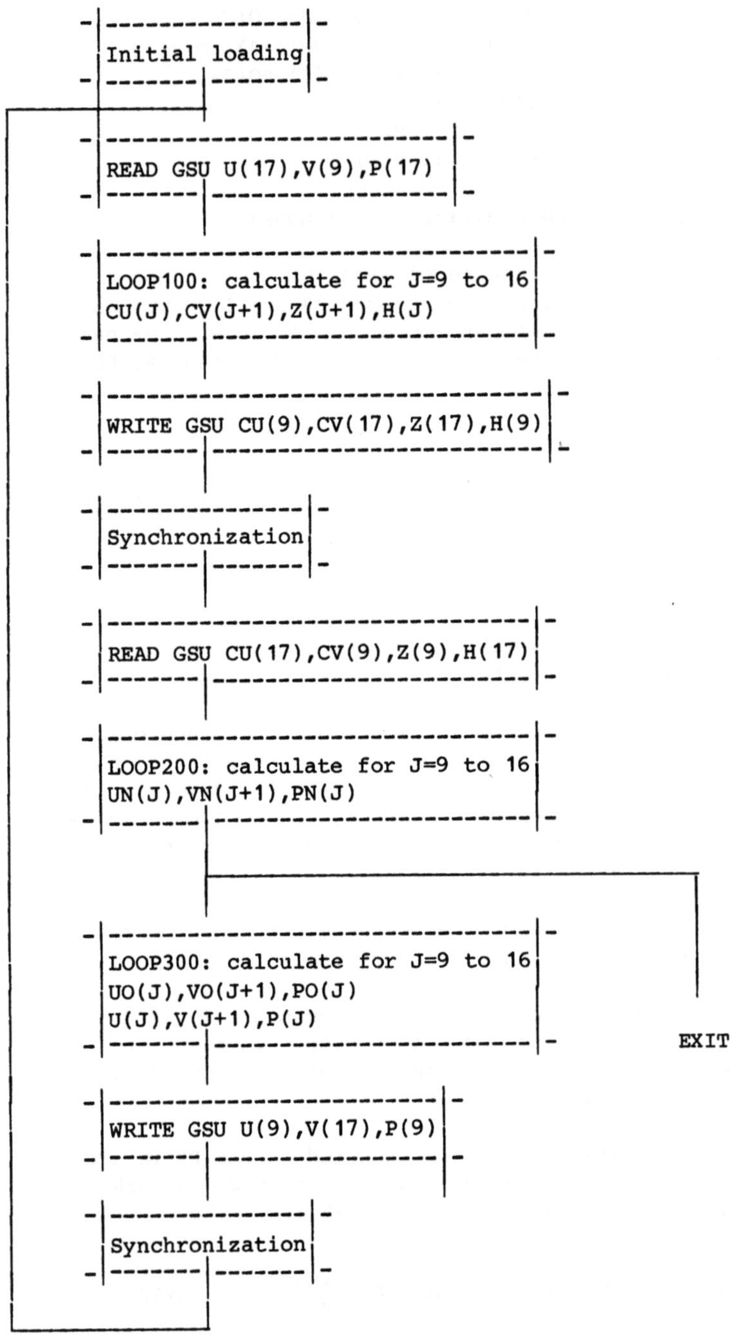

Fig. 10 Flow chart of modified code (E-4) on PE1

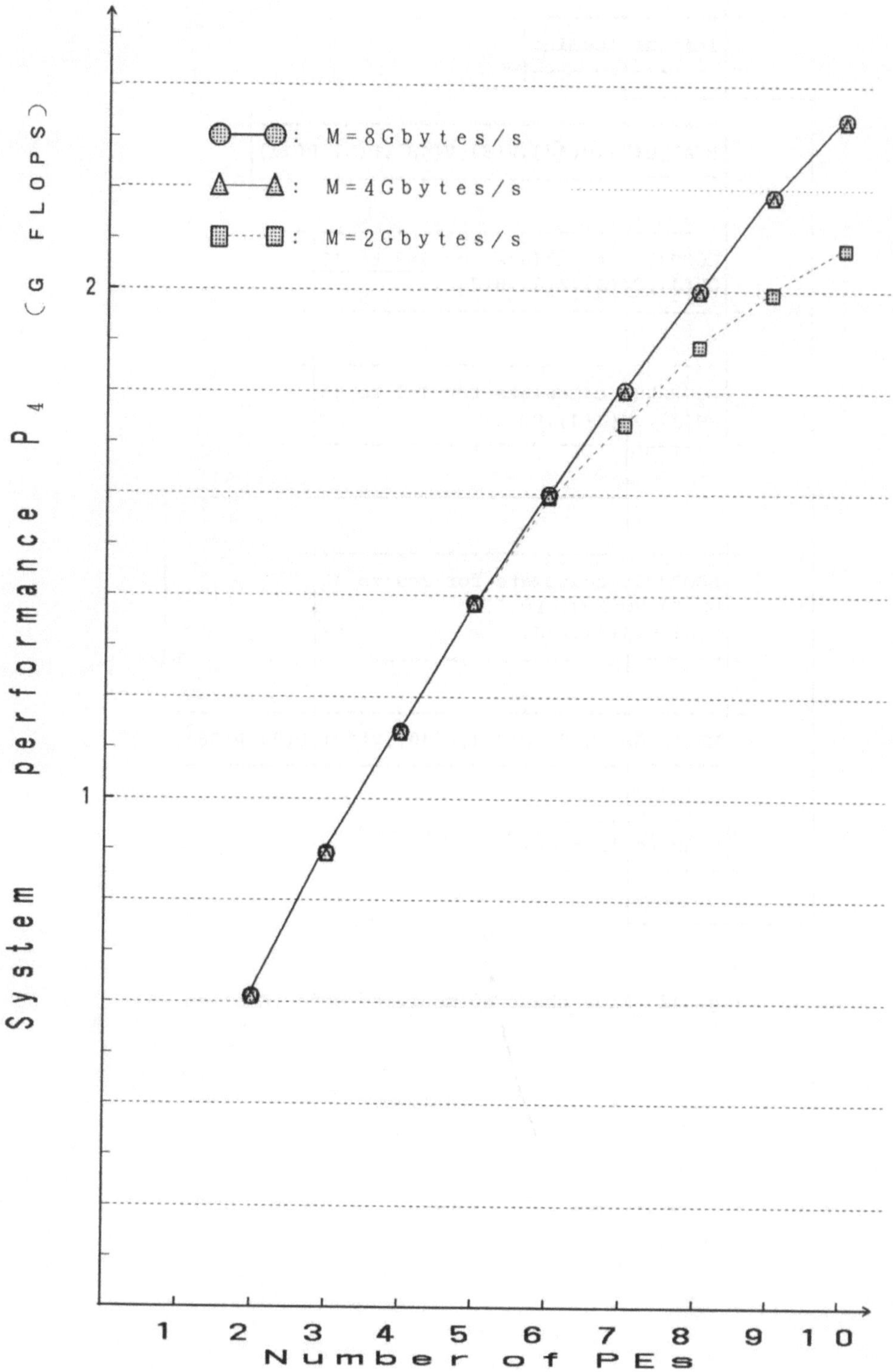

Fig. 11 Performance of multi-tasking
(Optimization of memory assignment (E-4))

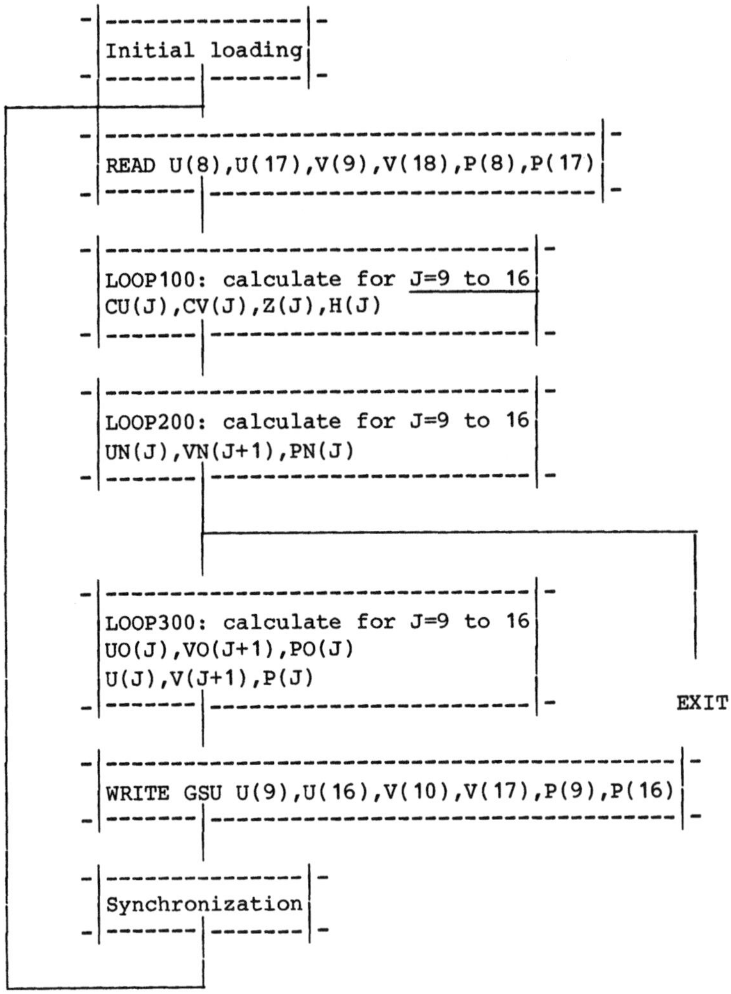

Fig. 12 Flow chart of modified code (E-5) on PE1

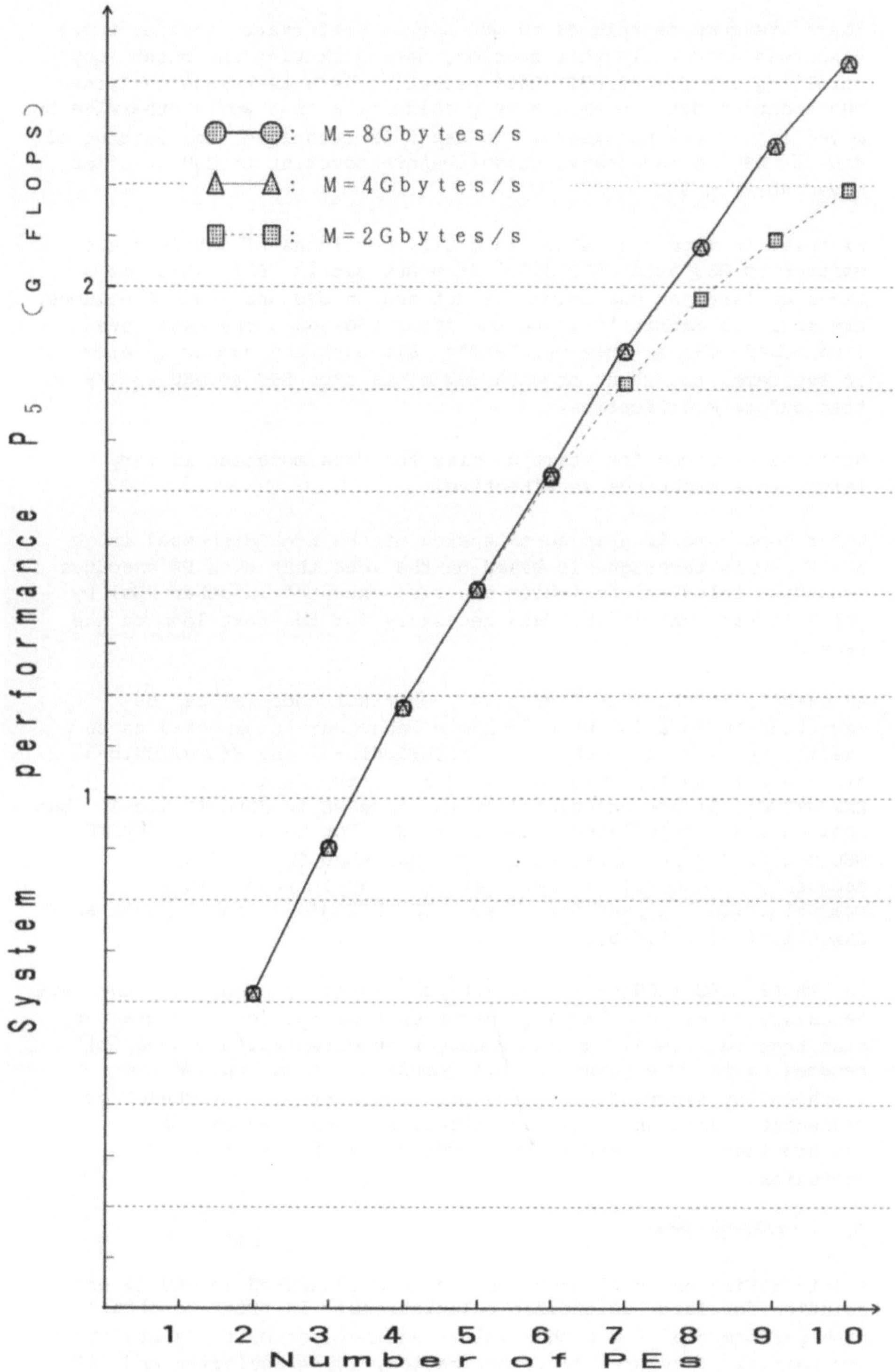

Fig. 13 Performance of multi-tasking
(Reduction of data movement (E-5)

6.3.3 Other techniques

There are some techniques to get higher performance besides those discussed above. In this section, data packeting and outer loop unrolling are discussed. Data packeting is a technique to lessen the count of data movements by packing data that would otherwise be moved in several movements. To use this technique, the editing of data on LSU is necessary, either before movement to GSU or after movement from GSU.

Example: In Figure 12 shown in 6.3.2, 6 columns of matrices are written to GSU after LOOP300. If a new matrix "PKT" which is 6 times as large as one column is defined on LSU and these 6 columns are moved to matrix "PKT" on LSU after LOOP300 , the data moved from LSU to GSU is only one "PKT". Although the number of elements is the same, the count of data movements from LSU to GSU is 1/6 of that before modification.

When the overhead for start up time for data movement is very large, this technique is effective.

Outer loop unrolling is an extension of the idea discussed in 6.3.2. This technique is based on the idea that each PE executes redundant calculations (which are also executed in other PEs) by which it can get all the data necessary for the next loop on the same PE.

An example is shown in Figure 14. Original code (a) can be rewritten to code (b) whose dynamic behaviour is expected to be exactly the same as code (a). EXECUTION(b-1) and EXECUTION(b-5) are the same as EXECUTION(a-1). Most of the data used in EXECUTION(B-5) are calculated by the same PE in EXECUTION(b-1), but soem data are calculated by another PE. For this reason "WRITE GSU(b-2)", "SYNCHRONIZATION(b-3)", and "READGSU(b-4)" are necessary. Code (c) is rewritten in order to reduce the communication overhead for these. EXECUTION(c-2) is the same as EXECUTION(a-1,b-1,b-5).

in "TASK2 EXECUTION(c-1)" each PE calculates by itself all the data necessary to execute TASK1. Execution time for (c-1) is greater than that for (b-1), but the count of synchronizations code (c) becomes half. The point of this technique is to replace the overhead for syncronization and data movement with overhead for redundant operation. This is effective, when overhead of synchronization is larger than performance of execution of operation.

7. CONCLUSIONS

Multi-tasking on multi vector-processors connected to GSU is one solution for large scientific calculations. In order to obtain high performance from such a multi-tasking system, there are three key points. They are (1) vectorization, (2) granularity and (3) memory localization. The performance in executing the benchmark code "SHALLOW" on multi-tasking system is estimated.

```
-|--------------------|-
 |                    |
 |          TIME STEP = TIME STEP + 1
 |               READ GSU
 |               TASK1 execution  -------------(a-1)
 |               WRITE GSU
 |               Synchronization
 |                    |
-|--------------------|-
```

(a) Original code

```
-|--------------------|-
 |                    |
 |          TIME STEP = TIME STEP + 1
 |               READ GSU
 |               TASK1 execution  -------------(b-1)
 |               WRITE GSU        -------------(b-2)
 |               Synchronization  -------------(b-3)
 |          TIME STEP = TIME STEP + 1
 |               READ GSU         -------------(b-4)
 |               TASK1 execution  -------------(b-5)
 |               WRITE GSU
 |               Synchronization
 |                    |
-|--------------------|-
```

(b) Simple expansion of outer loop

```
-|--------------------|-
 |                    |
 |          TIME STEP = TIME STEP + 1
 |               READ' GSU
 |               TASK2 execution  -------------(c-1)
 |          TIME STEP = TIME STEP + 1
 |               TASK1 execution  -------------(c-2)
 |               WRITE' GSU
 |               Synchronization
 |                    |
-|--------------------|-
```

(c) Outer loop unrolling

Fig 14 Technique of outer loop unrolling

It is shown that the code that is multi-tasking on ten vector-processors is from two to three times faster than on a single vector-processor. It is also shown that for bench mark code "SHALLOW", the modified code that is multi-tasked on ten vector-processors is about seven times as fast as original code on one vector-processor.

8. REFERENCE

G-R. Hoffman, P.N. Swarztrauber, and R.A. Sweet, "Aspects of Using Multiprocessors for Meteorological Modelling," WORKSHOP on Using Multiprocessor in Meteorological Models, (1984), PP270-358.

The Myth of Performance for Parallel Machines

CHRISTOPHER C. HSIUNG

Cray Research Inc., 900 Lowater Road, Chippewa Falls, WI 54729, USA

1. INTRODUCTION

It has been a decade since the first CRAY-1 computer was delivered to Los Alamos National Laboratories in the U.S. in 1976. Vector processing is becoming common place for large scale scientific and engineering computations. The fact that IBM introduced their own vector unit for the Sierra series in 1985 is the ultimate acknowledgement that vector processing is an accepted standard mode of operation in the conventional camp.

Even though 'MFLOPS', 'Livermore loops' and 'LINPACK numbers' have become jargon among financial analysts who follow this industry, how well is vector processing understood among the wider user community? How is performance being measured and perceived?

This paper will address performance issues surrounding parallel machines. The term 'parallel' primarily refers to multiprocessors. But, because of historical reasons, vector machines will also be addressed. Our attention will be directed toward the general purpose environment where more than one or two sets of code are used and where program development is an ongoing activity. However, dedicated usages will not be excluded. When performance is addressed for a general purpose environment, it is the belief of this author that more emphasis should be put on achieving a high throughput rate day in and day out. Moreover it is important to be able to achieve reasonable percentage of the potential speed with minimum amount of extra programming effort.

The second section reviews the status of vector processing. Several observations about user understanding will be made. The evolution of vector architecture in the past few years and its impact on performance will be discussed. The state of software technology and its impact on performance will only be briefly addressed. The impact of application algorithm development will also be discussed briefly.

Since the introduction of the CRAY X-MP, a multiprocessor vector computer in 1982, parallel processors and parallel processing have become more popular among computer vendors in the scientific and engineering market sector; namely, the supermini, minisuper and near supercomputer vendors. For the first time, users are beginning to seriously consider parallel processing as a viable approach to large scale computing on a daily basis.

In the third section, we are going to look into different machine features that are intimately related to performance. Several classes or architectures with commercial followings will be studied.

In the fourth section, multiprocessing experiences learned at Cray will be discussed. Major issues in parallel processing will be addressed, especially those affecting performance. The Cray software environment will be briefly mentioned. Multiprocessing exercises will then be illustrated with analyses of performance figures.

Even though there is a long way to go before parallel processing achieves its maturity, several major developments are under way. We will try to point out a few of the trends in the general purpose environment as a conclusion.

2. THE STATE OF VECTOR PROCESSING AFTER A DECADE

- Promises are not always promising -

At the time when CRAY-1 was introduced, even Seymour Cray, the founder and an architect of Cray Research, was quoted as saying "Nobody, I mean nobody knows how to program on the current generation of supercomputers". It was at a time that the ILLIAC IV with its multiple PE's met with only limited success. Pipeline architecture was still new and not quite proven. The compiler technology to exploit vector capability was limited to the research circle. It was still at a time when unstructured programming was more prevailing. The new Fortran 77 and its structured programming style was not common. But, vector processing has become a common place during the last ten years. It is interesting to see how far we have marched forward since the days of 1976.

2.1 THE UNDERSTANDING OF MACHINE PERFORMANCE

Vector and parallel machines typically have several modes of operations with various speeds. There is a tremendously wide spectrum of performance range. In many instances machine performances change drastically even under different operating environments. It can also vary as the software evolves. These, among other things, make it very difficult to compare machines.

One of the more popular and rather naive ways to compare machine performance is to look at the peak rate. The fallacy of this view is obvious. It assumes that most codes, with some effort, should execute close to its peak rate most of the time. It assumes that the performance of one simple loop can be generalised without qualifications. It assumes that it is very easy to scale up the performance of a few processors to that of, say, thousands of processors in a linear fashion. It also implies that over all machine performance increases linearly with the peak speed. The higher the peak rate, the better the machine should perform. This obviously goes against the well known argument of the Amdahl's Law [AE]. Another irony is: most people don't even make the distinction between the peak rate and the asymptotic rate.

Another popular way to compare machines is to compare their timings on a few well publicised kernel loops. Among them the (several versions of) Livermore Loops, the LINPACK dense system solver, the BLAS kernels, simple vector adds and multiplies, etc. Those measurements do shed some light on how machines perform under certain popular loop structures. The question is: Do those popular loops, with their given structures and vector lengths, represent the majority of the work load? Very likely, time consuming routines at a given site may have different loop structures. Besides, a full blown production code sometimes may have totally different timing characteristics than any of those kernel loops.

However, kernels representative of the major CPU work can be good measuring sticks, provided that problems with I/O and data communications are also included in the testbed.

As users have become more sophisticated, more models have been developed to assess the performance of different vector/parallel machines. A popular model, Hockney[HR], uses an additional gauge to measure machines. The Hockney model measures the vector length achieving half of the peak rate as an indicator of short vector performance. For multiprocessors, it uses the granularity achieving half of the peak rate as an additional gauge. The Hockney model does give more insight for memory-to-memory machines. However, it does not model the register-to-register machines very well as pointed out by Bucher and Simmons[BS2] in their modified model.

The old Amdahl's Law and the Ware's equation, with their associated speedup curves (hyperbolic in nature), are still good guidelines to understand performance. It highlights the importance of scalar performance as well as the compiler technology that helps to vectorise loops. For a general purpose environment, a more comprehensive approach to measuring machine performance is required in order to gauge machines in a more sensible way.

A recent experiment done at a major national lab showed that, on a CRAY X-MP, the mean MFLOPS(Millions of FLOating Points operations per Second) rate is about 25 MFLOPS per processor. At the same time, the mean vector length is less than 30 in a 24 hour span. This number may seem low to most people. But with compilers, loaders, debuggers, other utilities and the interactive operating system itself running on the supercomputer, this low number is quite understandable. The mean MFLOPS numbers could be a little higher if a site has a concentrated application area that contains highly tuned codes and with minimal program development.

To evaluate machine overall performance more objectively, several efforts have been attempted[BS1,HJ,WJ2]. I propose that the following points should be considered.

a) A set of kernel routines (not just loops) that are characteristic and representative of the CPU intensive part of the work load. A wide range of applications and code structures should be included.

b) A set of full blown benchmark codes. Pre- and post processing, file management and memory management should also be included.

c) Software impact should be distinguished from hardware impact. This gives a better understanding of the situation. Software related scenarios can change with time, sometimes very fast. Compiler capability, intrinsic functions and I/O utilities should be measured independently.

d) Efforts should be made to understand the behaviour of vector dominated code apart from scalar dominated ones. For vector dominated code, various access patterns for multi-dimensional arrays, complex number calculations and non-unit strides should all be included. One case in point for scalar performance: the effect of subroutine calling overhead is frequently overlooked, even though it may have tremendous bearing on performance.

e) The amount of conversion effort from a dusty deck version to a relatively efficient version should somehow be considered. Although it is quite possible, by putting in enough time and effort, to make any given code run faster, the cold reality is that time and effort may be better spent in improving the physics or in developing new numerical methods.

2.2 PERFORMANCE AND THE SYSTEM ARCHITECTURE

Subtle differences in architecture may have tremendous impact on machine performance. It is apparent that all vector/parallel machines are not equal. We'd like to address several key architectural issues that are crucial to performance. Even though the discussion is on vector machines in this section, comparison between multiple pipes and multiple processors will be discussed too.

a) Vector registers versus memory-to-memory organisation: The invention of vector registers on the CRAY-1 is a major advance in vector architecture. It was a breakthrough from the memory-to-memory STAR-100 organisation. Looking at the marketplace today, the vector register approach is almost unanimous except for the ETA-10, a distant cousin of the STAR. The major advantage of a memory-to-memory machine for vector loops is the reduced loop overhead. For register-to-register machines, because of the fixed length of vector registers, the compiler has to build in strip mining logic at the preamble. Memory-to-memory machines, however, pays the loop overhead in hardware through micro-coding.

The advantage of vector registers are many. They reduce the memory traffic by keeping intermediates and common expressions in registers. They allow the chaining of many operations together in one chime. Moreover, because of the strip mining approach, they eliminate the need to allocate huge space for storing intermediate results. The most overlooked advantage of register-to-register machines is the ability to mask pipe startup time for the second chime on in a vector loop. This is because the register architecture can allow better overlapping of unrelated instructions.

As to the number or registers, I feel that it is more important to have more registers than reconfigurable registers (such as the FUJITSU VP). Moreover, it is my observation that the number of elements per vector register does not really matter much as long as it is reasonably bigger than the vector startup time. Since in most cases the loop count is unknown at compile time, the compiler strip mining logic has to be put in anyway.

b) Real versus virtual memory: Users move up from mini and mainframe computers, where central memories are traditionally small, are used to virtual memories. Virtual memory obviously does make it easier for the user to expand the size of the problem. However, its impact on performance is very tricky. For large vector codes, it is no easy job for the user to figure out the optimal page size at run time. It also requires programmers to change the data and code structure to force stride one access pattern. For multi-dimensional problems, this is not always easy. The merit of virtual memory is supposed to be the ease of use for large problems, but the amount of tuning effort involved is as involved as physical I/O operations. Often times, for lack of tools, there is very little that the programmer can do.

Most well designed vector processors today (with virtual memories) either provide a bypass or disable virtual memory for vector operations. Otherwise, the vector speed, without performing a gathering operation, could even be slower than the scalar one. For scalar operations, the provision of cache memory and/or more scratch registers are far more useful alternatives than virtual memory.

In all, virtual memory is a useful approach to compensate a small central memory in small size problems. With the current generation of large memory vector processors, its usefulness is rather questionable for performance reasons. The benefit of paging in a multiprogramming environment is understandable. But for machines with real memories, either to swap the job to a fast extended memory or to do a partial swap to disk can certainly amend this shortcoming.

Another rather useful hardware feature in facilitating memory allocation is to support segmented memory or to support multiple base and limit registers.

c) Memory hierarchy: Obviously, the best solution for memory organisation is to have an one level fast large memory. Since this is generally not possible, memory hierarchy seems to be a necessary evil. In fact, no matter how big a central memory a machine has, bandwidth of secondary memory is always a serious issue.

In order to bridge the gap between the slow secondary memory and the fast CPU power, Cray uses the SSD(Solid-state Storage Device) as an additional memory level. The current SSD has a transfer speed 500 times that of a conventional disk (2 GB vs 4MB per second). It is capable of reaching an access time 800 times faster than that of a disk. It is conceivable that the SSD can be used in the following four ways. It can be used as an extended memory under user control (to store scratch files or used as pseudo virtual arrays). It can be used to store copies of frequently used permanent files, e.g. compilers, loaders, assemblers, etc. It can be used as job swapping and staging devices. It can also be used by the operating system as disk cache for large dataset residing on disks. Many vendors (IBM, FUJITSU, NEC and HITACHI) also start to offer solid state devices on their new vector processors. The majority of those offerings is to support memory paging in a virtual memory environment. Since this use is mostly for short transfers, the main benefit here is the faster access time.

d) Multiple pipes versus multiprocessors: The use of multiple pipes is becoming more commonplace recently. Multiple pipes works well for highly vectorised and long vector codes. It is especially impressive for single job performance. However, multiple pipe design put strain on the memory bandwidth. All multiple pipe machines today are forced to penalise non-unit strides. Consequently, it adds complications to program tuning if optimal performance is sought. Moreover, because of the Amdahl's Law, unless the code is highly vectorised, the benefit of extra pipes cannot be recognised at run time. To better utilise those expensive pipes, many machines (FUJITSU VP and the NEC SX) are forced to use a fixed frontend machine or coprocessor for program development and compilations. Typically those codes are very scalar in nature. Small number of powerful multiple processors, on the other hand, seems to be a better throughput machine in a multiprogramming environment. To exploit parallelism for a single job, it may require quite a bit of extra work (to be elaborated later). However, this added capability, if properly utilised, can offer outer loop parallelisation unattainable by vector machines.

This is only a brief account of the key differences between multipipe machines and multiprocessors. More discussions on multiprocessors can be found in section three.

e) Multiple function units versus multiple pipes: The benefit of multiple pipes is that it is transparent to the user. One not so obvious benefit of multiple function units, e.g. HITACHI S810 (it has multiple pipes as well as multiple function units), is that it allows chaining into the same operations. Multiple functional units are useful in reduction sums, fast Fourier transforms, multiple vector sums and vector products.

There is some confusion about the need to have separate floating point functional units for scalar and vector operations. Since scalar operations and vector operations do not occur together, practically we see very little advantage for separate units.

2.3 PERFORMANCE AND COMPILER TECHNOLOGY

The compiler/optimiser technology has advanced so much in the commercial world in the past decade that it is inconceivable for any major vendor to stay competitive with an inferior technology. Most vector/parallel processor vendors have their own development teams to keep pace with the advances in technology. At Cray Research, the improvements made in compiler techniques have resulted in an overall performance gain of 20-40% in the past three years (by comparing CFT1.11 with CFT77).

Except a few items, most of the standard techniques as mentioned in [KKLW] and [AK] are well known today. People even begin to realise that vectorisation is not everything. There is a lot more to code optimisation than just vectorisation. For example, for some loops with IF statements of very low true ratio, scalar code could outperform the vector one. Even if a loop is recognised as vectorisable, a lot more can be gained by reorganising a complex loop structure and by scheduling instructions intelligently. It is also very essential to let the user interact with and to direct the compiler.

2.4 PERFORMANCE AND APPLICATIONS DEVELOPMENT

The advancement in numerical algorithms and computational mathematics during the past ten years has been most instrumental in exploiting the new architectural features. It is so significant that most scientific codes running today on vector computers have pretty high vectorisation rates. Codes once thought not vectorisable, such as the Monte Carlo codes, are now vectorised with hardware gather/scatter operations. In fact, partly because of the ability to tackle large scale problems through the use of supercomputers such as the Cray, computational methods are gaining momentum and are fastly becoming at least as important as theoretical and experimental methods in many scientific fields. Many scientific findings were made through computations first and verified later through experimental means. The importance of algorithm development in exploiting vector/parallel machines cannot be emphasised more strongly especially since the problems are so complex and the computing powers are still so limited relatively.

3. THE EMERGING BRAVE NEW WORLD OF PARALLEL PROCESSORS

- How much can we defy the Ware's equation? -

Ever since the introduction of the first CRAY X-MP in 1982, parallel processors are gaining more popularity in large scale computations. With up to 65,536 processors available on a single machine in one case and a minimum configuration of 4000 processors in another, people are quite puzzled about how much parallelism we can stretch and even about their own common sense.

The main issues that affect performance and user interface for parallel machines (multiprocessors) are memory hierarchy and processor hierarchy. Observations for shared memory models and distributed memory models will be made. Comments on processor hierarchy will be made as is pertaining to effective system utilisations.

3.1 SHARED MEMORY MODELS

Among all models of parallel processors, the shared memory models are the most often employed. The term 'shared memory' refers to the fact that no data movement is involved in addressing shared data in the parallel processing mode. There are two major types of shared memory models. The Cray multiprocessor machines represent one kind in which the memory is interleaved and shared among all processors. The IBM RP3 and the Illinois CEDAR projects represent the shuffle network approach with memory modules and processor nodes situated at opposite side of the network.

The interleaved memory approach, because of the physical limitation on the number of banks, puts a ceiling on the number of processors that can effectively be supported. The shuffle network, on the other hand, has the advantage of dynamically adding processor nodes and/or memory modules by adding more switching packets and stages. It is suitable, with enough paths and/or queues built in, to support large number of scalar processors[MBW,PG]. To offer reasonable response for processor synchronisation, some low level indivisible operations, e.g. FETCH-AND-ADD, has to be implemented with the functional units close to the memory. The CEDAR approach, however, with its multi-level memory hierarchy, may create complications in its software development for memory sharing and allocation. The IBM RP3 may have to face similar issues with its two speed memories (local and global). However, it is still too early to tell how things will evolve.

The main issue over a shared memory model is the contention on memory. There are really two issues related to memory speed. The first issue is memory latency. The more stages in the memory resolution logic the farther away the memory will be. The other issue is about the memory bandwidth and the associated penalty when conflicts do occur.

On a vector machine such as the CRAY X-MP, the latency issue is less important for vector codes as long as the latency is reasonably small compared to 64, the length of vector registers. The bandwidth issue, however, demands more attention. With carefully designed memory resolution logic and small number of processors, the problem is manageable. The experience at Cray shows that, even with heavy memory traffic as is typical in computational fluid dynamics codes, the memory degradation is less than 10%. On the other hand, for scalar dominated codes, the bandwidth issue is rarely a problem. The latency issue for scalar code is more serious. However, it can be eased significantly through smart use, by compiler, of registers or through some (programmable or true) cache memory usages.

For machines with shuffle networks, the latency problem can be helped greatly if a much faster switching network is available. The bandwidth issue is a little tougher to solve. The well known tree saturation problem has the nasty cascading effect for hot spots. It can be helped through redundant paths and by combining memory requests in the network[PN,KP,LKK].

Another model that can be remotely considered as shared memory is the ETA-10. Processors share data in the Common Memory. But those data have to be specifically declared and their windows moved in and out of the local memory explicitly. The implication in programming effort is something to be considered seriously. This kind of sharing that appears in hierarchical central memory structures seems to create synchronisation overhead in data access. Moreover, for performance reasons, read only data need to be duplicated, again could result in complications in programming.

3.2 DISTRIBUTED MEMORY MODELS

This kind of machine employ message passing schemes for data sharing. The architecture that caught the most fancy is the hypercube family. There are merits to this kind of distributed approach if the links between processor nodes are fast enough and if there are high speed I/O channels to and from the external world[HMSCP]. For algorithms capable of exploiting data locality and with minimal data movement, the potential is there. However, it is one thing to be able to identify parallel work, it is something else if the user has to worry about how to place and move data in a message passing environment. The network communication architecture is another level of added complexity that most users don't have time for. Therefore, the serious issues facing hypercube architecture is the algorithm development and the software environment. It is a good research tool but not mature enough to be a viable commercial product in the general purpose environment.

Besides, the limited size of the local memory dictates the problem size. Hence, for a given problem, one may be able to find a hypercube configuration and devise an algorithm to solve it. But it is not always possible, for a fixed hypercube configuration, to find an algorithm to solve any given problem efficiently. (Of course, if the nodes have disks attached, things can be different.) This is precisely why it is important to have reconfigurable cubes. But even reconfigurable hypercube adds complexity in the processor hierarchy that is difficult to solve. For example, two freed 32-node subcubes don't necessarily make up a 64-node cube.

The problem with programming probably will be more serious once each node possesses vector capability[PTLP]. Take the 2D-FFT as an example, costly memory transposition has to be performed in order to use this architecture effectively. Besides, the faster vector speed on each node will put tremendous pressure on the link speed. Those are serious issues that hypercube developers and users have to contend with.

3.3 PROCESSOR HIERARCHY

Since shared memory multiprocessors put strain on the memory bandwidth, it is logical to tie clusters of multiprocessors together. Those clusters can then either share an extended memory or share their local memories through some kind of shuffle network, such as the approach taken by CEDAR. No matter what the processor-memory connections look like, this approach inevitably adds one more level of processor hierarchy in the architecture. The impact on performance is secondary, however, compared to the problems created in the software environment. This architecture put additional constraints in scheduling processors for multiprocessing jobs. The hierarchical concept has to be embedded in the support software. However, the added opportunity in algorithm development with multi-levels of parallelism can definitely inspire interesting research.

4. CRAY'S EXPERIENCE WITH PARALLEL PROCESSING

- Parallel processing is an overhead -

The Cray multitasking facility and its use[BH,CDH,CHLS,EHKR,RKEH, FP,FS] has sparkled a lot of interests in parallel processing in the past few years. The CRAY X-MP and the Cray 2 architecture introduced for the first time a multi-level parallelism as a commercially viable product.

Basic multitasking concepts and their usages at Cray can be found in [CR,LJ1]. A practical discussion of main programming considerations can be found in [LJ2]. In this section we will focus only on the philosophical differences between vector processing and parallel processing. The CRAY multiprocessing environment will only be mentioned briefly for the sake of completeness. Based on our own experience, an often overlooked issue with processor wait will be addressed.

4.1 MAJOR CHALLENGES IN PARALLEL PROCESSING

It appears that both vectorisation and parallelisation are based on dependency analysis. Does that mean that they are fundamentally the same? Well, conceptually they are, but in reality the two are very different. It is not only that parallelisation can exploit outer loop parallelism while vectorisation exploits only inner loop parallelism. The differences are rather subtle and create challenges for users.

a) The asynchronous nature of concurrent programming:

Concurrent execution streams in parallel processing significantly change the underlining programming environment. Moreover, unlike vector processing, the order of operations is indeterministic in parallel processing. This asynchronous nature adds complexity to the programming effort. As a result of the two factors, stack based compiler and calling convention are required to allow reentrancy.

It also necessitates synchronisation among tasks and the provision for the protection of critical regions. Indeterminancy makes it almost impossible to reproduce exactly the same execution behaviour and hence difficult to debug.

b) The issue with data scopes:

In the conventional single processing environment, there are only two data types, local and global. In parallel processing, however, there are two new concepts, private and shared data. Private data are local to the task that declared them. It may be local to a subroutine or it may be global across subroutine boundaries. At Cray, the latter is named TASK COMMON, a new data type. Shared data, on the other hand, are shared across tasks. The traditional Fortran COMMON data type belongs to this category. Moreover, data could be local to a subroutine but shared among tasks participating in the parallel work inside the routine.

If a child task is allowed to spawn other tasks, the sharing situation is even more complicated. How would data be shared between child and grandchild but not with the parent? What happens if a task is terminated before its offspring is? Data sharing creates even more serious challenges for compiler/optimiser or even human to do dependency analysis if the scope of the analysis has to go beyond the parallel block itself.

c) The issue with user interface:

The situation with concurrent programming and data scope intensifies for the lack of a proper parallel language and the lack of programming aids. The mechanism to implement parallelism could be the major factor between success and frustration. In vector processing, users can count on the underlining software to exploit the parallelism. In parallel processing, obviously the best solution is to let the software do all the analysis. Since the technology is not as advanced and because of the many pitfalls in concurrent programming, a simple user interface is deemed necessary. The ideal situation is to free the programmers from worrying about the implementation mechanism, but rather concentrate in identifying and exploiting parallelism.

4.2 THE SOFTWARE ENVIRONMENT AT CRAY

To support concurrent programming, the Fortran environment has to be able to support code reentrancy. The stack based compiler and calling convention are designed just to handle that. On top of this infrastructure, there are three kinds of multiprocessing support at Cray. At the base level is the multitasking library facility[see CR and LJ1 for details]. The multitasking library basically provides three major functions: spawning of tasks, synchronisation of tasks and protection of critical regions.

4.2 THE SOFTWARE ENVIRONMENT AT CRAY

To support concurrent programming, the Fortran environment has to be able to support code reentrancy. The stack based compiler and calling convention are designed just to handle that. On top of this infrastructure, there are three kinds of multiprocessing support at Cray. At the base level is the multitasking library facility [see CR and LJ1 for details]. The multitasking library basically provides three major functions: spawning of tasks, synchronisation of tasks and protection of critical regions.

The second kind of support is the microtasking facility. This facility allows the user to insert directives in the Fortran code to take advantage of the multiprocessing capability. It is an attempt to free the user of the drudging details of implementation and focus instead on identifying parallelism implicit in the code or problem. The microtasking approach has three unique points. It employs a concept termed 'fray' [CR], a subroutine in scope, which serves as an entry point for companion processors to join the execution. Actual parallel work is imbedded within the fray and marked by DO-ALL type of directives inserted by the programmer. The second unique point is, by marking parallel loops, there is no notion of tasks apparent to the user. This allows the user to think more in terms of parallel work instead of handling of tasks.

The third unique point is in its implementation on the CRAY X-MP, since everything was implemented through (shared and private) registers without having to go through memory at all, the scheme is extremely efficient, a key factor in handling fine granularity parallelism.

The third kind of support is the compiler automatic partitioning. The CFT77 is able to identify and perform FORK and JOIN functions at the loop level. It also allows the user to insert directives to aid the compiler to implement parallel structures. Depending on the intelligence of the compiler, this is probably the ultimate carefree way of exploiting parallelism from user stand. This approach obviously allows for accommodation of future parallel languages.

The operating systems at Cray are able to support multiple execution streams for a single user job. The multiprocessing software concepts and support are still emerging. We anticipate that the next few years will be very crucial in bringing up a matured and coherent user interface and friendly programming environment.

4.3 THE ISSUE WITH TASK SYNCHRONIZATION

There are two kinds of waits in a concurrent programming environment. The first kind occurs at critical regions. Tasks (processes) wait for some other task to exit from a small code region, typically containing the updating of some shared information. The second kind of wait occurs when tasks reached a point that requires indefinite wait time.

The first type of wait should be implemented in as efficient a way as possible. The only problem is when the processor gets yanked away midway through the critical region. When that happens, it changes the nature of wait from the first kind to the second kind. Hopefully, when the critical region is small, this does not happen too often.

The second kind of wait, however, is very sensitive to the implementation. It should not be implemented in a spin wait; namely, waiting indefinitely for some status flag to be cleared in a tight loop. The worst spin wait is to check the status of a memory cell. Because it puts so much strain on the memory bank which contains the particular memory cell, it could create tremendous memory degradation. The mechanism that is often used at Cray is the busy wait; namely, waiting for a binary semaphore to be cleared. This implementation basically puts the processor to sleep until some other task comes along and clears the semaphore. This mechanism is very efficient. However, it does not allow processors to voluntarily release themselves. (Even though it is safeguarded by the hardware interrupt mechanism against possible deadlocks). The third implementation is what the Cray multitasking library uses. It allows the user to set a timeout count. As soon as the time expires, control is passed back to the operating system. This seems to be the most flexible way but not necessarily the most efficient because of the built-in software queueing overhead.

It is important that appropriate task waiting scheme be chosen. We see examples of performance differences of up to 30% when a wrong one is used. Some of the work done elsewhere by employing the memory spin wait scheme [CSZ] failed to take into account the issues mentioned here. Its use may lead to serious drawbacks.

4.4 SOME PERFORMANCE EXAMPLES

Several experiments have been performed at Cray. Key issues in performance have been studied [CDH, EHKR, LJ2]; namely, granularity, load balancing, memory contention, synchronization mechanism, Ware's equation and the speedup curve, overlapping of I/O with CPU computations, etc. In this paper, we would like to illustrate the performance impacts through two examples.

A few remarks on speedups and overheads will be made first. To measure speedups, the best single processing wall time is compared against the best multiprocessing wall time.

Overhead is a very loosely used term. Let us denote by P the number of processors, R the percentage of time (of the single processing run) that can be parallel processed, T(1) the single processing time, T(P) the multiprocessing time, SP the theoretic speedup, Sp the realised speedup and OV the overhead. The overhead can then be measured in many different ways; namely

$$Sp(P) = 1 / (R/P + (1-R) + OV(P)) \quad (1)$$

$$Sp(P) = 1 / ((1+OV(P)) R/P + (1-R)) \quad (2)$$

$$OV(P) = (SP(P) - Sp(P)) / P \quad (3)$$

The overhead in (1) roughly measures the percentage related to the original single processor run. For example, for P = 1000 and R = 100%, SP(1000 = 1000. However, a 1% overhead in the sense of (1) means a speedup of only 90.9, a degradation of ten fold. The overhead in (2) roughly measures the percentage related to the multiprocessor run. Eq. (3) is similar to Eq. (2) only with a different scaling factor. In our experiments, we applied Eq. (2) to the parallel regions individually to measure overheads.

a) The first example is a large scale seismic 3-D migration code. The algorithm chosen is an ADI method with a Crank-Nicolson type discretization scheme [please see HW for details]. The dimension of the particular problem is a 200 x 200 grid with 1024 time samples per trace. The parallelism is identified at the frequency domain as the whole problem is nicely decoupled. The amount of central memory used is roughly 3.7 MW's and the SSD space used is 40 MW's. The percentage of time that can be processed in parallel is 98.7%, which translates into a theoretical speedup of 3.85 for a CRAY X-MP/48.

The realised speedup is 3.45, with a granularity of 50 milliseconds, very large in our standard. The observed memory contention overhead is around 8.7%, other overheads caused by I/O, synchronisation and other operating system functions account for about 1.3%. Through the use of the SSD, I/O wait time is totally masked by the CPU computations. In comparison, without the SSD the job is totally dominated by I/O. It takes 24 hours to migrate 1000 depth levels with a single DD-29 disk. The same problem, by using SSD, will only take roughly one hour with four processors.

b) The second example is a computational fluid dynamics code that solves a 3-D Reynolds-Averaged Navier-Stokes equation. An approximate factorisation numerical scheme is used which lends itself very well for parallel processing. Parallelism is identified by integrating individual plans independently [see BH for details]. For a 30 x 30 x 30 problem, the percentage of parallelism is 98.8%, which amounts to a theoretical speedup of 1.98 by using two processors and 3.86 by using four processors.

The realised speedup is 1.87 for two processors and 3.50 for four processors on an CRAY X-MP/48. The granularity varies between 2 and 78 milliseconds for different sections of the code. With all the data fit in core, the main overhead seems to be from memory contention. The observed contention differs for different parts of the code. It varies between 3% to 11%, not too bad for a code with heavy vector memory traffic on all processors.

This is a benchmark code originally developed for the ILLIAC IV. The amount of coding changes involved is about 50 lines, for a code of 3500 lines, in the form of directives without having to change the Fortran source at all.

5. CONCLUSIONS

Industry trend suggests that multiprocessor vector computers are here to stay. Our experiments indicate that shared memory multiprocessors are very amenable for parallel processing. It can help to take the mind of the user from data movement and manipulation. The user can, instead, focus his/her effort on creating and exploiting parallelism.

With parallel processing exploiting outer loop level parallelism, which is inaccessible to vector processors, the potential to be gained through parallel processing is very promising. We anticipate the driving force again to be scientists and applied mathematicians instead of computer scientists. Their insatiable appetite for computing power and the pressure to advance sciences will provide the kind of impetus needed to shape future directions.

As parallel processing matures, there will definitely be an industry shakeout. We foresee the coming of two level parallel processing on top of vectorization, with inner level parallel processing automatically handled by compilers. We foresee the standardization of user interface and the emergence of higher level language.

On the other hand, parallel processing may be easier to grasp conceptually than vectorization, but it is definitely a very complex problem in terms of implementation and debugging. For that matter, vendors will be hard pressed to provide very friendly programming environment in order to attract serious users and to keep market shares.

REFERENCES

[AE] Amdahl E.M. (1967) The Validity of the Single-Processor Approach to Achieving Large-Scale Computing Capabilities. AFIPS Conf. Proc. 30:483-485.

[AK] Allen J., Kennedy K. (1982) PFC: A Program to Convert Fortran to Parallel Form. Proc. of the IBM Conf. on Parallel Computers and Scientific Computations, IBM, Rome, Italy.

[BH] Barton J.T., Hsiung C.C. (1985) Multitasking the Code ARC3D. Proc. of the GAMM Workshop on The Efficient Use of Vector Computers with Emphasis on Computational Fluid Dynamics, 204-208.

[BS1] Bucher I.Y., Simmons M.L. (to be published by North Holland). Performance Assessment of Supercomputers, Vector and Parallel Processors: Architecture, Applications, and Performance Evaluation, M. Ginsberg, Ed.

[BS2] Bucher I.Y., Simmons M.L. (1987) A Close Look at Vector Performance of Register-to-register Vector Computers and a New Model. ACM SIGMETRICS Conf. on Modelling and Measurement of Computer Systems, Banff, Canada.

[CDH] Chen S.S. Dongarra J.J. Hsiung C.C. (1984) Multiprocessing Linear Algebra Algorithms on the CRAY X-MP/2 Experiences with Small Granularity. J. Parallel ^ Distributed Computing 1:1,22-31

[CHLS] Chen S.S. Hsiung C.C. Larson J.L. Somdahl E.R. (to be published by North Holland). CRAY X-MP: A Multiprocessor Supercomputer, Vector and Parallel Processors: Architecture, Applications, and Performance Evaluation. M. Ginsberg, Ed.

[CSZ] Camevali P. Sguazzero P. Zecca V. (1986) Microtasking on IBM Multiprocessors. IBM J. Res. Develop. 30:6, 574-582.

[CR] Multitasking User Guide (1985). Cray Computer Systems Technical Note, SN-0222.

[EHKR] Edwards M. Hsiung C.C. Kosloff D., Reshef M. (1985) Elastic 3-D Forward Modelling by the Fourier Method, presented at the 55th Annual meeting of the SEG, Washington, DC, Oct. 7-11.

[FP] Forcrand P. de (1986) QCD from Chippewa Falls. J. Stat. Phys. 43: 1077-1094.

[FS] Forcrand P. de Stack J.D. (1985) Spin-dependent Potentials in SU(3) Lattice Gauge Theory. Phys. Rev. Lett. 55:1254-1257.

[HJ] Hack J.J. (1986) Peak vs. Sustained Performance in Highly Concurrent Vector Machines. IEEE COMPUTER 19:9, 11-19.

[HMSCP] Hayes J.P. et al. (1986) Architecture of a Hypercube Super-Computer. Proc. Int. Conf. on Parallel Processing, 653-660.

[HR] Hockney R.W. (1985) (r, n, s) Measurements on the 2-CPU CRAY X-MP, Parallel Computing 1:1-14.

[HW] Hsiung C.C. Butscher W. (1984) A Numerical Seismic 3-D Migration Model for Vector Multiprocessors. Parallel Computing 1:2, 113-120.

[KKLW] Kuck D.J. Kuhn R.H. Leasure B. Wolfe M. (1980) The Structure of an Advanced Retargetable Vectorizer. Proc. COMPSAC'80, IEEE.

[KP] Kumar M. Pfister G.F. (1986) The Onset of Hot Spot Contention. Proc. Internat. Conf. Parallel Processing 28-32.

[LJ1] Larson J.L. (1984) Multitasking on the CRAY X-MP/2 Multiprocessor. IEEE COMPUTER 17:7.

[LJ2] Larson J.L. (1984) Practical Concerns in Multitasking on the CRAY X-MP. Workshop on Using Multiprocessors in Meteorological Models.

[LKK] Lee G. Kruskal C.P. Kuck D.J. (1986) The Effectiveness of Combining in Shared Memory Parallel Computers in the Presence of 'Hot Spots'. Proc. Internat. Conf. on Parallel Processing, 35-41.

[MBW] McAuliffe K.P. Brantley W.C. Weiss J. (1985) The RP3 Processor/Memory Element. Proc. Int. Conf. on Parallel Processors.

[PG] Pfister G.F. et al. (1985) The IBM Research Parallel Processor Prototype (RP3): Introduction and Architecture. Proc. Int. Conf. on Parallel Processors.

[PN] Pfister G.F. Norton V.A. (1985) "Hot Spot" Contention and Bombining in Multistage Interconnection Networks. Proc. Internat. Conf. on Parallel Processing, 934-948.

[PTLP] Peterson J.C. et al. (1985) The Mark III Hypercube-Ensemble Concurrent Computer. Proc. Int. Conf. on Parallel Processing, 71-73.

[RKEH] Reshef M. Kosloff D. Edwards M. Hsiung C.C. (subm. to GEOPHYICS) 3-Dimensional Forward Modelling by the Fourier Method, Part I, Acoustic Case.

[WJ1] Worlton J. (1986) Toward a Science of Parallel Computation. Symposium on Future Directions of Computational Mechanics. Dec 10-11, Anaheim, CA.

[WJ2] Worlton J. (1984) Understanding Supercomputer Benchmarks. Datamation, 121-130.

The ECMWF Model: Past, Present and Future

D. Dent

European Centre for Medium-Range Weather Forecasts, Shinfield Park, Reading,
Berkshire RG2 9AX, United Kingdom

INTRODUCTION

ECMWF has been involved in multitasking development ever since the first decision to acquire a Cray-XMP with more than one processor. Until recently, this has been a single-minded effort concentrating solely on the operational weather model. Within the last 6 months, the Analysis code, has also become a multitasked operational application. This presentation will however be limited to discussion of the spectral model.

HISTORY OF DEVELOPMENT

The current ECMWF production weather model first went into daily production use in 1983 executing on a Cray-1A. It uses the spectral technique and the first version had a wave resolution of 63 (known as T63) with 16 vertical levels. In 1985 the model was moved to a Cray-X22 and multitasked, thus allowing a resolution increase to T106 within operational time constraints. By the beginning of 1986 further development had taken place to allow efficient use of the 4 processors of a Cray-X48. The vertical levels were increased to 19 and additional physical parameterisations were added. This represents the current state of the model in production.

CODE CHARACTERISTICS

The application is written in Fortran and constitutes about 100000 lines. A few important functions make use of assembler coded routines such as Fast Fourier Transform and matrix multiply. The Fortran is highly vectorised by the complier and makes extensive use of the Cray vector merge functions (CVMGx).

Since the model is in daily production use it must run as fast as possible. However, it is also the vehicle for research into model improvements and must therefore be flexible and easy to change, not only for staff familiar with the code but also for visitors and consultants.

Flexibility in handling memory is achieved by use of a memory manager which is based on the Cray Fortran POINTER extension. This has the disadvantage of making the code non-portable but the advantage of allowing flexibility to the extent that the model is resolution independent and can be run without source change using a variety of initial conditions at differing horizontal and vertical resolutions. For research purposes, it is straightforward to allow space for additional prognostic variables.

Model resolution is normally such that data must be held on backing store in work files. In order to achieve adequate wall-clock times, heavy use is made of the Solid state Storage Device (SSD) to hold these work files. In fact, I/O rates are so fast between memory and SSD that the I/O can be made synchronous, thus saving substantially in memory requirements and code simplicity by abandoning double buffering techniques. The penalty for using synchronous I/O is about 3% of the model elapsed time.

For reference purposes, the costs of various significant parts of the model computation have been separated so that the relative expense of the dynamics, spectral tranforms, FFTs, physics parameterisation, and memory manager overheads can be compared. These have been measured with the SPY facility and are illustrated in figure 1.

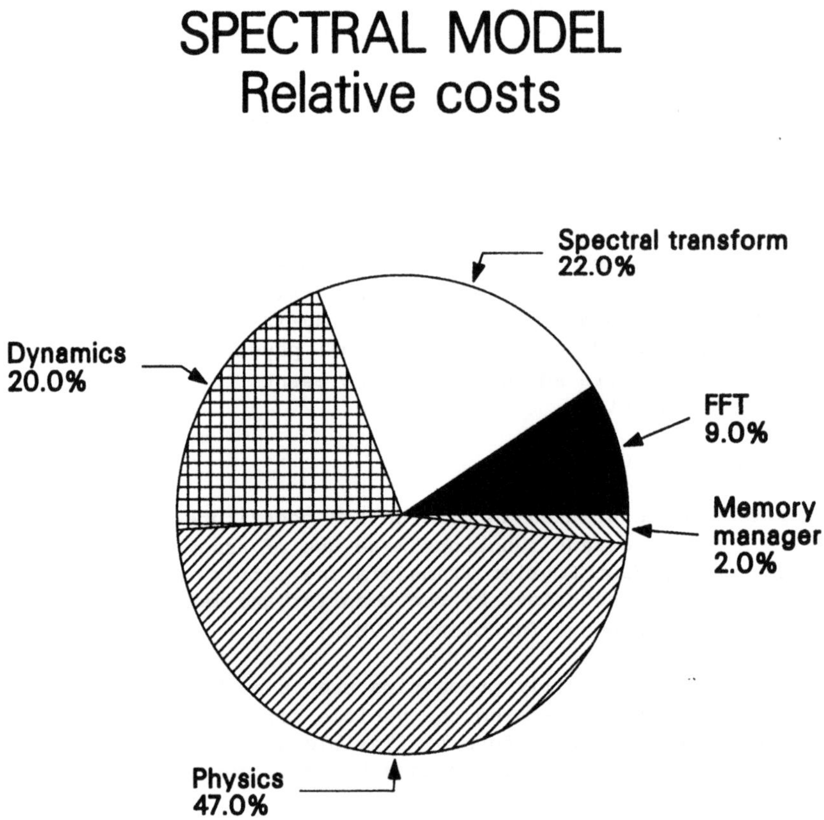

Figure 1

The elapsed time of the complete operational 10 day forecast with its accompanying postprocessing tasks is shown in figure 2. The high and low values correspond to changes in the supercomputer configuration and to changes in model resolution.

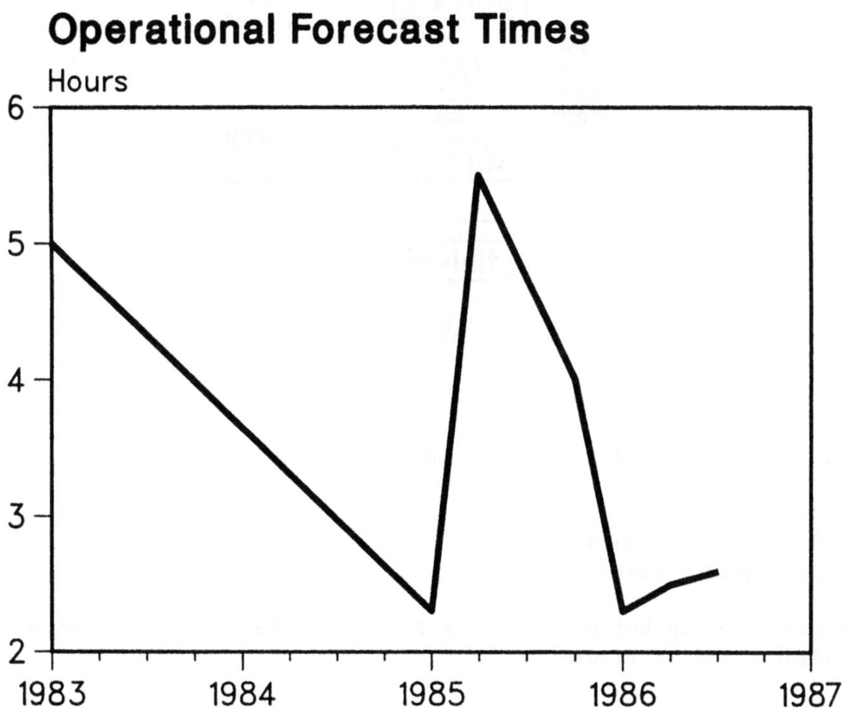

Figure 2

MODEL STRUCTURE

This section deals with the data organisation, since this is crucial to the multitasking strategy.

The work file data is processed in 2 scans for each model timestep (see figure 3). Within each scan, the I/O and computation is inside a loop over lines of latitude. That is, data belonging to all vertical levels for one line of latitude are collected together into one record and stored in the work files. At the operational resolution(T106), there are 160 rows between poles and 960 timesteps in a 10 day forecast. The workfiles occupy about 16 million words of SSD storage.

Between the scans (marked D) is some computation in spectral space for diffusion and semi-implicit timestepping.

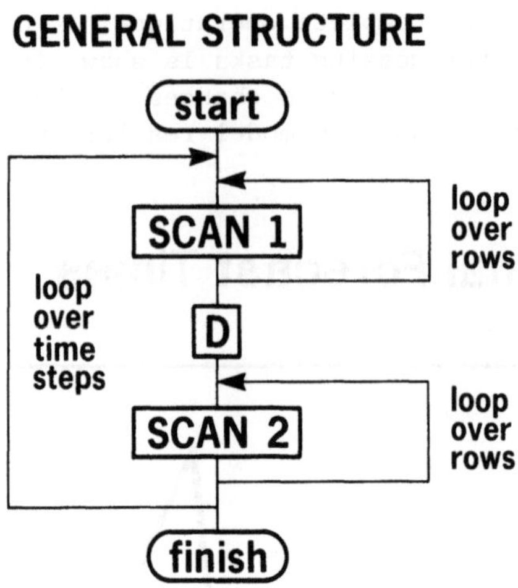

Figure 3

Data in the work files is split into 3 types:

(1) Fourier data
(2) Grid point data
(3) Legendre coefficients

The relationship between the work files and the 2 scans of the model is illustrated in figure 4.

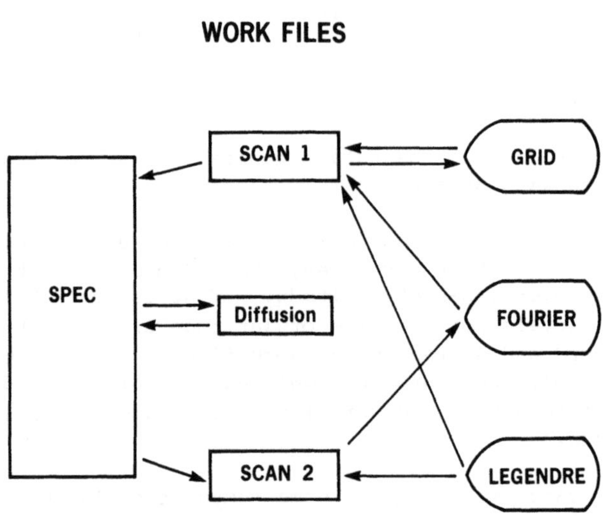

Figure 4

Scan 1 commences with dynamical calculations in Fourier space and requires input from the Fourier work file. Conversion to grid point space is followed by all the physical parameterisation and results in the gridpoint work file being read and rewritten. Another Fourier transform then generates contributions for a Legendre transform. When all lines of latitude have been processed, the scan is complete and the Legendre transforms have generated the fields in spectral form (the memory space labelled SPEC).

Scan 2 is relatively simple and computes the inverse Legendre transform. Output is in the form of the rewritten Fourier file.

MULTITASKING STRATEGY

Since the first multitasking was carried out for a Cray-X22, memory usage was a dominant consideration. Hence a strategy was chosen which provided efficient computation combined with minimum memory requirement. This strategy involves splitting scan 1 into 2 processes. The first includes the dynamics, the physics and the subsequent FFT. The second consists of the direct Legendre transform.

Process 1 can be applied simultaneously to a Northern hemisphere row and to its equivalent Southern hemisphere row with a synchronising point to ensure that the next process does not proceed until the data from process 1 is ready. Each north-south pair of Fourier transforms provides components which are symmetric and antisymmetric about the equator. These are combined in different ways to provide contributions to the Legendre transform. This becomes process 2 which can be multitasked so that the combinations are carried out simultaneously. Each copy of process 2 updates one half of the spectral arrays so that there is no danger of writing to the same memory simultaneously and therefore no need for locks.

Process 3 becomes the inverse Legendre transform executed in scan 2. Again, this work can be split across 2 processors and the resulting data is the Fourier record representing a North-South latitude pair.

The small amount of work between the scans can also be multitasked but will not be considered here since it is not inside a latitude loop and is relatively unimportant from the timing point of view.

The multitasking structure for two processors is illustrated in figure 5.

Generalisation of this scheme to any even number of processors is straightforward and involves a logical splitting of the work at a higher level. Each pair of processors is given a North-South latitude pair of data to work on independently. The loop over latitude rows is now shared by N processors. For scan 1, this is illustrated in figure 6. The second scan is trivial to generalise.

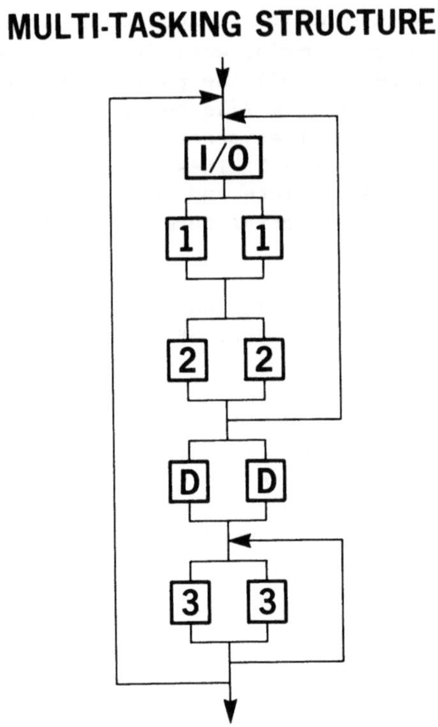

Figure 5

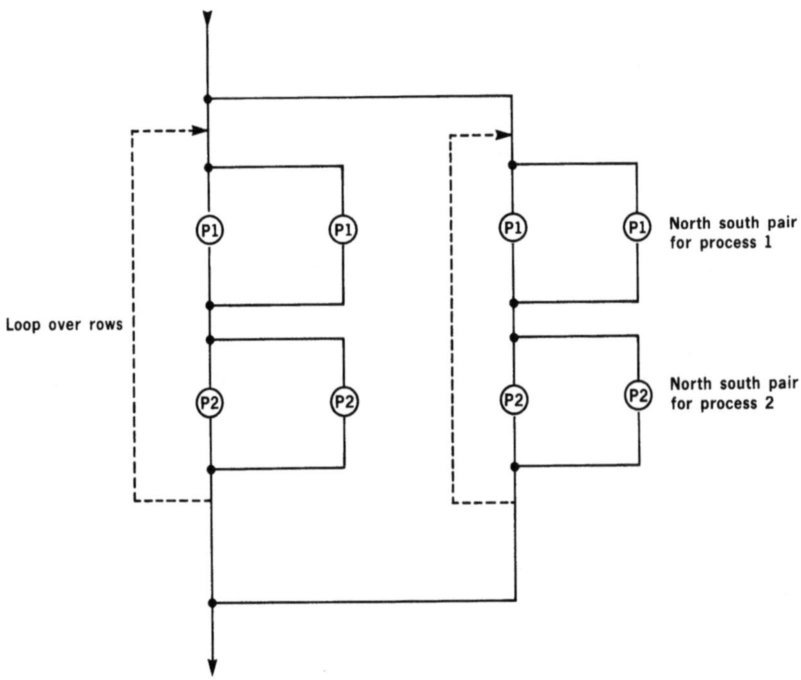

Schematic of multitasking strategy for scan 1 of 4 processor model

Figure 6

A potential hazard arises in the direct Legendre transform since 2 northern rows may update the same part of the spectral array simultaneously. Similarly, 2 southern rows may cause conflict with each other (see figure 7). The danger may be removed by the use of LOCKS although this must be done with care to avoid either

(a) excessive multitasking overhead from too much lock control

or (b) substantial idle processor waste when one processor is prevented from executing while a lock is held by another task.

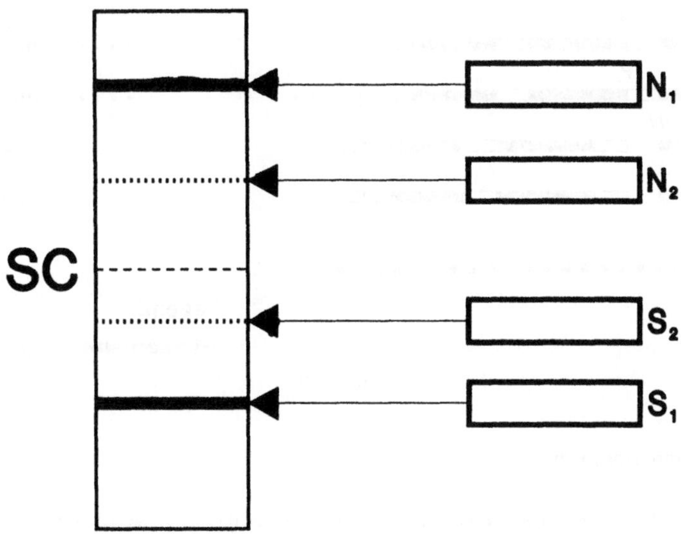

$$SC(I,*) = SC(I,*) + CINC(*)$$

Figure 7

Unfortunately, there is now an indeterminacy in the results since the order in which latitude rows update the spectral array is no longer predetermined by the code. Since on any finite word length computer

A + B + C is not identical to A + C + B

the output from the Legendre transform is no longer unique. The differences are of course small and physically insignificant, but nevertheless extremely irritating from the testing point of view. EVENT logic may be used to force the computation into a certain order and hence re-establish reproducability. This turns out to be quite cheap in an operational environment and is therefore implemented as default although it can be easily disabled. Figure 8 illustrates processor utilisation for scan 1. 2 processors are idle waiting for an event (W) which is posted (P) when process 2 is completed by the first latitude pair. Assuming no other disturbances, this stagger is then maintained for the rest of the scan with a correspondingly opposite delay at the end while waiting for the last pair to complete. Normally, the presence of postprocessing jobs (PP) allows this idle time to be utilised without unacceptable waste.

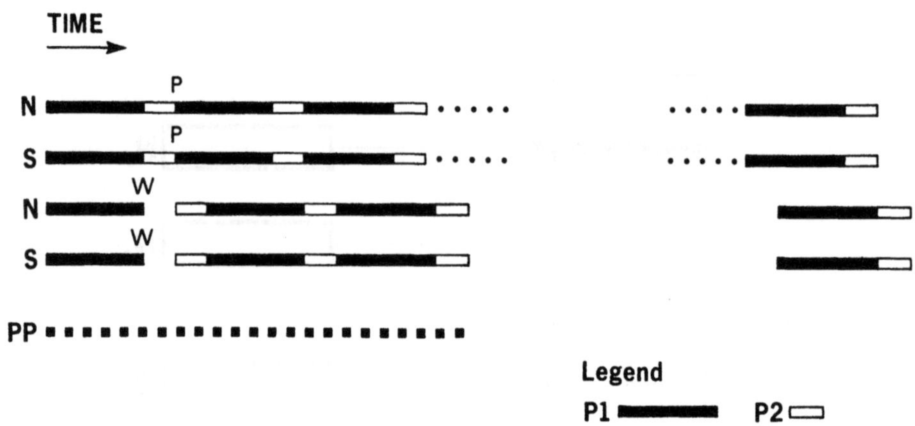

Figure 8

CURRENT PERFORMANCE

Using the Cray XMP hardware performance monitor, measurements of the production model environment reveal the following statistics:

Multitasking efficiency = 3.6
execution rate = 335 Megaflops/sec
vectorisation = 99% (of all floating point operations)

Time stamps inserted into the model code at suitable points allow the granularity and the variability of task sizes to be measured (see figure 9). Variations in processes 2 and 3 are small and due to memory bank conflicts. Process 1 shows variations of 15% and these are mainly due to inequalities in work content (see figure 10). Particularly in the physical parameterisation of convection, the work load for different latitude lines is not equal. The graph clearly shows greater cost in equatorial latitudes compared to polar areas where convection is virtually nonexistent. Fortunately, the pairing of north-south latitude lines helps to minimise this variation but figure 11 shows that it is still the largest inefficiency (out of balance). 'Reproducibility' is the waste due to EVENT logic in the direct Legendre transforms, while 'MT overheads' are the costs of task management and locks. 'Single-tasked' identifies that part of the code which has not yet been multitasked.

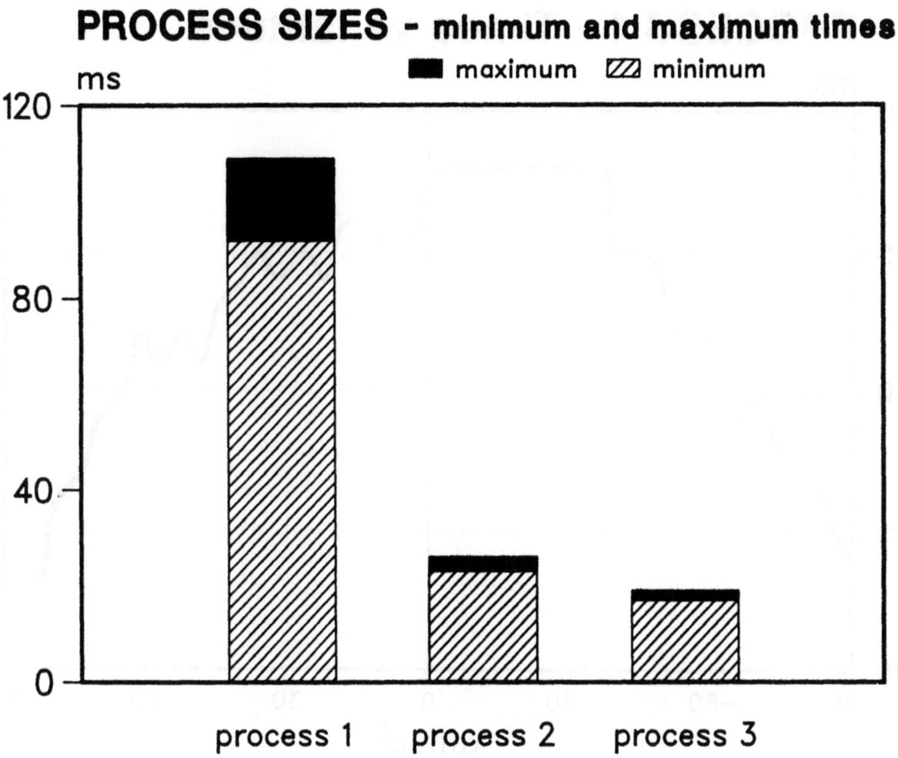

Figure 9

REVISED STRATEGY

Having identified a number of places where inefficiencies exist, the multitasking strategy can be revised. Given the more relaxed limitations of main memory compared to Cray X22, a dynamic strategy becomes attractive. This revised approach has been labelled the DSC technique for the following reasons:

(a) it has been designed and implemented by
Dent Snelling and Carver

(b) it can be described as
Dynamic SCheduling

The revised strategy retains the North-South pairings of latitude rows for computational economy but computes them sequentially, i.e. North followed by South. To be more precise, Process 1 is sequentially executed, first for the Northern row and then for the equivalent Southern row. All data is then available for the direct Legendre transform (process 2) which can be executed twice to complete the contributions to the spectral array from the pair of latitude rows. Once this series of calculations is complete, it can be repeated for another (essentially independent) pair of rows. This is illustrated in figure 12. The important advantage is that no synchronisation is necessary after each pair of rows since the

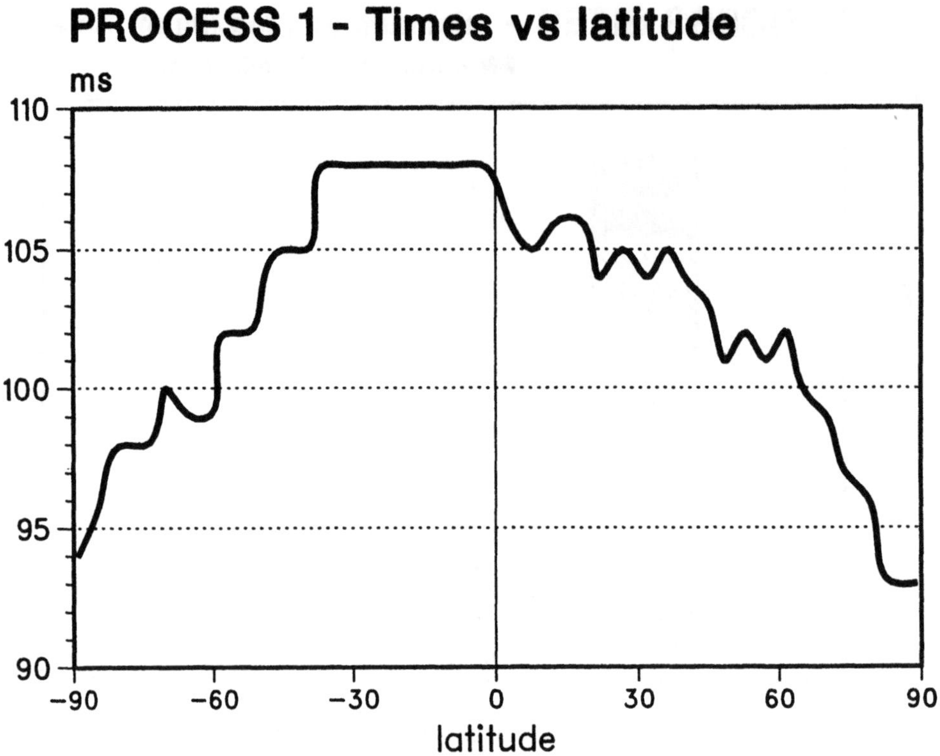

Figure 10

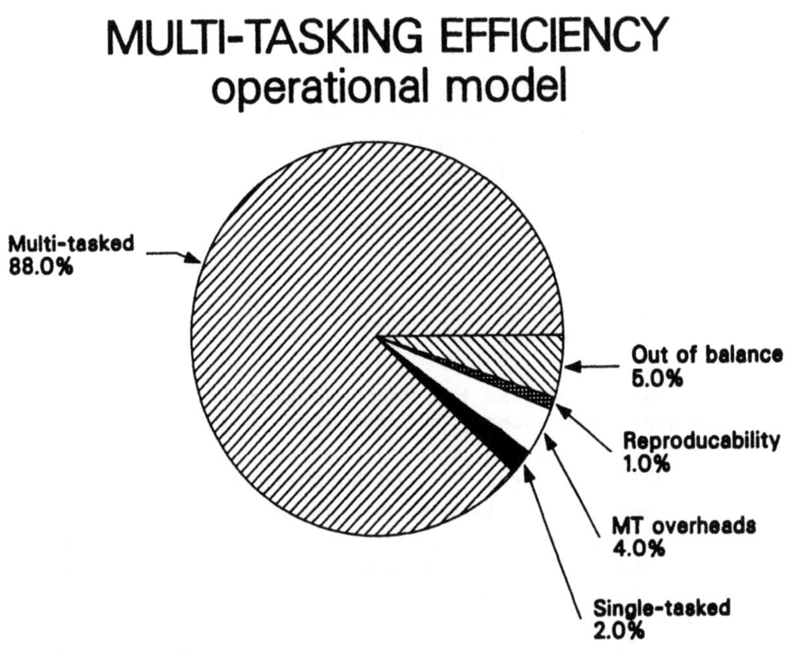

Figure 11

necessary data must be available. Since individual processors can be given independent pairs of rows to compute, the only synchronisation necessary is at the end of the scan to ensure that all rows are complete. The decision as to which row pair to compute can be taken at run time and hence the strategy is completely flexible, allowing for any number of available processors.

There is a modest penalty of increased main memory since each task must retain a Fourier record and a pair of symmetric-antisymmetric components from the Fourier transforms. The static strategy requires this data only for each pair of tasks.

The I/O scheme for the model workfiles is based on the static multitasking strategy and must therefore be revised. Since COS 1.15 offers an alternative access method which achieves savings by queueing I/O requests in order to minimise system overheads, the revision was planned to make maximum possible use of this facility. It is particularly attractive for random I/O requests to SSD storage in circumstances when several requests may be outstanding at a time. The revised scheme therefore always attempts to keep the available buffer space full by issuing (possibly in advance of requests) as many READs as possible. If sufficient buffer space is allocated (by means of a run time parameter) double buffering is therefore achieved.

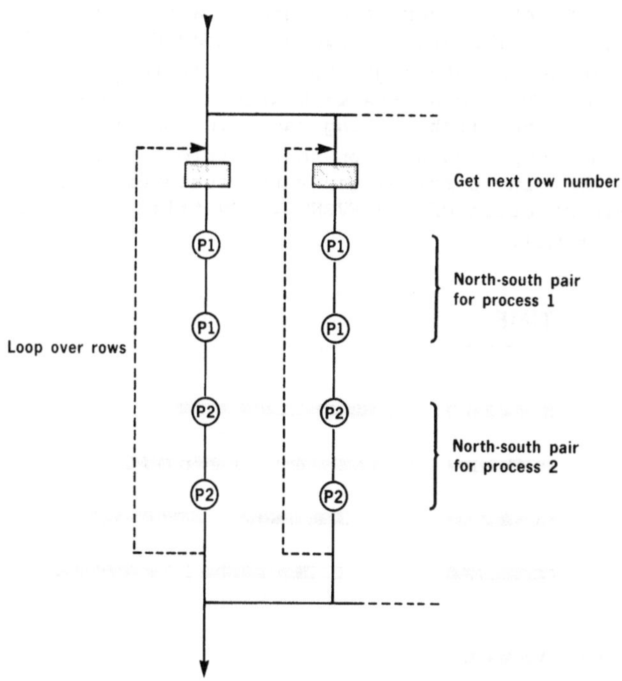

Dynamic multitasking strategy for scan 1 any number of processors

Figure 12

Since the I/O master routines are called from within the multitasked code, locks are used to enforce single-threading. However, this does not cause much inefficiency since the current execution of the I/O routine may do work on behalf of subsequent calls to the same routine from other tasks. These other calls are delayed by the lock but normally find no work left to do when the lock is finally released. By queueing as many requests as possible, maximum advantage may be taken of the economy in operating system overheads.

It also becomes possible to offer a new option for research into physical parameterisation techniques. The current production code allows access to all data on the current line of latitude only. For any grid point, its neighbours to the East and West are available as well as vertically below and above it. However, because of the I/O structure, it is not possible to access data to the North and South. As model grid resolution becomes finer, this limitation is likely to become more serious. The revised I/O scheme allows records and hence lines of latitude to be retained in buffer space until signalled by the running tasks that they are no longer needed. Providing that there is sufficient buffer space, data to the North and South for any desired number of latitude lines can remain available. This option is controlled by a run time parameter.

The inefficiencies introduced by enforcing reproducability become larger with the DSC technique. This is because only one task can be executing the direct Legendre transform at a time compared to 2 in the static scheme. The startup overhead needed to establish the stagger in time is therefore greater (see Figure 13). The loss may be minimised by subdividing the work within the transform into portions so that the first waiting task can commence as soon as the active task has completed work on that portion. Excessive subdivision into tiny portions generate counterproductive overheads from the administration of the EVENTS. By trial, subdivision into 8 seems to be optimal.

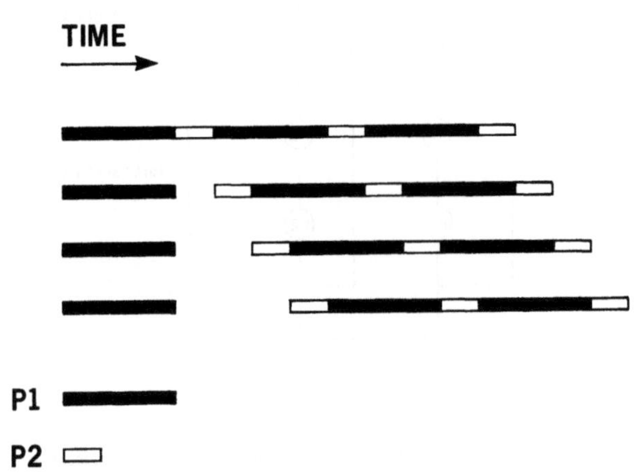

Figure 13 Processor utilisation for DSC with reproducability

The performance analysis has been repeated on short tests using the DSC method. However, the queued I/O access method is not yet available at ECMWF so that simulation had to be made using standard I/O techniques. The performance gains so far are small but are expected to improve with further enhancement of the code. The multitasking efficiency has increased to 3.7. Figure 14 shows a reduction of the out of balance overhead to 1% and the multitasking overhead to 3%. The multitasked portion has increased to 92%.

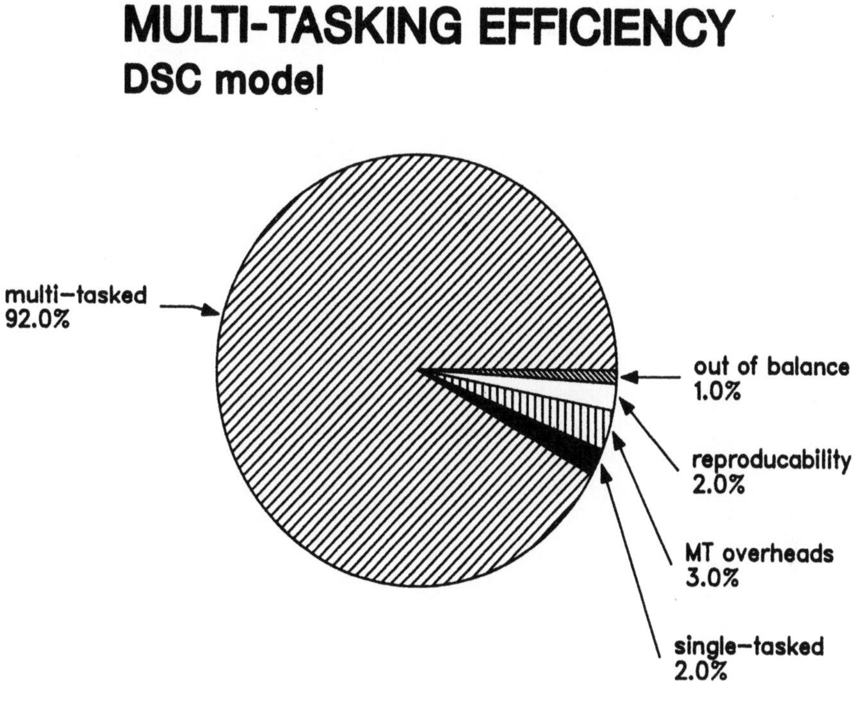

Figure 14

SUMMARY

Inefficiencies in the current operational code have been identified. While these are relatively small, they are likely to become more objectionable with future hardware containing larger numbers of processors. Hence the revised DSC scheme offers hope that a reasonably high multitasking efficiency can be achieved on future multi-processor configurations. Current performance of the DSC scheme will be improved when the queued I/O access method becomes available. Further gains may be made by eliminating many of the locks currently present only to enable statistics gathering. Finally, the residual single tasked code can be tackled either by the conventional macrotasking techniques or by use of microtasking in cases where the granularity is small.

Microtasking an Atmospheric Model at Environment Canada

ANDREW STANIFORTH and DAVID PARKS

Environment Canada/Cray Canada, Dorval, Quebec, Canada

ABSTRACT

Environment Canada has just replaced its Cray 1-S computer with a two-processor Cray XMP. Work has started on adapting the operational finite-element regional model to run in a multi-processing environment. The adopted strategy is outlined and some preliminary timings are given of a simplified version of the model.

1. INTRODUCTION

1.1 User Environment

The Atmospheric Environment Service (AES) of Environment Canada has a mandate to provide comprehensive information to Canadians concerning past, present, future and possible conditions of the atmosphere. Numerical models of the atmosphere have been developed for both research and real-time operational applications in the broad areas of weather prediction, climate and air quality. Approximately half of the resources of our principal computer are dedicated to the running of an operational suite of real-time weather analysis and prediction models. The remainder are used for R and D, principally in numerical weather prediction, climate and air quality, but some time is also made available to university researchers in several disciplines.

The conflicting demands of operational and R and D jobs create a challenging task for the scheduling of jobs. Real-time operational jobs have to meet strict completion deadlines and considerably fragment the available resources, even during prime time, while good turnaround is also important for a vigorous R and D environment. The ability of the principal computer to efficiently multiprogram jobs is therefore very important.

1.2 Hardware Environment

The principal computer is a recently-installed (December 1986) two-processor Cray XMP having 8 Mw of MOS main memory, 8Mw of buffer memory, an 8.5 ns clock cycle and has hardware gather/scatter and compressed index instructions; a 32 Mw SSD is scheduled to be installed during the coming year. The Cray XMP replaces a single-processor Cray 1-S, having 1 Mw of bipolar main memory, 4 Mw of buffer memory and a 12.5 ns clock cycle. It is currently front-ended by two Control Data Cyber 830's. The memory-banking of the XMP is such that transfers between the two processors and main memory are limited to 2 words/processor/clock cycle, whereas 3 words/processor/ clock cycle would be required to fully exploit the arithmetic capabilities of the processors. Jobs may thus be expected to be memory bound from time to time.

2. THE MODEL

2.1 Attributes

The model code is a hydrodynamics code that:

- is a two-dimensional (2-D) version of the 3-D operational regional forecast model code;
- is representative of approximately half of the computations that are performed during execution of the 3-D model;
- uses a variable-resolution finite-element discretization in space coupled with a semi-implicit time integration scheme and a semi-Lagrangian treatment of advection; and
- uses algorithms that are well-suited to parallel processing.

2.2 Governing equations

The governing equations of the model code describe the motion of a thin layer of fluid on a rotating sphere. Using Cartesian coordinates on a polar stereographic projection, true at 60°N, they are:

$$\frac{dU}{dt} + gh_x - fV = -\frac{1}{2} S_x (U^2 + V^2) \tag{1}$$

$$\frac{dV}{dt} + gh_y + fU = -\frac{1}{2} S_y (U^2 + V^2) \tag{2}$$

$$\frac{d}{dt} (\ln h) + D = 0 \tag{3}$$

where

$$D = S (U_x + V_y), \tag{4}$$

$$\frac{d}{dt} = \frac{\partial}{\partial t} + S(U \frac{\partial}{\partial x} + V \frac{\partial}{\partial y}), \tag{5}$$

$U = u/m$, $V = v/m$ and $S = m^2$.

Here, x and y are the coordinates of the projection, u and v are the components of the wind vector along the axes of the coordinate system, g is the acceleration due to gravity, h is the height of the free surface, m is the map-scale factor, f is the Coriolis parameter and U and V are termed the wind images.

3. MULTIPROCESSING

Multi-processor machines such as the Cray XMP potentially offer:

- better throughput across all users, or
- better real-time execution for a single application, or
- both of the above but to a lesser degree.

The needs of the Atmospheric Environment Service of Environment Canada lie between the two extremes. There is a need to multi-program user R and D jobs with operational ones, while at the same time exploiting the availability of two processors in order to speed up time-sensitive operational jobs such as the regional forecast model. The balancing of the computational load of the model code between processors need only therefore be approximate, since lower-priority user jobs can be expected to mop up free time on an available processor provided sufficient main memory is available for such a job to be simultaneously memory resident with the operational one.

A fundamental question that needs to be addressed in designing programs for multi-processor execution is at what hierarchy level should the user(s) multi-task, that is partition the work-load into a set of more or less independent tasks for independent execution. Should it be at the level of:

(i) the job (different jobs executing on different processors),

(ii) the job-step (different programs of a single job executing on different processors),

(iii) the program (different subroutines of a single program executing on different processors),

(iv) the subroutine (different loops of a subroutine executing on different processors), or

(v) the loop (different parts of a loop executing on different processors)?

At one extreme (i.e. level (i) of the above hierarchy) the user relies on the operating system to optimize throughput and helps the system by minimizing resource requirements, such as memory: this extreme would perhaps be appropriate in an environment where turnaround is not an issue. At the other extreme (i.e. level (v) of the above hierarchy) a single user tries to squeeze the maximum real-time juice out of the proverbial lemon.

The Cray microtasking software employed in the present study and described in the next section, may be used at levels (iii) - (v) of the above hierarchy.

4. CRAY MICROTASKING

4.1 Overview

A good introduction to the Cray microtasking software is given in an article by Mike Booth and Kent Misegades that appeared in the Summer 1986 issue of Cray Channels, a quarterly publication of Cray Research Incorporated. It consists of a preprocessor, PREMULT, that interprets user-supplied directives embedded as comment cards in a Fortran code, and inserts appropriate microtasking code. The resulting code is then compiled by the Fortran compiler and executed in the normal way.

From a user standpoint the principal attributes of the software are:

- it is useful for codes of both small and large granularity (i.e. task size), and is targeted particularly at codes of relatively small granularity having task sizes of the order of several milliseconds;

- code changes are generally localised in the code which simplifies debugging;

- it has fairly low overhead;

- it works well in an environment where the number of available processors is unknown or varies during execution; and

- the code does not lose portability since user directives are seen as comment lines (starting with CMICS) if the preprocessor is not executed.

4.2 Directives and Code Structures

All microtasking are preceded by "CMICS", starting in column 1. The following directives and code structures are supported by the software.

(a) **CMICS GETCPUS**

* inserted once to make multiple processors available.

(b) **CMICS MICRO**

```
CMICS MICRO
      SUBROUTINE name (arguments)
    - directives interleaved with code -
      RETURN
      END
```

* tells the processor to process microtasking directives in the following subroutine.

(c)
```
CMICS DO GLOBAL
          DO 10 I = 1, LIMIT
      - code -
        10 CONTINUE
```

* tells the processor to spin off a task to an available CPU as the loop index varies.

(d)
```
CMICS PROCESS
      - code -
CMICS END PROCESS
```

* ensures that the intervening code is only executed by a single CPU.

(e)
```
CMICS PROCESS
      - code A -
CMICS ALSO PROCESS
      - code B
..................
CMICS ALSO PROCESS
      - code Z -
CMICS END PROCESS
```

* tells the preprocessor that Code A, Code B, ... are independent of one another and to spin them off as separate tasks for execution by available CPU's.

(f)
```
CMICS GUARD
      - code -
CMICS END GUARD
```

* tells preprocessor that intervening code can be executed in parallel but only by one processor at a time.

4.3 Scope of Variables

Variables are either global or local in scope. Global variables are available to all processors and are statically assigned. Examples of global variables are those that appear in COMMON blocks, DATA statements and in a subroutine's argument list. For local variables, each processor has its own which are unknown to other processors, and they are dynamically assigned to a stack. DIMENSIONed variables that do not appear in the subroutine's argument list are local variables, and they provide a convenient way of providing scratch work arrays.

4.4 Example

To illustrate the use of the Cray microtasking software, we consider a simple example, the microtasking of a subroutine that multiplies the transpose of a matrix by another matrix. The original code and two different ways of microtasking it are given in Fig. 1.

The original code (Fig. 1a) is simply a call to a library routine MXMA that post-multiplies the transpose of a matrix by another matrix. For square matrices of dimension 101, this code takes 9.69 millisecs when using a single processor of our two-processor XMP.

In a first attempt at microtasking, the single call to MXMA is first replaced (Fig. 1b) by IC calls (loop 10) to another library routine MXVA that post-multiplies the transpose of a matrix by a vector. Looping over calls to MXVA is mathematically equivalent to the single call to MXMA, and does not change the number of arithmetic operations. However, each of the calls to MXVA may now be executed in parallel, and the subroutine is in a form suitable for the insertion of microtasking directives. Note however that this alternate way of writing the subroutine takes a little longer to execute on a single-processor (11.59 millisecs compared to 9.69 millisecs for the original).

To microtask the alternate form, we first insert the directive "CMICS MICRO" just before the SUBROUTINE statement (Fig. 1c). This is a flag to tell the preprocessor to process the directives embedded in subroutine MATPR, of which there is only one, namely "CMICS DO GLOBAL". This last directive states that the code inside the 10 loop may be executed in parallel for each value of the loop index. IC tasks, each one a call to MXVA, are thus spun off to processors as they become available. On a two-processor XMP, this reduces the wall-clock time to 5.62 millisecs, which is 1.72 times faster than the original. That this is not closer to the optimum speed-up of a factor of two is due primarily to the overhead of rewriting the code to define independent tasks (Fig. 1b).

Another way of microtasking the code is shown in Fig. 1d. The strategy here is to send half the calculations (namely for the first IC/2 rows of the matrix A) to one processor, and the remainder to the other, using the PROCESS/ALSO PROCESS/END PROCESS structure. The wall-clock time is then reduced to 4.97 millisecs (410 megaflops), a speed-up of 1.95 when compared to the original; this is within 2.5% of the optimum speed-up.

The strategy of Fig. 1c is a little more flexible than that of Fig. 1d, since the code is independent of the number of available processors. It should, for example, work well without modification on a 4 or 8 processor machine. On the other hand, the code of Fig. 1d would have to be modified (by chopping up the problem into 4 or 8 processes of corresponding smaller size), and performance would be severely degraded if one of the processors were not available for any reason. The flexibility of the strategy of Fig. 1c however has to be traded off against the increased performance (under optimum circumstances) of that of Fig. 1d.

```
      SUBROUTINE MATPR (A,B,C,IA,IB,IC)
C
C  MATRIX PRODUCT, A = TRANSPOSE (B) * C
C
      DIMENSION A(IA,IC), B(IB,IA), C(IB,IC)
C
      CALL MXMA(B,IB,1,C,1,IB,A,1,IA,IA,IB,IC)
C
      RETURN
      END
```

Fig. 1a Original code; takes 9.69 millisecs on single-processor xmp for IA=IB=IC=101.

```
      SUBROUTINE MATPR (A,B,C,IA,IB,IC)
C
C  MATRIX PRODUCT, A = TRANSPOSE (B) * C
C
      DIMENSION A(IA,IC), B(IB,IA), C(IB,IC)
C
C  REWRITE AS LOOP OVER MATRIX/VECTOR PRODUCTS
C
      DO 10 I = 1, IC
      CALL MXVA(B,I,1,C,1,A,1,IA,IB)
   10 CONTINUE
C
      RETURN
      END
```

Fig. 1b Code rewritten as independent processes; takes 11.59 millisecs on single-processor XMP for IA=IB=IC=101.

```
CMICS MICRO
      SUBROUTINE MATPR (A,B,C,IA,IB,IC)
C
C  MATRIX PRODUCT, A = TRANSPOSE (B) * C
C
      DIMENSION A(IA,IC), B(IB,IA), C(IB,IC)
C
C  REWRITE AS LOOP OVER MATRIX/VECTOR PRODUCTS,
C  SEND TO PROCESSORS AS AVAILABLE
C
CMICS DO GLOBAL
      DO 10 I=1,IC
      CALL MXVA(B,I,1,C,1,A,1,IA,IB)
   10 CONTINUE
C
      RETURN
      END
```

Fig. 1c Microtasked version of the code of Fig. 1b; takes 5.62 millisecs on two-processor XMP for IA=IB=IC=101

```
CMICS MICRO
      SUBROUTINE MATPR (A,B,C,IA,IB,IC)
C
C  MATRIX PRODUCT, A=TRANSPOSE(B) * C
C
      DIMENSION A(IA,IC), B(IB,IA), C(IB,IC)
C
C  REWRITE AS 2 TASKS OF HALF THE SIZE
C
CMICS PROCESS
      CALL MXMA (B,IB,1,C,1,IB,A,1,IA,IA,IB,IC/2)
C
CMICS ALSO PROCESS
      CALL MXMA (B,IB,1,C(1,IC/2+1),1,IB,A(1,IC/2+1),1,IA,IA,IB,IC-IC-/2)
CMICS END PROCESS
C
      RETURN
      END
```

Fig. 1d Microtrasked version using PROCESS/ALSO PROCESS/END PROCESS structure; code takes 4.97 millisecs on two processor XMP.

5. RESULTS AND CONCLUSIONS

For the shallow-water code of Section 2, 82% of the total single-processor time is expended in just 3 subroutines, each one of which was easily microtasked. Speeding-up each of these subroutines by a factor of 1.95 (as in the matrix multiply example of the previous section) would theoretically result in an overall speed-up of 100/60 = 1.7. This is because 82 time units per 100 units of the single-processor time would be reduced to 82/1.95 = 42 units when using two processors, the remaining 18 units would remain unchanged, and the two-processor time would thus be expected to be 42 + 18 = 60 units of time (compared to the 100 units using a single-processor). Experimentally on a dedicated machine we obtained a speedup within a couple of percent of that expected. These preliminary results indicate that the microtasking overhead is indeed small for a microtasked code where the task sizes vary between 5 and 42 millisecs.

In conclusion, we found the Cray microtasking software quite easy to use and were able to gain an appreciable speed-up in real-time performance for a relatively modeset effort. Most of the remaining code should be amenable to microtasking, although proportionally more programming effort will be required because the time spent in non-microtasked subroutines is spread fairly uniformly over a dozen or so routines. Because of the need in the AES environment to multi-program user R and D jobs with operational ones, it may only be necessary to microtask a relatively small number of the most time-consuming subroutines in the operational regional forecast model to ensure that operational deadlines are met.

6. ACKNOWLEDGEMENT

The authors gratefully acknowledge the expert typing of Christine Yim.

Application of Multi-Processors to Operational Analysis and Forecasts at the French Weather Service

J. CLOCHARD

Direction de la Météorologie Nationale, Paris, France

1. INTRODUCTION

The aim of the French Weather Service is, for a major part, to make and deliver products about the current meteorological situation and its expected evolution up to a few days, with an update twice daily. These features have always required a lot of computing facilities, and especially a high-level computing power. For these reasons, the French operational weather models have been running on a Cray-1S since January 1985. This machine is a vector and remote computer, which is shared with other French scientists in the framework of an economic group (G.C.C.V.R.). However, this computer has shown its limits (!) and will be replaced in the first quarter of 1987 by a Cray-2, which is also a vector computer, quicker in this mode, including a much larger core memory (256 MW instead of 1) and 4 processors, with the possibility of running multitasked jobs. As this feature is especially well-suited to run time-critical and big codes, we are particularly concerned with use of multi-tasking for our operational suite.

Use of the Cray-2 for French NWP models will consist of 3 phases:

1) Make the current suite run and be operational on the Cray-2; this phase must be over no later than March '87, when the Cray-1S goes away.

2) Use the current codes with minimum changes (none in the logical structure) to take appreciable advantages from the machine without waiting for new adapted codes to be developed; only the spatial resolution and/or domain sizes will be concerned. Use of multitasking seems the best way to achieve that.

3) Develop and use new codes involving "new" numerical techniques, adapted to the large core memory (such as semi-lagrangian technique, that has a higher numerical stability than semi-implicit schemes).

This presentation will emphasise the second phase, which is our first direct contact with multi-tasking.

2. DESCRIPTION OF THE OPERATIONAL SUITE ON THE CRAY-1S AND PLANNED SUITE FOR THE SECOND PHASE

2.1 Current Operational Suite on Cray-1S

It mainly consists of two systems:

- a large-scale one, on a hemispheric domain, called Emeraude;

- a meso-scale one, on a small domain centered over France, called Peridot.

Each of these systems is split in two parts:

- an analysis model which gives a snapshot of the meteorological situation at a given moment, on a three dimensional regular grid;

- a forecast model which computes the future state of the atmosphere at different ranges starting from the analysis result.

The observations data sets of 00Z and 12Z are used as soon as possible.

The suite may be described as 3 sets of jobs, nearly sequential:

- the 00Z chain: it consists of an Emeraude analysis, followed by an Emeraude forecast up to 96 hours (split in 2 jobs to let Peridot run) and of a Peridot analysis, followed by a 36 hours range forecast.

- the 12Z chain: consists of an Emeraude analysis, followed by a 48 hours range Emeraude forecast, and of a Peridot analysis.

- the data assimilation: when the observation data sets for a given time are complete (or nearly), cycles of analysis followed by a forecast are made, with a 6-hour cycling for Emeraude, starting from 00Z the previous day, and with a 12-hour cycling for Peridot.

Details of jobs flow are given in figure 1.

The Emeraude analysis is made on an almost hemispheric, regular latitude-longitude grid, with a 1.5° resolution in latitude and 2° in longitude (63 latitudes, 180 longitudes), on 16 pressure levels (standard levels + 950, 900, 800, 600, 70, 50 hPa). Height, wind, temperature, humidity and some ground-level fields are analysed. It takes about 8 to 9 minutes to run, depending on the number of observations available.

The Emeraude forecast model is a spectral hemispheric model based on primitive equations. The horizontal fields are described on a truncated series of functions (spherical harmonics); the truncation is of triangular type at wave-number 79. The vertical discretisation is made with finite differences with 15 hybrid levels. The time-stepping is of semi-implicit leap-frog type, with a 24 minutes time step. The prognostic variables are temperature, wind components, specific humidity and ground-level pressure. Subgrid-scale processes are treated with an extensive physical package. A 6-hour range forecast takes about 4.5 minutes to run.

The Peridot analysis is made on a polar stereographic grid which resolution is about 35 km (51 x 51 grid-points), with 15 sigma levels. Wind, temperature, humidity and some ground-level fields are analysed. It takes about 5 minutes to run, all interfaces included.

The Peridot forecast model is of grid-point type, using finite differences on both horizontal and vertical, using the same grid as the Peridot analysis. The time-stepping is of semi-implicit leap-frog type, with a 4 minutes time step. Prognostic variables are the same as for Emeraude, the physical package being nearly the same. Lateral boundary conditions are provided by the Emeraude forecast. A 6-hour range forecast takes about 4 minutes to run.

2.2 Planned suite on the Cray-2 for second phase

As written before, no changes will be made in the general structure; there will be the same 2 systems (large-scale and meso-scale), but intended modifications are:

* Global data assimilation for Emeraude, that implies a global forecast to be made up to 24 hours range from 00Z for safety reasons (to avoid climatological starts in the analysis in case of failure). The analysis grid will be at the same resolution but extended to the whole sphere (120 latitudes, 180 longitudes). The forecast model will then be run in its global version, with a T95 truncation, maybe with extra levels.

* Extension of the Peridot domain, mainly towards the Atlantic: a 95 x 95 grid may be expected, maybe with extra levels too. The larger horizontal domain should lead to less problems with lateral boundary conditions over the area which is of direct interest for us.

* Range of 00Z Peridot forecast increased up to 48 hours.

* 12Z Peridot forecast, 36 hours range.

All these features, especially horizontal and vertical resolution, will be tuned depending on performance of the adapted codes on the Cray-2. However, we will keep enough computing time for our own development and research tests.

3. GENERAL PHILOSOPHY FOR MULTI-TASKING IMPLEMENTATION

Use of multi-tasking in our codes should have a certain number of properties:

- efficiency: for a given number of processors available, a good ratio CP time/elapsed time is expected; overhead coming from multi-tasking codes running in batch mode for development tests should have minimum dependencies between simultaneous tasks.

- flexibility: the codes must be able to run with any number of processors available, without much overhead; this number should be known only at the run time. This feature is very important in the G.C.C.V.R. context, because some programs will run on a different number of processors, both in operational mode (4 or 2 expected), and in batch mode (2 or 1 for tests).

- numerical results should be reproduced from one run to another run, even if different numbers of CPUs are being used: because of rounding, many arithmetic operations are not associative on a computer (for instance, sum of floating-point values). But, even if it should lead to few numerical differences, results should not depend on the context of the run, mostly for testing and debugging.

- possibility of running the multi-tasked codes both on Cray-2 and on Cray-XMP48 of ECMWF: this is also important to provide some kind of compatibility for development tests on the XMP (we have some computer resources on this machine).

- on the other hand, the work load to reach these goals must not be too big with respect to human time spent on adapting the codes. So only the codes which "are worth of" will be multi-tasked: only those spending a lot of elapsed time and with enough internal parallelism.

Some general guidelines may be given to satisfy the previous requirements.

First, efficiency and flexibility lead us towards dynamic partitioning of work rather than static; it implies that the internal order of work done by the different tasks will not be deterministic.

As the most evident parallelism in codes is often at outer loops level (inner loops are vectorised or should be), one example of dynamic partitioning may be as follows:

Let us suppose that we have a DO loop within a Fortran program which requires a big part of CP time, with independent iterations as regards numerical results, without I/O, and that NBCPU processors are available.

Original code:

```
      DO 1000 K=1,KMAX
      CALL DOK (K)           [work for item K]
1000  CONTINUE
```

"straight-forward" multi-tasked code (for simplification, KMAX is supposed to be a multiple of NBCPU):

```
      DO 1000 K=1,KMAX,NBCPU
C
      DO 100 ITASK=1,NBCPU
      start task executing DOK for item (K+ITASK-1)
 100  CONTINUE
      DO 200 ITASK=1,NBCPU
      wait for end of task executing DOK for item (K+ITASK-1)
 200  CONTINUE
C
1000  CONTINUE
```

This last way of coding may lead to non-optimum use of the processors, especially if work load for a given iteration depends upon the loop index, and also if the NBCPU processors are not always available (batch mode). There are as many tasks as loop iterations, leading to some overhead due to the "task starting" management.

Dynamic partitioning of this loop:

```
      COMMON /PARALL/ K, LOCKK
      ...
      define LOCKK as a lock to protect use of K
      K=1
C
      DO 100 ITASK=1,NBCPU
      start task executing DOK, logical number ITASK
 100  CONTINUE
      DO 200 ITASK=1,NBCPU
      wait for end of task executing DOK, logical number ITASK
 200  CONTINUE
C
      release lock LOCKK
```

the code of subroutine DOK being modified as follows:

```
      SUBROUTINE DOK
      COMMON /PARALL/ K,LOCKK
      ...
  10  CONTINUE [first executable statement]
      set lock LOCKK (and if already set, wait for it to be cleared)
      KLOC=K
      K=K+1
      clear lock LOCKK
      IF(KLOC.GT.KMAX) RETURN
C
      work for item KLOC
C
      GOTO 10 [ready to process another item]
      END
```

This leads in each task to a dynamic search for next iteration to be made, if any; a task ready to process another item will not wait for other tasks to finish their work (if the time spent for an iteration is large enough, the risk of waiting for lock to be cleared is very low). There is a dynamic balance of work; and you only use NBCPU tasks to compute KMAX iterations, so the overhead from task starting is much slighter than in the previous case.

However, if the overhead coming form the lock-protected zone of code is too big, you can compute several following iterations instead of one before updating the "loop index", with increments of K larger than one, as described by John Larson, 1984.

Another aspect concerns the ability to reproduce computations. In some cases, it involves that some order-dependent computations be removed from the "multi-taskable" zone of code, and put afterwards (for instance, some statistics). In other cases, you cannot change the place of these order-dependent computations as easily because of huge additional storage implied. An example will be given in the next paragraph, and a solution will be proposed.

Flexibility may be achieved by the definition of a minimum number of arrays with the highest number of processors available, and by taking the number of processors to use at run time in the program environment (reading it from the control card loading the program or from a directive file).

The last general guidelines are about I/O. First, with dynamical strategies of multi-tasking you do not know exactly in which order the operations will be done. So if file manipulation is involved, you will have to change the sequential files into random access files, if possible. Then if you want your codes using working files (or using files for reading and/or writing almost twice) to run faster on the Cray-2 but with the possibility of running them on an XMP, you will have to provide interfaces using rather memory as big "buffers" than files on the Cray-2, and using only files on XMP.

4. DETAILS OF CODES AND SEARCH FOR INTERNAL PARALLELISM

4.1 Large-scale Emeraude analysis

This analysis is split in 6 programs, 4 of which being overlapped with observations files transfer from front-end computer. A description of them, mean CP time and elapsed time on the Cray-1S in seconds are given below.

		CP / real
A1:	setting up of the guess-field for the analysis	5 / 30
A2:	humidity field analysis	35 / 40
A3:	height and wind field analysis	240 /310
A3B:	statistics on guess-field errors	2 / 4
A4:	temperature analysis	34 / 42
A5:	makes final results	4 / 9

A1 must run before all other programs; A5 must run after all previous ones. A2B must run after A3, A4 must run after A2 and A3. So A2 and A3 could run in parallel, but we would lose the benefit of a big file transfer overlap; no merge of programs will be made then.

Looking at times, A3 is worth of being multi-tasked; its internal structure is:

- reading observations file (this file is and will stay sequential because it comes from front-end)
- compute deviations from guess-field
- check these deviations
- evaluate statistics on deviations from guess-field
- merge some observations into "super-observations"
- search for observations in the neighbourhood of each grid-point
- compute matrix elements
- solve linear systems

The first 5 steps take about 25 seconds of CP and 50 wall-clock seconds: they are order-dependent and no parallelism may be seen in them. The 3 following steps use about 210 seconds of CP and 260 wall-clock seconds, and these 3 steps show a good parallelism, because they consist of a single outer loop over latitudes. Dynamic partitioning as shown in the previous paragraph may be achieved.

Two other programs, A2 and A4 are worth of being multi-tasked and show a common structure:

- reading observation file
- checking of deviations from guess-field
- analysis itself (humidity or temperature)

The same kind of remarks as for A3 can be made for these 2 programs: the most part of time is taken by the analysis step itself, that can be easily dynamically multi-tasked over latitudes.

The 3 other programs have too many order-dependent steps for convenient multi-tasking.

Estimated proportion of "multi-taskable" CP time, and task granularity estimated with the hypothesis of a single processor of the Cray-2 twice as fast as the Cray-1S CPU, 2 physical CPUs being available:

- A2: 85% - 15 seconds
- A3: 85% - 105 seconds
- A4: 85% - 15 seconds

4.2 Large-scale Emeraude Model

It consists of 6 programs:	CP / real on Cray-1S
Y1: preprocessing (makes data coming from the analysis suitable for the forecast model)	25 / 50

Y2:	initialization (eliminating nasty gravity waves)	85 /140
Y3:	forecast model itself (6 hours)	190 /205
Y4:	post-processing (makes results used later on Cray)	13 / 19
Y5:	zonal diagnostics	1 / 1
Y6:	post-processing for front-end	10 / 12

Let us suppose we have a program running on a Cray-1 that we want to be multi-tasked with all requirements described in para. 3. This programs contains a loop spending a major part of time, with no sequential I/O, this loop being split in 3 steps:

1°) independent computations over loop iterations, storing results in arrays
2°) order-dependent computations over loop iterations, using array results from the previous step
3°) independent computations over loop iterations, not using the same results from 1° than 2° - this last step may be empty

This loop may be multi-tasked with numerical results that can be reproduced and with a good efficiency as follows.

```
     Split the original code into:
     ...
     DO 1000 K=1,KMAX
     step 1 --- results to be used by 2 in array R
     step 2 --- using array R
     [step 3, if any - not using R]
1000 CONTINUE
     ...

     Multi-task it as follows (NBCPU is the number of available
     processors):
     ...
     COMMON /PARALL/ K, LOCKK
     ...
     define LOCKK as a lock, and a pool of memory to be managed
     ...
     K=1
     start tasks 1,2,...NBCPU (subroutine STEP1A3)
C
     DO 1000 KPOST=1,KMAX
     wait for KPOST to be "preprocessed" (step 1 to be done for this
     item)
     compute step 2 for KPOST
     disallocate memory allocated for steps 1 and 2 of this item
1000 CONTINUE
C
     wait for end of tasks 1,2,...NBCPU
     ...
     release lock LOCKK and the pool of managed memory
     ...
     SUBROUTINE STEP1A3
     COMMON /PARALL/ K,LOCKK
     ...
```

```
10 CONTINUE (first executable statement)
   get next item KLOC to process and increment the "loop index" K
   (zone of code protected by LOCKK)
   if all done, RETURN, otherwise for KLOC:

   begin step 1
   allocate memory (contiguous zone) in managed pool for R
   end step 1 (allocated memory content definition, called
   "preprocessing")

   "set the allocated memory free": post an event to say item KLOC
   is ready for step 2
   [compute step 3, if any, but without reference to memory set
   free]

   GOTO 10 (go back for a next item)
   END
```

In this scheme, the order-dependent computations are being made by the main task that has started the tasks computing the loop iterations in parallel, and which afterwards would be asleep until the end of these tasks. The results computed in the first step are "sent back" to the main task as a piece of managed memory.

Advantages:
- numerical results may be reproduced
- efficiency gained from potential "continuous" use of available CPUs if the pool of managed memory is big enough
- dynamic load balance
- possibility of running on XMP without additional I/O with a reduced amount of managed memory

Possible limitations are overheads due to:
- management of memory (including synchronisation of the main task), that must use locks and events (however, the managed memory may be statically allocated in a common block to avoid extra memory expanding/compressing);
- multi-tasking scheduler, because of the management of one more active task than available CPUs. But for the Emeraude model, this additional task should not be active so often if only 2 CPUs were used;
- a large number of available CPUs, so that the order-dependent computations will not be made as quickly as the production of the other tasks. However, 2 answers may be given in this case: start one less task with dynamical partitioning, and/or try to split work of order-dependent computations in 2 tasks.

Estimation of "multi-taskable" CP time and task granularity with the same hypothesis as those made for the Emeraude analysis:

- about 98% multi-taskable
- first loop over latitudes: 82 seconds for a task computing order-independent steps, while the main task will compute about 33 seconds for the order-dependent step.
- computations on spectral coefficients: 1.8 seconds
- second loop over latitudes: 43 seconds

4.3 Meso-scale Peridot Analysis

This analysis is made of a single program surrounded with interfaces programs, shared with the forecast model; these makes essentially I/O and may not be multi-tasked. It takes about 110 CP seconds and 140 wall-clock seconds to run. Its internal steps may be split into:

- 10 cp seconds of control, deviation from guess-field, statistics computations
- 90 cp seconds of "pure" analysis of temperature and wind fields
- 10 cp seconds of "pure" analysis of humidity

The 2 pure analysis steps may be dynamically partitioned as the Emeraude analysis. But here, in this meso-scale analysis, the last step may here be multi-tasked over the 2 types of observations being used: classical ones and satellite radiances, coming from front-end on 2 different files.

Estimations as for the Emeraude analysis, but with 4 physical CPUs available:
- 91% multi-taskable with any number of CPUs, 9% with only 2 CPUs
- temperature and wind analysis: 40 seconds per task
- humidity analysis: 4 seconds per task but 2 only

4.4 Meso-scale Peridot Forecast Model

The same kind of preliminary remarks as for Emeraude can be made, so we will concentrate on the internal structure of one time step iteration of the model itself. All internal steps consist of outer loops over horizontal or vertial plans:

- explicit dynamics: 1 over horizontal plans + 1 over vertical plans
- physical parametrizations: over vertical plans
- horizontal diffusion: over horizontal plans
- explicit time integration: over horizontal plans (including ground level)
- semi-implicit computations: 1 over vertical + 1 over horizontal plans
- final evolution: over vertical plans
- lateral boundary conditions: application of Davies's method, looping over horizontal plans (including ground level)

For each of these steps, dynamical partitioning may be applied over external loops, but here we will not be far from the estimated minimum task granularity necessary to gain real-time from multi-tasking on the Cray-2.

Estimations with the same hypothesis as those made for the Peridot analysis:
- 98% multi-taskable

Granularity per task in CP seconds, in the order of external loops given above:

- 0.06	- 0.06	- 0.06
- 0.11	- 0.05	- 0.07
- 0.49	- 0.04	- 0.04

These estimated granularities show that task starting use may have to be restricted in this model. This could be achieved by starting tasks only before (or at the beginning of) the loop over time steps, instead of at each time step, and by using events to synchronize sub-tasks (the different loops described above, each loop being dynamically partitioned).

5. PRELIMINARY RESULTS

Up to now, the maximum effort has been made for the benchmark and the first phase of our use of the Cray-2. So only some very preliminary tests have been made on the XMP48 of E.C.M.W.F., in batch mode. A summary is given below:

The management scheme proposed in para. 4.2 has been written and has run correctly in test cases, but no implementation in the Emeraude model has been done yet.

A program from the Emeraude analysis (the humidity one) has been multi-tasked with dynamic partitioning over latitudes and has run giving the same results as the rearranged original code, with exactly the same results using 1, 2 or 4 processors. The only problems encountered came from arguments given to subroutines through common blocks. The timings in terms of CP were:

- about 30 seconds for original rearranged code
- no appreciable overhead form multi-tasking use
- nearly 25 seconds multi-tasked, with no differences between tasks more than 0.2 seconds when using the 4 CPUs.

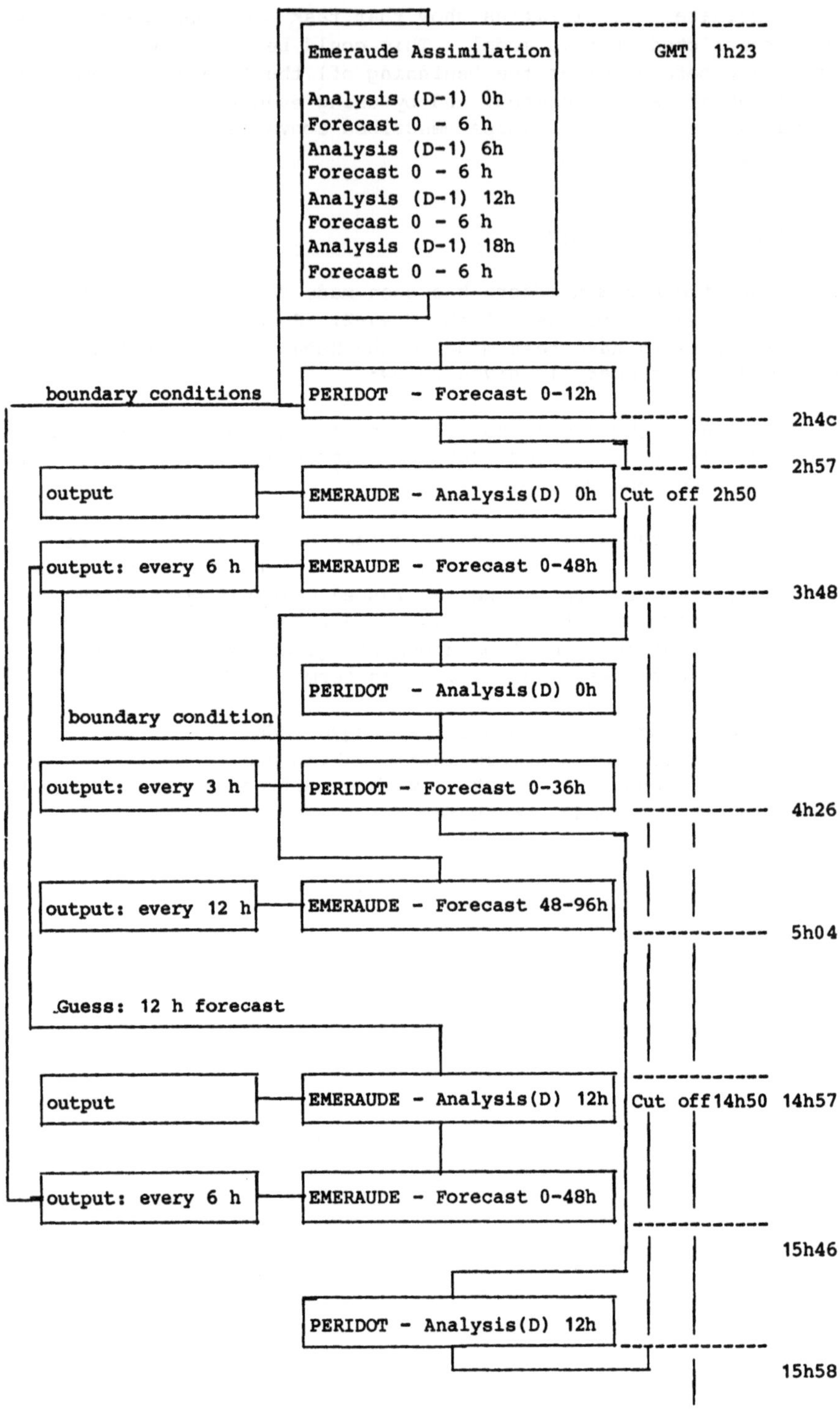

Figure 1 Operational Suite on Cray-1S

Fig. 2 Internal Details of Time Step for Emeraude Forecast Model

Loop iteration initialisations

Reset spectral coefficients

First loop over latitudes:

Time filtering (second step, for t- t variables)

Loop over hemispheres:
Computations on Fourier coefficients (longitude derivatives)
Reverse Fourier transform -- grid-point variables
Explicit dynamics
Physical parametrisations
Direct Fourier transform -- Fourier coefficients
Computations on Fourier coefficients(semi-implicit corrections)

Time filtering (first step, for variables at t)

Direct Legendre transform: from Fourier to spectral coefficients

Computations on spectral coefficients: semi-implicit resolution, horizontal diffusion

Second loop over latitudes:

Reverse Legendre transform: from spectral to Fourier coefficients

REFERENCES

Cray Research Multitasking User Guide, 1985, CRAY Computer Systems Technical Note no. SN-0222, Cray Research Inc., Mendota Heights, Minnesota.

Larson J., 1984: Practical Concerns in Multitasking on the CRAY-XMP, ECMWF Workshop on Using Multiprocessors in Meteorological Models.

Parallel Integration of the Norwegian Prediction Model

SIBGJØRN GRØNAAS

The Norwegian Meteorological Institute, P.O. Box 320, Blinder, 0314 Oslo 3, Norway

1. INTRODUCTION

In May 1985 a mesoscale model (LAM50) was made operational at the Norwegian Meteorological Institute. The horizontal mesh is 50 km and 24 hour integrations are made three times a day. Lateral boundary values are applied by prediction from a more coarse mesh routine called LAM150. The prediction model used is essentially the same in both systems, and described by Grønås and Hellevik (1982) and in a catalog of existing modules of limited area NWP systems, WMO (1984). The assimilation systems are described by Grønås and Midtbø (1986), and the physical parameterization, which includes radiation, by Nordeng (1986). The integration areas are shown in Figure 1.

Our experiences with the mesoscale model are very promising. For the first time the Scandinavian mountains are reasonably well resolved in an operational numerical model. The prediction of precipitation and wind in the boundary layer are significantly better than in LAM150. However, due to lack of computer capacity the integration area coves 3000 grid-points only, and meteorological disturbances are often advected from the boundary and into the middle of the area in less than 24 hours. A substantial increase of the integration area is desirable, but this demands a similar increase in computation speed.

The present routine is run on a FPS-164 attached processor, Charlesworth and Gustafson (1981). Fortran programming is used and the speed of computation is about 4 MFLOPS. Acquisition of a supercomputer is probably too expensive for our institution. However, the new technology with parallel processors seem to be an interesting, and less expensive alternative which have to be considered. A parallel vector supercomputer (Hawkinson, 1986) is now manufactured as the FPS T Series. This system consists of processor nodes which each incorporate vector-oriented floating point arithmetic (64-bit) and a local memory design. Peak Vector performance of one node is 12 MFLOPS and there is fast communication links between the nodes. The FPS T Series is built from modules containing eight nodes. In this paper the potential of the system is investigated in relation to our prediction model. The time integration method is explicit and very suitable for a parallel processing where each processor computes on sub-areas of the total integration area.

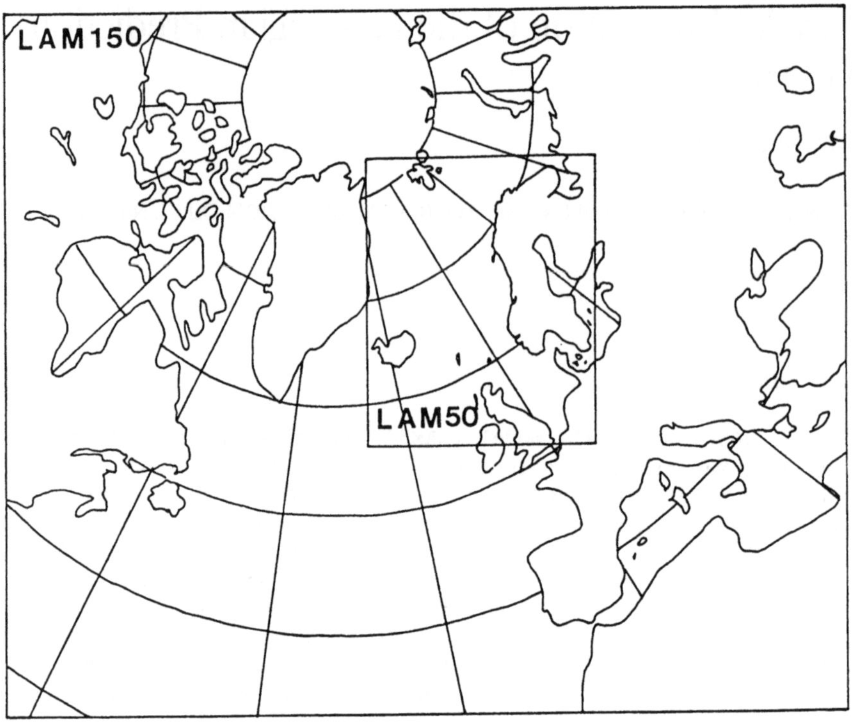

Fig. 1 The operational integration areas.

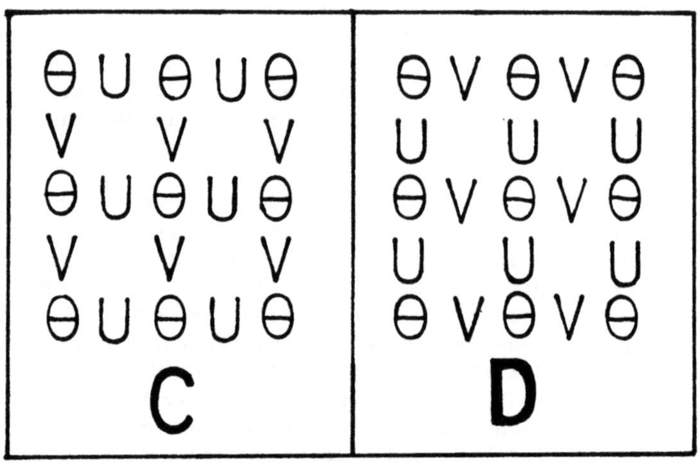

Fig. 2 The C and D grid.

2. TIME INTEGRATION IN PARALLEL

2.1 The time integration method

The limited area model is integrated on a stereographic map. There are four dependent variables: The wind components u and v, potential temperature θ, and specific humidity q. In addition the surface pressure p_s is a dependent variable. Horizontally the variables are staggered in space according to the Arakawa D-grid, (see Fig. 2).

The basis for the time integration method is the time-staggered leapfrog scheme. The spatial derivatives of a variable are then computed at the centres of the squares defined by the points carrying the variable. To make finite difference approximations with a grid length d we define:

$$\delta_x \alpha = (\alpha(x+d/2) - \alpha(x-d/2))/d$$
$$\overline{\alpha}^x = (\alpha(x+d/2) + \alpha(x-d/2))/2 \qquad (1)$$

Second-order centred differences are used and the derivatives are approximated as follows:

$$\delta_x \overline{\alpha}^y \qquad \delta_y \overline{\alpha}^x \qquad (2)$$

The accuracy of this scheme could be compared to the accuracy using the Arakawa C-grid (see figure 2) where the derivatives are approximated by:

$$\delta_x \overline{\alpha}^x \qquad \delta_y \overline{a}^y \qquad (3)$$

The two methods of approximation give the same level of accuracy if the grid intervals are in proportion $\sqrt{2}$: 1. In this respect a distance of d = 50 km in our scheme corresponds to d = 35 km in the C-grid.

The computation of time derivatives can be written as follows:

$$\partial \alpha_{i+1/2,j+1/2}/\partial t = F \qquad (4)$$

where F is the finite difference approximation of the right hand side of the equation involved.

The next step in the time integration is according to Bratseth (1983). The time derivatives are interpolated to the positions where the corresponding variables are defined by the following formula:

$$\langle \partial \alpha_{i,j}/\partial t \rangle = (1+2a)/4 \; \partial \overline{\alpha}^{xy}/\partial t \qquad (5)$$
$$- a/2(\partial \overline{\alpha}^{xy}_{i+1,j}/\partial t + \partial \overline{\alpha}^{xy}_{i-1,j}/\partial t + \partial \overline{\alpha}^{xy}_{i,j+1}/\partial t + \partial \overline{\alpha}^{xy}_{i,j-1}/\partial t)$$

The interpolation is correct to fourth order in d for a=1/8. By this operation the time step can be almost doubled, and according to Bratseth (1983) the accuracy is virtually unchanged.

In the third step an intermediate value are computed at time step n+1 with the leapfrog scheme:

$$\alpha_*^{n+1} = \alpha^{n-1} + 2\Delta t\langle\partial\alpha_{i,j}/\partial t\rangle \qquad (6)$$

At last a pressure averaging technique is used in combination with a time filter according to Brown and Campana (1978):

$$\alpha^{n} = (1-c)\alpha_*^{n} + c/2(\alpha_*^{n+1} + \alpha^{n-1}) \qquad (7)$$

Also this operation allows for almost a double time step. In this way the time step is nearly four times longer than in the original time-staggered scheme. In our operational routine LAM50 the time step is 2 ½ minutes. A similar mesoscale routine is operating at the French Weather Service (Juvanon Du Vachat et al., 1986). They use an implicit integration method on the C-grid. The time step is 4 minutes with a grid distance of 35 km. However, their scheme is more complicated and involve solutions of elliptic equations which might be more difficult to solve in parallel.

Our integration method can easily be applied on sub-areas of the total integration area. Information from the neighbouring areas are then needed for the operations contained in (4) and (5).

2.2 One-dimensional and two-dimensional slicing

The partitioning of the integration area could be made in several ways. We are only considering rectangular sub-areas and assume that one processor (node) performs the integration for such an area.

When the integration area is divided into rectangular slices across the total area, we have one-dimensional slices, Fig. 3. Each slice will in this case be connected to the two neighbouring slices only. There will be data transfer in both directions between two nodes, and we assume that two channels are needed to do this simultaneously. We further assume that data transfer between both neighbouring nodes also can be done simultaneously. Altogether four channels are then needed (see Fig. 3). The transfer time will be proportional to the number of grid-points across the slice.

When the slices are done in two dimensions, we get rectangular sub-areas with m*n points for computation (see Fig. 3). Such a sub-area will have four main neighbour nodes and four corner nodes and there have to be data transfer to all of them. Altogether either channels are used for transfer between the main neighbours. The amount of data transfer between corner nodes will be much smaller than between main nodes, and we assume that the needed

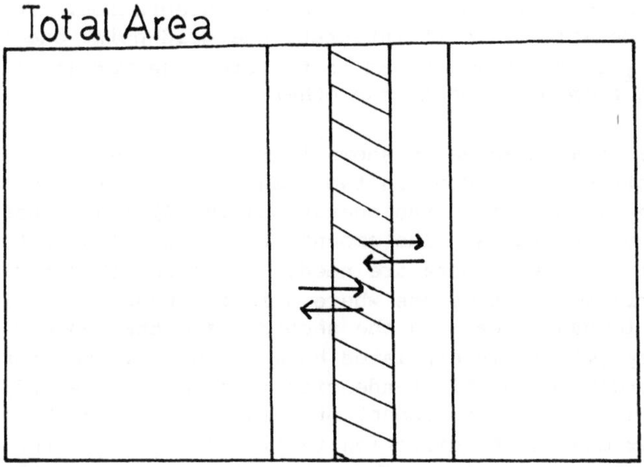

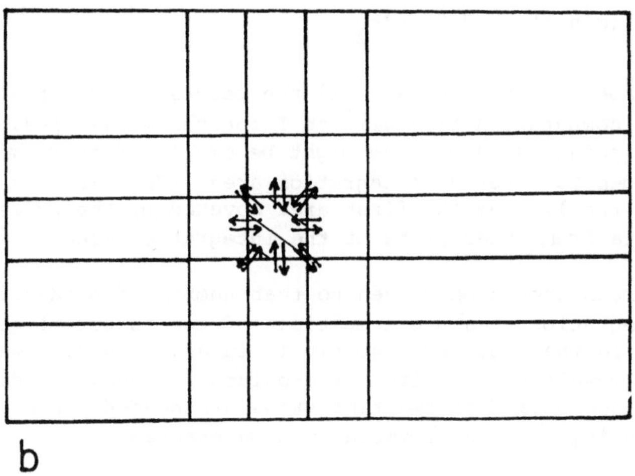

Fig. 3 a) One-dimensional slicing. b) Two-dimensional slicing. The arrows show direction of transfer of boundary values between one slice and the neighbouring slices.

sequential transfer in both directions is less than maximum transfer in one direction between main nodes. In this case another four channels are needed. Again, when the data transfer between nodes takes place simultaneously, the transfer times will be proportional to the number of grid-points along the longest side of the sub-area.

2.3 Integration of a sub-area

The number of variables contained vertically in a grid-point, or group of gridpoints since the grid is staggered, are 4*k + 1 where k is the number of layers. For short we call these variables the variables in a grid-point. We now introduce two times needed to evaluate the efficiency of the parallel integration. The time t_1 is the computation time to integrate one grid-point one time-step. The time t_2 is the time needed to transfer the variables at one grid-point from one node to the other.

An example of a sub-area is shown in Fig. 4. The computation within a time-step is done in two stages. First the boundary values needed to perform the operations in (4) are transferred from the neighbouring nodes. The amount of data is illustrated in Fig. 5. Two time steps of data are needed and altogether maximum 2*.75 of the variables at n points where n is the number of grid-points along the longest side. In the second stage the tendencies computed by (4) are interpolated back to the original positions by (5). In order to do this tendencies must be transferred from neighbouring nodes. The amount of this transfer is shown in Fig. 6 and the maximum in one direction is 1.75*n times the information contained in a grid-point. The total time T to integrate one time step is as follows:

$$T = (m*n)*t_1 + 3.25*n*t_2 \qquad (8)$$

If the number of nodes covers all the points in the integration area, the computation time will be T for the whole area. The number of points at each node might be small and there might not be enough nodes to cover an integration area. In this case we have to store the result from the first area covered by the nodes and read in new data from other parts of the integration area.

The sub-areas should be chosen so that number of grid-points in the total integration area are a multiple of the number of grid-points contained in the nodes available. In such a situation we have to store the result of the first grid-points and read new data to the nodes. Since data from two time steps are needed, this time will be $2.*(m*n)*t_1$, and the formula must be changed to:

$$T = (m*n)*t_1 + 3.25*n*t_2 + 2.*(m*n)*t_2 \qquad (9)$$

For a n by n square the additional term is squared in n and introduces a serious decrease in efficiency. If this transfer could be done asynchronously while the computation takes place, the additional term would vanish. There is nothing in the integration method to prevent this, so such a possibility will be welcomed. If asynchronous transfer is possible, there might be problems with buffer sizes. On the other hand, there is then less to lose by decreasing the number of points in the sub-areas.

3. TIME INTEGRATION ON FPS T SERIES

3.1 FPS T Series specifications

The formulas (8) and (9) are now applied on the FPS T Series. A model with 12 vertical layers is investigated. The number of variables contained vertically in a grid-point is 4 * 12 + 1 = 49.

The computation time t_1 is estimated from actual computation on the FPS-164. The total area computer time is here set to 6.0 seconds a time step. There are (61-1)*(49-1) = 2880 internal grid-points. The computation time will not vary much from grid point to grid-point, and we let

$$t_1 = 6./2880 = 2.1 \text{ milliseconds}$$

This computer speed is approximately 4 Megaflops. The computer speed of the FPS T-Series node might be faster than at FPS-164, but in the following this has not been taken into consideration. The time t_1 might of course vary with the size of the area according to length of the loops etc., however, it is believed that such variations are small for actual areas.

The transfer speed at channels from node to node, t_t, is measured by Mbytes^{-1}. For FPS T-Series $c_t = 0.69$ when a small set-up time is included. If 8 bytes in each word is assumed, the transfer time for the variables in a grid-point is:

$$t_2 = (49*8)/c_t = .60 \text{ milliseconds}$$

A node in the FPS T-Series contains local memory of 1 Mbytes only. The code for the model is equal to or less than 0.1 Mbytes. We assume that .6 Mbytes are available for data storage. The model needs storage for approximately 5 sets of fields for an "all in core version". We can then store 600000/(50*5*8) = 300 grid-points in the memory of a node (8 bytes a variable).

3.2 Results

The computation times from the two formulas for small squares are found in Table 1. It is seen that parallel processing is in our case very efficient when no startup time is involved. For n^2 = 256, the transfer time is only 6% of the pure computation time. When startup times have to be included, the efficiency is significantly less.

An actual area for computation is planned to have at least four times as many horizontal points as today and the desired total integration time should be reduced to one third of the present integration time. Such a model needs at least 12 times more computation speed that we have today. In other words, the desired speedup is at least 12. We now investigate computation on one such area with the following size:

$$(16*8)*(16*6) = 128*96 = 12288 \text{ grid-points}$$

For curiosity, the computer time and speedup is computed in case a large number of nodes are available. The results are given in Table 2. It is indicated that a large number of FPS nodes gives

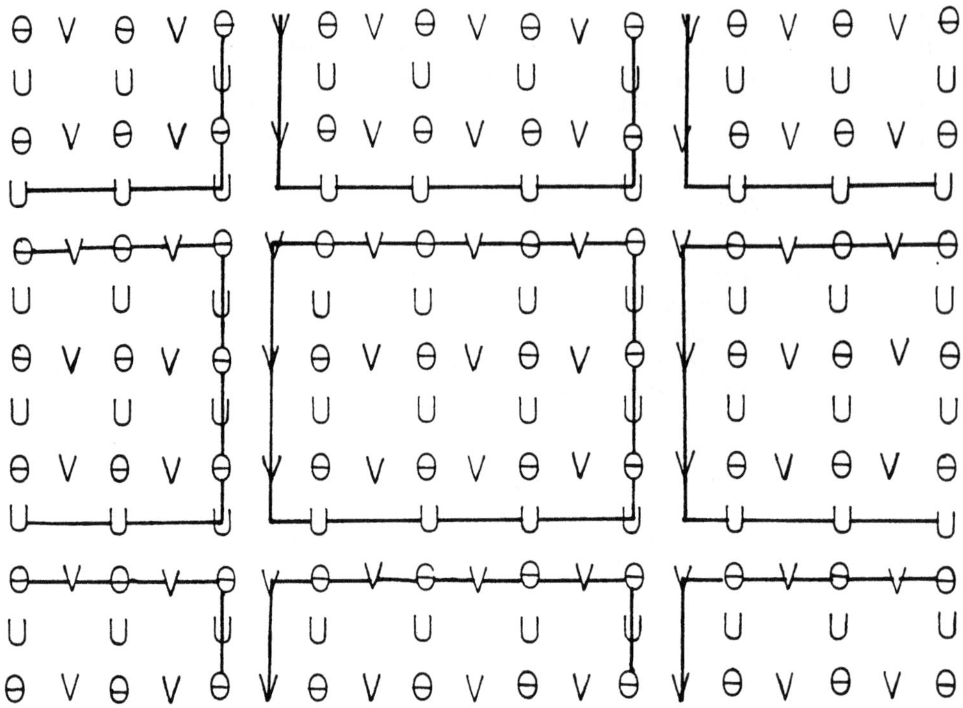

Fig. 4 Example of partitioning of the integration area, (two-dimensional slicing).

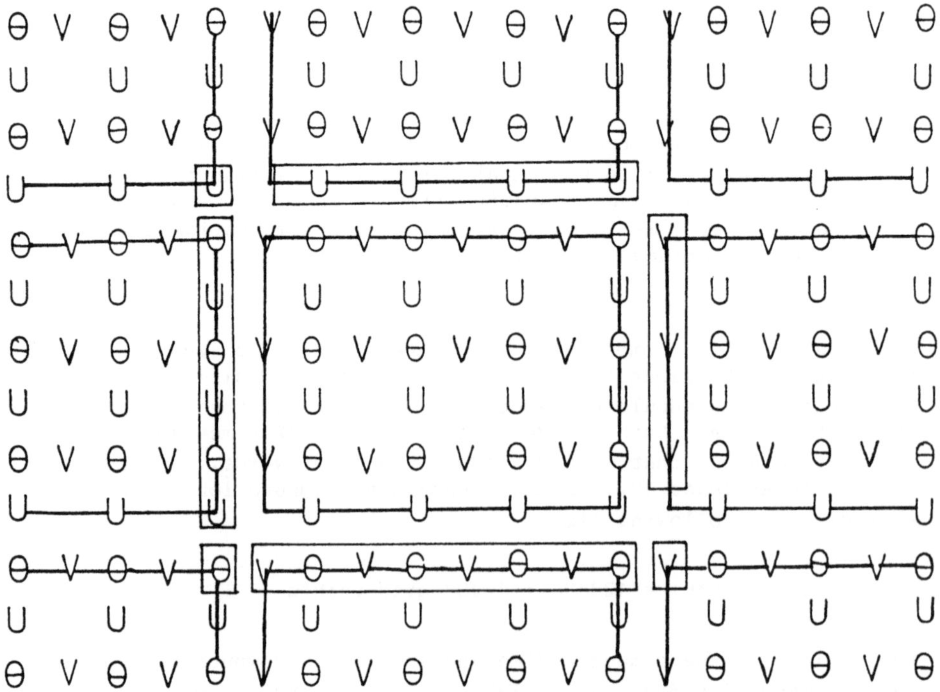

Fig. 5 Boundary values needed in the first stage.

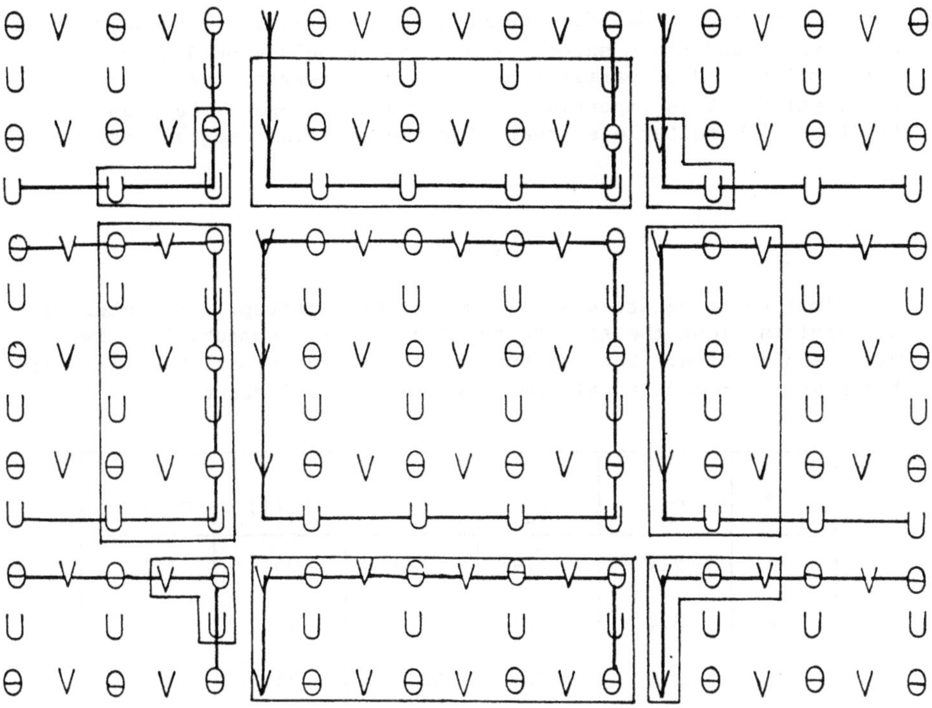

Fig. 6 Boundary values needed for interpolation in the second stage.

very efficient computation. In case of 48 nodes the speedup is 45 relative to today computation. This must be considered to be an exceptional good result and represent an actual computer speed close to 200 MFLOPS.

We assume now that only a relative small number of FPS-nodes are available. In Table 3 and 4 computation time and speedup is computed in case of 8 16 and 48 FPS nodes. Table 3 is for squares of 16*16 points, and Table 4 for 96*3 points slices.

The tables show how two-dimensional slicing is more efficient than one-dimensional slicing. Further, the effect of startups is demonstrated. The desired speedup of 12 is achieved with 16 nodes by both one-dimensional and two-dimensional slicing. However, this is only the case when no startups are included.

5. CONCLUDING REMARKS

It is shown that parallel computation is very promising for the integration of our prediction model. It is now important to evaluate the resources needed to do the programming. The model can easily be reprogrammed to integrate sequentially on sub-areas on our present computer. In such a program there will be FORTRAN subroutines calls for each square (slice). The needed parallel programming (OCCAM) is then assumed to be a minimum. In our operational environment there are also other models to be considered, i.e. model for analyses of observations, models for ocean waves and storm surge. Also these models should be suitable for similar kind of parallel processing. However, for all our applications it is important that FORTRAN can easily be used together with suitable software for the parallel programming.

Table 1

Time (ms) to integrate a small square one timestep. n: number of grid-points along the side of the square. T_1: computation time. T_2: transfer time. T_3: startup time. F_1: additional time (%) for transfer. F_2: additional time for transfer and startup.

n^2	T_1	T_2	T_3	F_1(%)	F_2(%)
1	2.1	1.9	1.2	90	147
4	8.4	3.8	4.6	45	100
16	33.6	7.6	18.5	23	78
64	134.4	15.1	74.3	12	67
256	537.6	30.2	297.1	6	61

Table 2

Speedup and computation time each timestep in case of a large number of nodes covering 12288 points.

Nodes	points/node	speedup	time(ms)/step
12288	1	6451	4.0
3072	4	2112	12.2
768	16	626	41.2
192	64	173	149.5
48	256	45	567.8

Table 3

Computation time (s) and speedup in case of 8, 16 and 48 nodes, and 16*16 points squares covering 12288 points. T_1: computation time per timestep. T_2: computation time per timestep when startup is included. S_1 and S_2 are speedup in the two cases.

Nodes	Points	No of startups	T_1	T_2	S_1	S_2
8	2048	6	3.41	5.19	7.6	4.9
16	4096	3	1.70	2.59	15.1	9.9
48	12288	0	0.57	0.57	45.4	45.4

Table 4

Computation time and speedup in case of 8, 16 and 48 nodes, and 96*3 points slices covering 12288 points. T_1: computation time per timestep (s). T_2: computation time per timestep when startup is included. S_1 and S_2 are speedup in the two cases.

Nodes	Points	No of startups	T_1	T_2	S_1	S_2
8	2304	6	4.42	5.90	6.1	4.2
16	4608	3	2.21	2.95	12.2	8.5
48	13824	0	0.79	0.79	36.6	36.6

References

Bratseth, A., 1983: Some economical, explicit finite-difference schemes for the primitive equations. Mon.Wea.Rev., 111, 663-668.

Brown, J.A. and Campana, K.A., 1978: An economical time-differencing system for numerical weather prediction. Mon.Wea.Rev., 106, 1125-1136.

Charlesowrth, A.E. and Gustafson, J.L., 1986: Introducing replicated VLSI to Supercomputing: the FPS-164/Max scientific computer. IEEE Computer, 19,3.

Grønås, S. and Hellevik, O.E., 1982: A limited are prediction model at the Norwegian Meteorological Institute, Oslo, Norway.

Grønås, S. and Midtbø, K.H., 1986: Four dimensional data assimilation at the Norwegian Meteorological Institute. Tech. Rep. No. 66, The Norwegian Meteorological Institute, Oslo, Norway.

Hawkinson, S., 1986: The FPS T Series: A parallel Vector Supercomputer. Proceedings from the international conference on vector and parallel computing. Juen 2-6, 1986, Loen, Norway. Chr. Michelsen Institute, Bergen, Norway.

Juvanon Du Vachat, R., Imbard, M., Joly, A., Audoin, J.M., Marie, N., 1986: The Peridot fine-mesh numerical weather prediction system - Description and evaluation. WMO PSMP report series No. 19, Tokyo 1986.

Nordeng, T.E., 1986: Parameterization of physical processes in a three-dimensional numerical weather prediction model. Tech. Rep. No. 65, The Norwegian Meteorological Institute, Oslo, Norway.

WMO, 1984: Catalog of existing modules of limited-area NWP systems. Short- and medium range weather prediction research publication No. 6.

A Comparative Study of the ECMWF Weather Model on Several Multiprocessor Architectures

GEERD-R. HOFFMANN and DAVID F. SNELLING

European Centre for Medium-Range Weather Forecasts, Shinfield Park, Reading, Berkshire RG2 9AX, United Kingdom

INTRODUCTION

The European Centre for Medium-Range Weather Forecasts (ECMWF) first begun using multiprocessors in the early 1980s for the production of their weather service products. In the past few years, with the increasing number of and variety of multiprocessor systems on the market, it has become necessary to broaden the evaluation of multiprocessing. Toward this end, ECMWF has developed a suite of four benchmarks for the evaluation of multiprocessing. It should be noted that this benchmark suite is strictly for the evaluation of multiprocessing and is not in itself a complete benchmark suite. Every effort was made to allow for the study of multiprocessing and not raw performance. As a result, this study focuses on the efficiency of the multiprocessing systems available. Of the machines available on the market, only four were chosen for detailed study, partially because of their performance characteristics and partially because of the availability of hardware. The study will focus on the Cray X/MP 4-processor system, the ETA 10 (only simulations were performed), the IBM 3090/400 with vector facility, and the Floating Point Systems T series. The four benchmarks developed at ECMWF are MANDEL, MATRIX, SPEC and MODEL. Each of these is described below in terms of their structure and what aspect of multiprocessing they address.

BENCHMARKS

MANDEL

MANDEL is a graphics program which draws images of the Mandelbrot set, a mathematical structure generated through a series of transformations in the complex plane. For some values in the complex plane the iterative algorithm z = z**2+c will eventually cause the magnitude of z to become arbitrarily large. This happens for a variable number of iterations depending on the initial point c. In this benchmark the graphics screen is assumed to be 640 pixels by 480 pixels, with a color resolution of a 1000 different colors. Values for c were chosen from the domain x = -2 to 1 and y = -2 to 2. MANDEL has no vector calculations, although it is register intensive. Each of the pixels can be computed independently and by grouping them together in lines or sections of lines, very large grain parallelism can be achieved. MANDEL was chosen essentially to demonstrate the best possible case for multiprocessing. It requires little or no synchronization and as a result may be used to demonstrate the best possible multiprocessing performance on given computer hardware. The fact that it is register intensive also avoids measurement of any memory bank conflicts that might occur on shared memory systems.

MATRIX

MATRIX is a very simple program which performs a square MATRIX multiply. The size of a MATRIX is 1024 elements x 1024 elements. The vector length is 1024, although other algorithms were implemented for some machines for which more parallelism could be achieved if a nested MATRIX multiply was used.

The code is highly vectorizable and memory intense. Because of the large size of the MATRIX, highly parallel large GRAIN algorithms can be developed. As a result, MATRIX may be used to evaluate the effect of global data requirements on a distributed memory system. Since each parallel process performing part of the matrix multiply must, in the course of the algorithm, access part of one of the matrices and all of the other. On a shared memory system, there are no major algorithmic difficulties, although memory bank conflicts may occur because of the intense memory access. As a result, MATRIX is a fairly general purpose program which addresses one of the more difficult issues in multiprocessing (globally shared data).

SPEC

SPEC performs the direct and inverse Legendre transforms, which are the kernel of the spectral model at ECMWF and consume about 5, 16, and 23 percent of the calculation time in operational models at T21, T63, and T106 resolutions respectively.[1] For the purposes of this study, a T63 resolution transform using four prognostic variables, was chosen. The vorticity, the divergence, the humidity and the temperature are transformed first from spectral space into Fourier space, then re-transformed back into spectral space. The program has a vector length of 32, is highly parallel and exhibits the same kind of global data requirement as the matrix multiply. SPEC has a fairly large memory requirement of approximately 1.2 million words. SPEC was also used to develop algorithms for use in the model benchmark. It allowed us to do a detailed analysis of various algorithms for performing the spectral transformations in parallel. In the weather model at ECMWF, the spectral transformations are the only place where there exists a global data requirement and as a result it is the one aspect of the weather model that requires special attention in terms of multiprocessing. In addition to its use as a development tool, SPEC provided an opportunity to do a detailed analysis of the worst part of the model, the portion where a global data memory requirement exists.

Even on a global memory system, where the parallel algorithms are much simpler, SPEC provides an interesting area for study. In the direct Legendre transform, variations in the order that operations occur cause an irreproducibility effect in the calculations. By studying SPEC, we were able to get a better understanding of different algorithms and approaches for performing the spectral transformations. The spectral transformations are, because of their structure, the fastest growing computational part of the model. As resolution is increased, the computational load in the Legendre transforms grows with the cube of the resolution. SPEC was extracted as an algorithm from a T63 version of the weather model.

The first issue that had to be dealt with was a reorganization of the data. In the original version, the total wave numbers are stored one after another in a large sequential array. Because each total wave number is composed of a varying number of zonal wave numbers, the triangular shape of the data structure was difficult to describe in terms of the square matrix notation used by SPLIB. In order to solve this problem, the data was reorganized so that the largest total wave number was paired off with the smallest total wave number. As a result, the spectral data structure was made to conform to the square matrix structure of SPLIB and allowed the data to be partitioned in several different ways.[2]

In particular, with a T63 version of the model the data may be divided into 1, 2, 4, 8, 16 and 32 parts. This is useful in evaluating the optimum grain size for this algorithm. Three different algorithms were developed for performing the Legendre transforms. The first, which is equivalent to the current production version at ECMWF, was a global memory version assuming that the spectral data resided in memory accessible directly by all parallel processes. Another version, referred to as the one part version, assumed that each parallel process had its own local copy of the entire spectral array. The third version was an n-part version, in which a small portion of the spectral array was local to each parallel process and a global copy of the entire spectral data structure was maintained in the shared matrix structure in SPLIB.

MODEL

MODEL is a benchmark based on the model section of the ECMWF operational forecast suite.[1] In this study a T63 version was run, although other resolutions may be parameterized and run as well. All the code was originally vectorized and already written for a parallel machine. Although some changes were made to the vector and parallel structure to allow for increased portability. There were five phases in the development of the model benchmark. The initial phase was to change the multiprocessing strategy, the second was the removal of Cray specific code, the third was the removal of the Cray pointer construct and the ECMWF memory management system, and in the fourth and fifth phases the model was converted so that it could be run in an SPLIB controlled parallel environment.[3,4,5]

---

[1]This benchmark is based on cycle 25 of the operational forecast model.

The multiprocessing strategy in the original model, in which north and south rows were run parallel, was altered in order to provide a simple parallel structure for dealing with a variety of computer systems. As a result, the code was restructured, so that a north row and its corresponding south row could run sequentially in one parallel process. At the same time, another north/south pair could be dealt with in another parallel process. In addition to this general restructuring of the multiprocessing, the multiprocessing control was centralized into one routine. In addition, all LOCKON and LOCKOFF calls to the Cray multitasking library were put under the control of a global logical variable. This would allow the program to ignore this synchronization structure when the code was running on a distributed memory system. Major changes were made to the I/O code to allow for the new parallel structure. These changes have subsequently been included in the production version of the operational weather model at ECMWF. For more details on this activity see [6].

Cray specific code was removed to allow for easier compilation on a variety of computer architectures. In addition, some 54,000 lines of operational code were removed. Most of this code was associated with the initialization of the model and with other aspects of running in an operational environment. It is the belief of the authors that this merely simplified the benchmarking effort and in no way would reflect on the ability of a particular machine to actually run in an operational environment. Since the resolution in the production model was determined by run time variables, the code was parameterized so that different resolutions could be selected at compile time. The run time approach used in the production version of the model does not conform to standard Fortran. There are two parameters that need to be changed in order to run different size models. Since all the model data files at ECMWF contain an initial data description record, which does not conform to uniform record size required by Fortran 77 I/O, the data description records were removed from the data files and the initial data converted from binary to ASCII. In addition, timing diagnostics were installed in the code to provide information on the parallel behavior of the code.

In addition to converting to an ASCII format for data, the I/O structure was changed from a buffer-in/buffer-out asynchronous I/O structure to standard Fortran random access. In addition to this type of change, changes were made to replace Cray assembly language code with Fortran. Most of the routines involved were simple system interfaces such as date and time, although two subroutines were changed that make a significant difference to performance. The assembly language fast Fourier transform and matrix times vector routines were removed and replaced with their Fortran equivalents. This was done with the understanding that any machine would have different versions of these algorithms that performed with peak capabilities.

The major change to the vector structure of the model was the removal of the vector merge functions. These functions, provided in the early days because of inadequacies in the compiler, have been replaced with block if statements, which today's compilers can vectorize. In spite of the advances in compiler technology there were quite a few loops, some in critical parts of the code, which the Cray compiler could not vectorize without the vector merge functions. As a result the balance of computation changed, and the Legendre transforms' contribution to the total computation was reduced. In this benchmark version of the model, the Legendre transforms constitute about 2.5 and 3.4 percent of the calculation in the T21 and T63 versions respectively.

Changes made to the ECMWF model, involving the memory manager and the Cray pointer construct, were necessary in order to transport the code to a variety of computer architectures. All data local to a process was moved from the dynamic memory management system into TASK COMMON on the Cray. (On the other machines studied all common is local to the process and therefore equivalent to Cray's TASK COMMON.) Other managed memory, such as the spectral data coefficients, were moved for the time being into a global common. In addition to removing these managed data structures, temporary data structures were moved from the memory manager into stack based local arrays.

Removal of managed memory required a significant rewrite of much of the radiation code. In the radiation code section, pointers were passed as arguments to subroutines, which had to be changed in order to allow for execution using variables instead of pointers. In large sections of radiation code pointers were assigned to each other in a manner which was functionally equivalent to Fortran equivalence statements. These statements had to be rewritten into 700 equivalence statements in various parts of the code. Two other minor changes had to be made in order to deal with the memory manager. Since the spectral calculation involved accessing data structures by pointer, these spectral calculations were changed in minor ways to allow variables to reside in common. In addition, the Legendre transform algorithms had to be rewritten. These transforms were rewritten again in the second phase of the SPLIB conversion effort. Once all these changes were made, the memory manager and all calls to it, could be removed.

In the first phase of the SPLIB development, the fork/join structure of the multiprocessing environment was changed to a barrier structure. This provided for efficient access to the multiprocessing facilities of a variety of computer systems. Since the code would no longer be running in a Cray multitasking environment, all calls to the Cray multitasking library were removed and replaced with calls to SPLIB.

Several changes involving the common block structure were made. Any common block that did not change during execution, was moved into TASK COMMON and copied from the initial process to the parallel process at the beginning of the first parallel section. Common blocks that were changed outside the parallel sections were treated similarly, except that the transfer was performed at the start of the parallel section each time step. All the remaining commons can be called "reduction" commons. In these cases, each process performed an associative reduction operation (sum, product, max, min...) into its own copy of the common block. Then at the end of a parallel section of code, all reduction commons were combined into the TASK COMMON of the main process. This was done for computing the global accumulative diagnostics. At the conclusion of this phase of the conversion there was only one global common block remaining in the code, the spectral data common block. All other commons had been converted to TASK COMMONs. Since this was the most complex shared data in the entire model, it was dealt with independently in phase 2 of the SPLIB conversion.

In the second phase of the SPLIB conversion effort, the spectral data had to be restructured from the triangular form to a square form as discussed earlier in the SPEC benchmark section of this paper. The spectral data space was moved from the global common block into the shared matrix structure provided by SPLIB. In addition, TASK COMMONs, containing all or part of the spectral data structure, were used. The size of these TASK COMMONs depended on the Legendre transform algorithm being used. Accordingly, the spectral calculations were rewritten to perform the one and n-part transforms described earlier.

RESULTS

The results presented below are preliminary, since the study is an ongoing activity. Not all the benchmarks have been run on all the equipment, but some effort has been made to provide an indication of how these programs might be used in evaluating parallel architectures. Since all the programs where developed on the Cray X/MP at ECMWF, the Cray runs provide a reference point for this study.

MANDEL

In figure 1 the bar graphs represent the speedup achieved using 2, 3, and 4 processors on the CRAY X-MP at ECMWF. The two different groups of results describe the performance achieved using the standard Cray multiprocessing library (Global) and the SPLIB library (Private). The latter represents the performance one would expect from a CRAY X-MP if all access to shared memory was via library calls. Since MANDEL has only a limited amount of memory activity, most of the computation being in registers, this program exhibits almost identical behavior using both approaches. In particular, since memory contention is avoided, in both cases the program runs with almost perfect efficiency and exhibits a linear improvement in performance as the number of processors is increased. Also the impact of the additional software required to handle the communication with a shared memory is insignificant. In short, the library over-heads for problems of this granularity are insignificant.

MATRIX

Unlike MANDEL, MATRIX is highly memory intensive, and therefore some degradation of performance is expected as the number of processors is increased on a shared memory machine. The results displayed in figure 2 represent the performance of 2, 3, and 4 processors on CRAY X-MP using FORTRAN and assembler versions of MATRIX, and of 16, 64, and 128 nodes on the FPS T-Series. Since the grain size is larger and the synchronization overhead smaller, in MATRIX than in MANDEL, efficiencies of nearly one would be

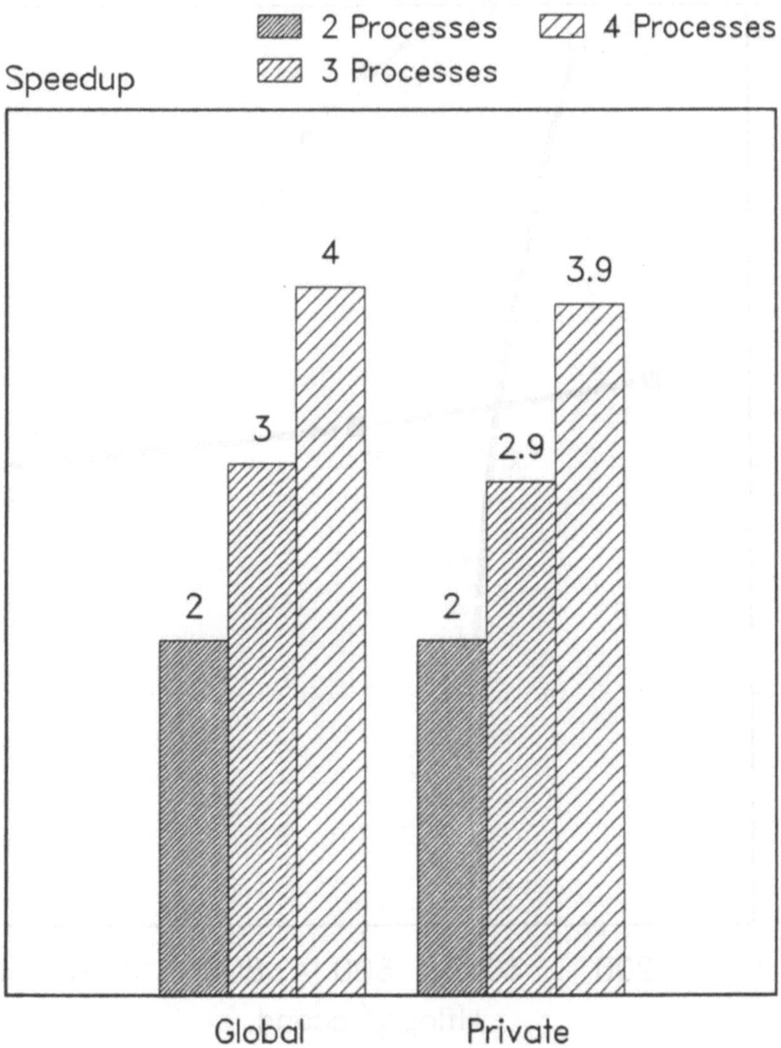

Figure 1 True Speedup for MANDEL Using Two Algorithms on the Cray X/MP. Speedup = Ts/Tp.

expected; except for the impact of memory contention. The timings show that the impact on performance in much greater in the shared memory plots than in the distributed memory plot, although the absolute efficiency on the shared memory machine is higher in all but the four processor FORTRAN version. The complexity required to implement even this simple algorithm on a hypercube machine causes the low efficiency found in this example. If the linear trend on the T-Series plot is continued, the efficiency at 2048 nodes is

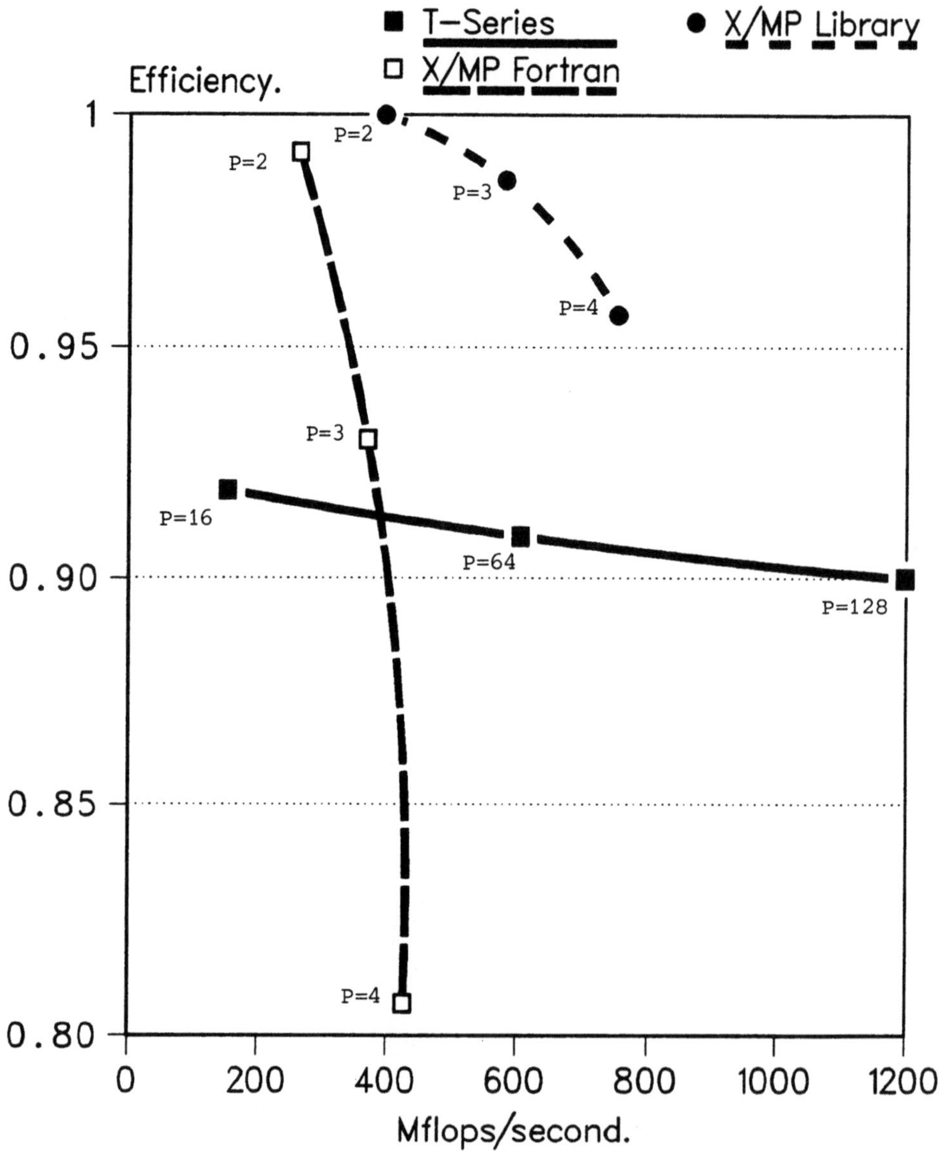

Figure 2 Efficiency vs. Performance for MATRIX on the Cray X/MP and the FPS T-Series. Efficiency = Ts/(Tp*p).

only 0.57. 2048 is well below the largest configuration of the T-Series, and therefore new algorithms, larger grain sizes, or more efficient software would be requited to match the 0.81 efficiency exhibited by the maximum configuration of the X-MP.

By comparing the slopes of the two X-MP runs it is clear that, although the raw performance increase can be accounted for by the reduction in instruction count, the change in slope is opposite to what is expected. Since the assembler version implies a reduction in grain size, the efficiency should drop accordingly. Instead the efficiency drops less rapidly in the assembler version. This is

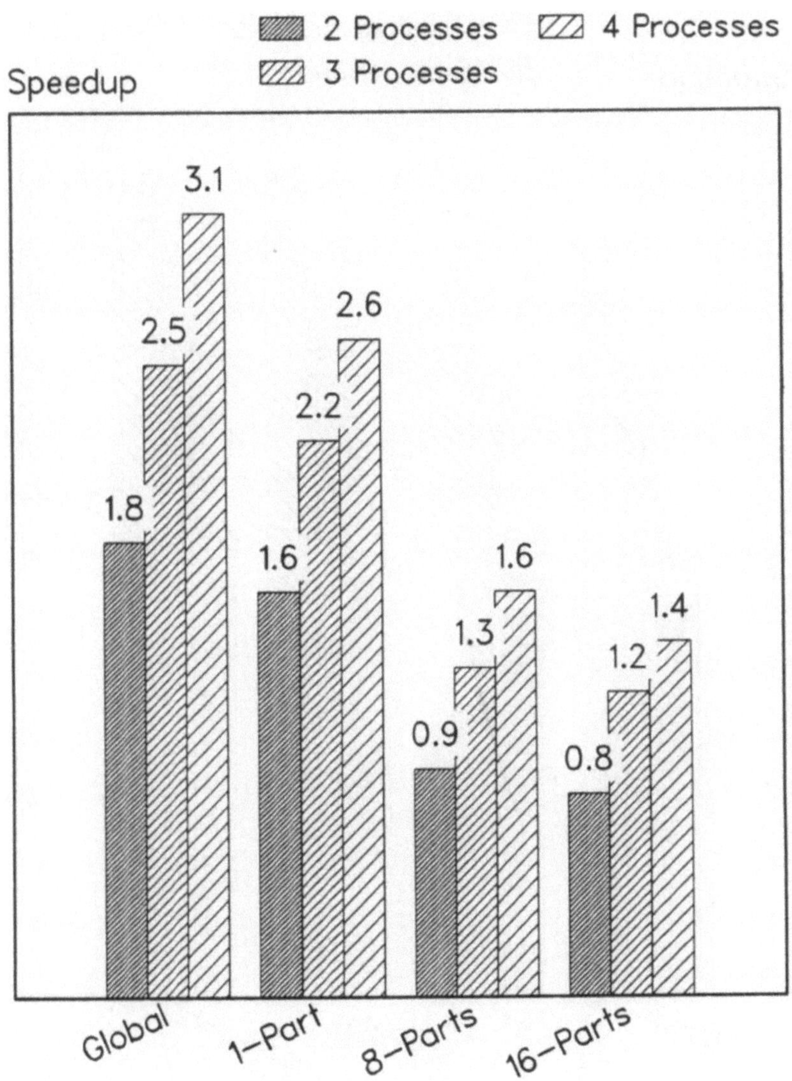

Figure 3 True Speedup for Four Legendre Transform Algorithms. True Speedup = Ts/Tp.

due to an algorithmic difference in the two approaches. The assembler routine holds a working group of vectors in registers for several stages of the calculation, and thereby reducing the memory reference to calculation ratio, which in turn reduces the memory contention.

SPEC

Figures 3 and 4 provide an interesting comparison between the two different approaches used in parallel Legendre transforms. Where there is support for global memory less total data motion is

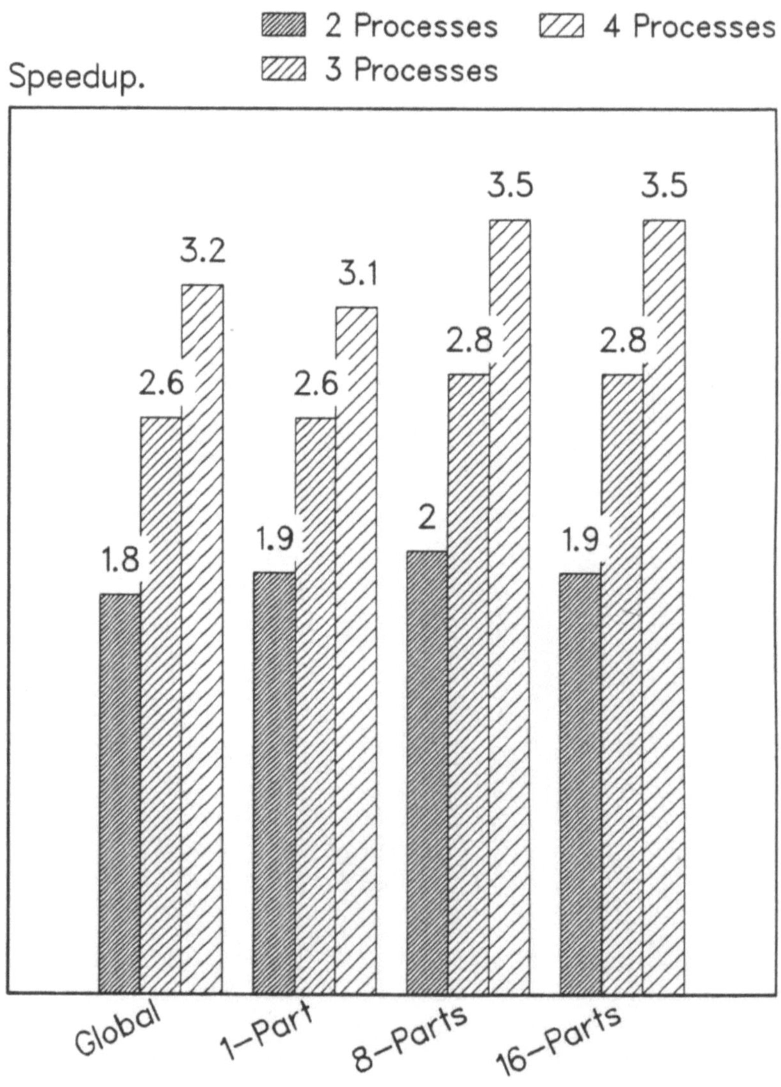

Figure 4 Relative Speedup for Four Legendre Transform Algorithms. Relative Speedup = T1/Tp.

required than in the hierarchal approaches, which require large quantities of data to be transmitted to and from a shared memory. The four examples in each figure are: 1) Global, in which the spectral data structure is updated in place, 2) 1-Part, in which a local copy of the spectral data structure is updated in place and then these copies are combined in parallel at the end of a time step, 3) and 4) 8-Parts and 16-Parts, in which the spectral data structure is kept in shared memory and one part at a time is brought into local memory, updated, and then written back out to

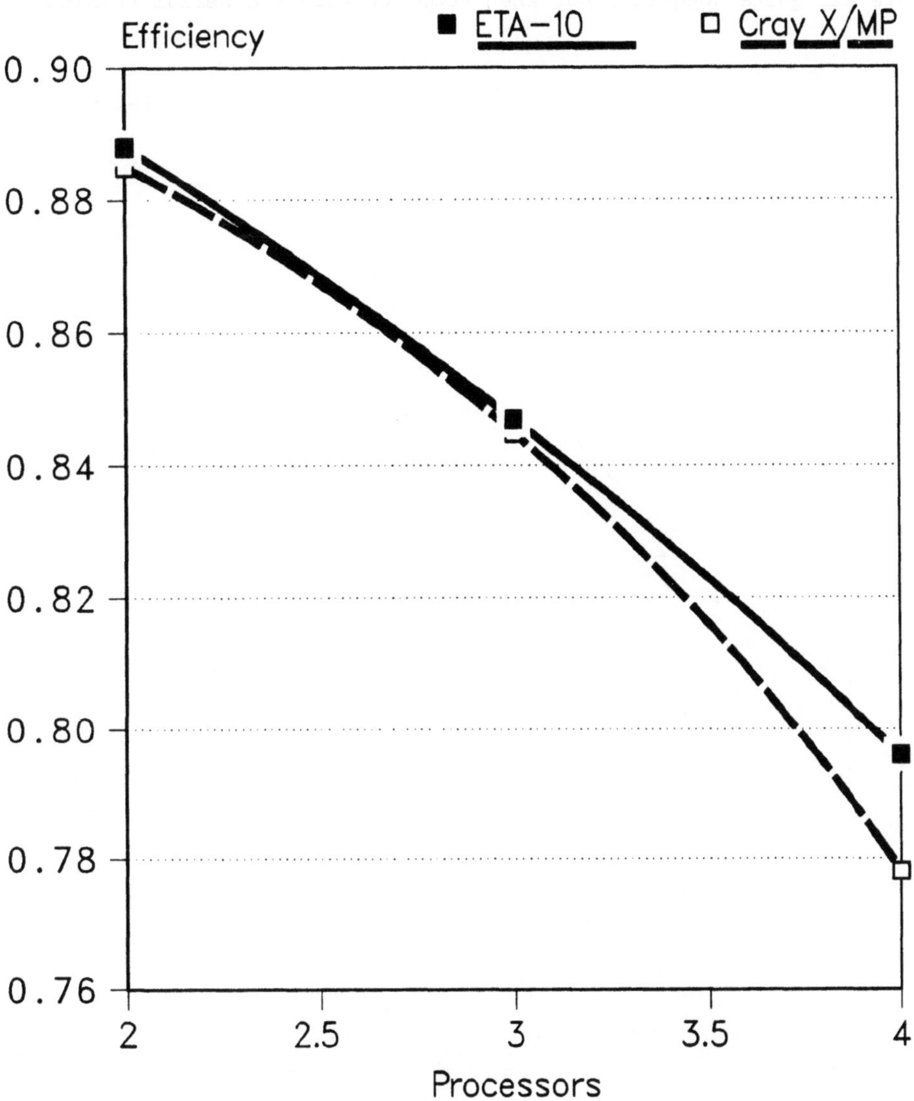

Figure 5 Efficiency of Parallel Legendre Transform Algorithms on the Cray X/MP (Global memory algorithm) and the ETA-10 (Private memory algorithm). Efficiency = Ts/(Tp*p).

shared memory. In figure 3 the quoted speedups are true speedups, that is they represent the speedup in relation to the original serial algorithm on the same machine. In this case it is clear that as the complexity of the algorithm increases the speedup drops. This drop is partially due to multiprocessing inefficiencies, but the bulk of the degradation is due to the change in algorithm. In figure 4 the given parallel algorithms were run using only one process, and then this time was used as the reference for the calculation of the relative speedup. It is therefore clear that as parallel implementations these algorithms are all quite adequate, but when compared with the serial version

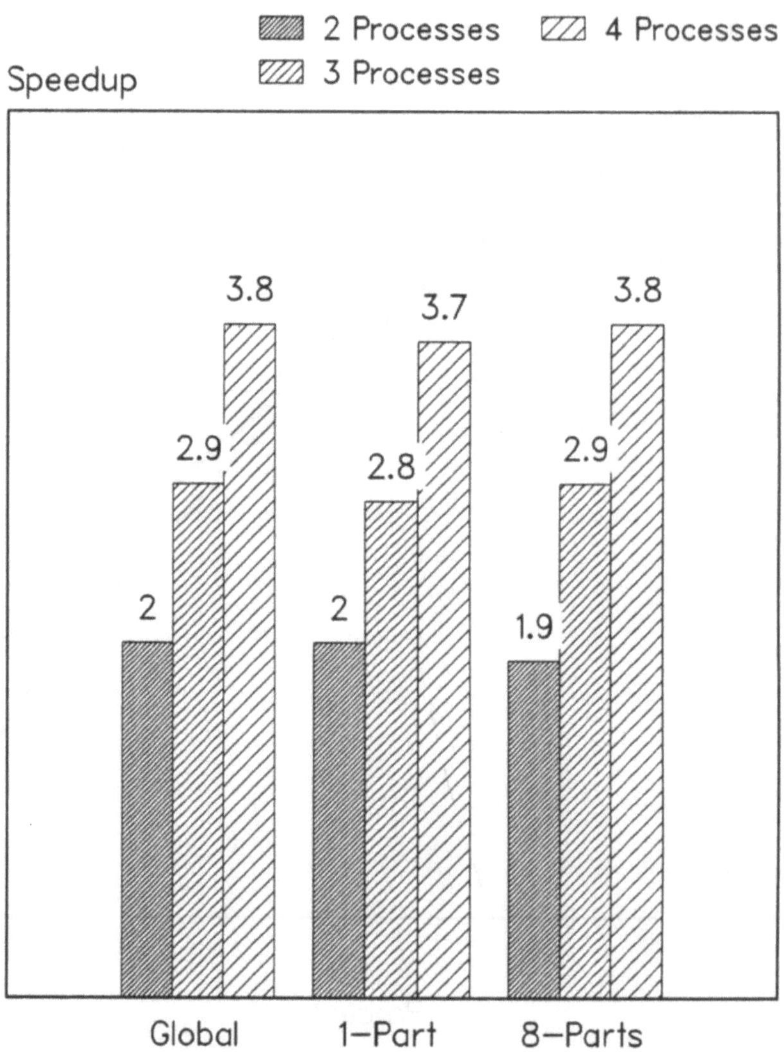

Figure 6 True Speedup for Full Model at T63 Resolution Using Three Legendre Transform Algorithms on the Cray X/MP. True Speedup = Ts/Tp.

some degree of improvement is lost, depending on the algorithm in question. It should be noted that, if the size of local memory on a hierarchal memory system or the size of primary memory on a shared memory system is limited, the entire spectral data structure might not fit into the memory closest to the processor. The only way this type of model could then be run is by using one of the multiple part algorithms and allowing the entire spectral data structure to reside on a secondary level of memory.

In both figures 3 and 4 all timings are from a Cray X-MP. As the number of processors is increased the portion of the potential speedup achieved drops slightly. This is due to the increased memory contention when more processors are active. Because the multiple part algorithms are less memory intensive, these two approaches show better relative efficiencies than the global algorithm.

In figure 5, the effect of memory contention becomes clear. The ETA version of this algorithm shows a slightly higher efficiency than the Cray version, in spite of the fact that the version of this program running on the ETA multitasking simulator was an inefficient implementation of the 1-Part algorithm. This is because in all stages of the calculation, including the accesses to shared memory, there is no memory contention in the ETA version. It is important to note that the improved efficiency in the ETA version is at the cost of increased memory usage. This is because in order to avoid extensive transfers to and from shared memory, a complete copy of the spectral data structure is kept in each local memory.

MODEL

In figure 6. the three basic algorithms for the Legendre transform were used in the model and the true speedups plotted. When the spectral transforms are combined with the rest of the benchmark model, their contribution is only about 3.4%. Notice that, compared to the results obtained using SPEC, the impact of the choice of algorithm on the achieved speedup is much less than would be expected. A straight forward application of these results, however, yields a maximum true speedup of 3.75 for the 8-Parts algorithm on a 4 processor system. This is compared to the measured true speedup of 3.76. In fact the difference between the global speedup (3.84) and the 8-part speedup (3.76) is so small as to be insignificant. Nonetheless, it is clear that further detailed studies into memory contention and instruction issue are required to fully understand these results.

CONCLUSION

These benchmarks allow the European Centre for Medium-Range Weather Forecasts to evaluate multiprocessors in a straight forward and controlled manner. They do not, however, provide an exhaustive analysis of the issues which arise in multiprocessing. Each of these programs addresses one or more of the issues which are important to anyone interested in multiprocessing, but, because of the specialized nature of the application from which they were extracted, they do not suffice as a universal tool for measuring the capabilities of a multiprocessor system. In particular, the surprising result, which was obtained using the full model - rather than a subset - as a benchmark tool, confirms the naive view that the only realistic benchmark for a machine, particularly a supercomputer, is the actual operational application. In light of this, it is particularly important to note that the task of evaluating multiprocessors can be addressed using operational quality benchmarks, if the programming strategy and techniques for parallel processing are chosen carefully.

[1] Steppeler, J., Galerkin and Finite Element Methods in Numerical Weather Prediction series: Bonner Meteorologische Abhandlungen (Dummler, Bonn WG) 1987.

[2] Snelling, D. F., "Tools for Accessing Multiprocessing," This publication.

[3] Cray Research Inc., Fortran Reference Manual, SR-0009, (Cray Research Inc.; Mendota Heights, Minnesota), 1986.

[4] Cray Research Inc., Programming Library Reference Manual, SR-0113, (Cray Research Inc.; Mendota Heights, Minnesota), 1986.

[5] Dent D. W. and J. K. Gibson, A Memory Manager for Single and Multitasking Applications, European Centre for Medium-Range Weather Forecasts, Technical Memorandum No. 104.

[6] Dent, D., "The ECMWF Model Past Present and Future," This publication.

Multiprocessors: Requirements and Perspectives Summary of the Discussion

DIMITRIS K. MARETIS

European Centre for Medium-Range Weather Forecasts, Shinfield Park, Reading, Berkshire RG2 9AX, United Kingdom

1. INTRODUCTION

At the first Workshop some requirements for future processors were discussed. Since then advances have taken place that have fulfilled some of these requirements, e.g. large memories of at least 100 MW seem to be in reach, faster I/O devices have been developed, etc.

However, there still remain areas where users have requirements, and where problems need to be resolved if supercomputers are to be used sensibly.

The following discussion report gives a brief summary of the needs and expectations of the users, as well as of the views and development plans of the manufacturers present during the Workshop.

2. HARDWARE STRUCTURE AND MEMORY

There seems to be general agreement on the fact that for meteorological models to be run reasonably, the machines should not only be able to provide 10^{12} - 10^{14} operations in a certain time frame, but also have large memories. These memory requirements will probably be only slightly affected by the planned introduction of new algorithms in the next few years. The current spectral models seem to have relatively high global memory requirements. Future schemes could improve the situation, or at least not make it worse. It is therefore anticipated that memories of at least 100 MW - probably more - will be needed for the meteorological models to come.

2.1 Memory Organization and Management

The question that immediately comes up is 'What kind of memory organization will be supported by the architectures of the future machines?'. Could it be, e.g., a one level global memory which is large and fast, and has enough bandwidth to support 8 to 16 processors?

Looking down the road, it looks as if cost and reliability issues make it unlikely that large (approx. 100 MW) and at the same time fast (i.e. supporting 4 nsec machine cycle time) global memories will be made available in the next 3 or 4 years. Although an increase of the chip density by a factor of 4 seems realistic, the increase in speed is unlikely to exceed 10-20% over the next few years.

ECL chips still remain too expensive to be used outside of the main memory, and plans by Cray to introduce GaAs chips to increase the speed of local memory have not yet been implemented. So, the situation remains that memories will be either fast but not big enough, or big but not fast enough. The need for architectures which provide a hierarchy of memories is, therefore, apparent.

However, since the bandwidth of secondary memory is almost always a serious issue, no matter how big a central memory a machine has, additional memory levels are needed to bridge the gap between the slow secondary memory and the fast CPU power. Here Cray uses the SSD as such an additional memory level, and other manufacturers will probably follow soon with similar concepts. But the fact remains that there is a general need for cost effective hierarchical memories in addition to local memory.

2.2 Memory Management

Related to this topic is another issue of great importance for the users; namely, what is the prevailing memory organization likely to be in future machines, and which concept seems to be feasible from the technical point of view. Will the trend be towards shared memories, as the X-MP has, or towards hierarchies of memories, as some vendors seem to advocate, or will the technical development lead towards completely distributed systems?

In shared memory models there are two main issues related to memory speed: latency and bandwidth. On vector machines the latency does not pose real problems, the bandwidth, however, demands more attention.

A major problem is still memory management. The manufacturers provide a fairly clear 2 level structure: a layer to accommodate local data structures and a second one that allows for communication of a large amount of data between levels of Fortran subroutine calls. However, for decent multitasking and for a sensible use of memory, at least 3 levels of data structure are essential, one of which should be user explicit. Because the user has a much better idea than the system can ever have of what may be removed from memory, or what should be fetched in. Furthermore, the layer at the top should be the one where all debug issues become most critical. There, the user should be given explicit control.

It is well known that a major part of the problems encountered in memory management stems from the deficiencies of the chosen programming language. Fortran gives only the choice between a local variable and a global one. There is nothing in between.

The deficiencies in memory management as provided by the manufacturers, and the need to devise means of addressing variables in a sensible way within the current constraints of the language, led ECMWF to opt for a third layer above the two in the form of the Memory Manager. It is based on the Cray Fortran Pointer extension and creates great flexibility in handling memory. The use of the

Pointer statement certainly reduces the portability of the code, but the Memory Manager provides the user with a great amount of flexibility, both for the analysis and the model.

With this in mind, the question arises as to whether it is necessary in the future to have some form of dynamic memory allocation within a high level language like Fortran in addition to the facilities that are available at the moment. If this were the case, then the users should make it clear to the programming standards bodies like X3J3 that two levels of data structures are not enough, and that there is a need for at least a third level and possibly a fourth.

However, it seems unlikely that anything in the form of a memory control structure will be provided by the standards bodies in the near future. It is also unlikely to expect the manufacturers to agree on 3 levels of memory hierarchy. ECMWF has therefore spent effort and manpower to produce SPLIB, a portable multitasking library, in order to have something that makes it possible to benchmark and compare multiprocessing machines with each other, e.g. for a future procurement exercise.

2.3 Addressing the Memory

The next question concerns the way of addressing memory. Should it be done by using syntactically correct Fortran subroutine calls, or by allowing for extensions of the Fortran syntax.

Users often find the flexibility they get by non-standard features very valuable. But when it comes to comparing multitasking environments from the point of view of measuring multitasking capability, then syntactically correct Fortran subroutine calls are crucial (the only allowed way of doing it). Benchmark code has to be syntactically correct Fortran, and has to be able to go through different compilers and at least compile, if the benchmarks are to provide meaningful results.

From the point of view of the manufacturers, there is very little difference in how the needed features for addressing memory are implemented. Features like TASK COMMON, for example, can be implemented either by extending the Fortran syntax - as Cray did it - or as a compiler directive. By extending the syntax, however, it is made clearer that this is going to result in differently generated code, whereas providing such features as a compiler directive means that the code would compile on different machines without error messages, although it would not behave as expected.

3. USER-FRIENDLINESS

Another important issue is the user-friendliness of the programming environment. It is felt that there remains still much to be done in this respect, and the users would like to see the manufacturers addressing this issue more seriously over the next few years.

3.1 Deficiencies of Programming Languages

User-friendliness is connected first of all with the language used. It is obvious that there are some problems with standard Fortran in this respect. So it is only natural that many users do not really like Fortran, but still use it because they see no viable alternative. Unfortunately it is not true that if the right algorithm has been chosen, any language could be used to implement it. There are indeed differences between programming languages, and so it is not only the algorithm that matters. The probability of taking a good algorithm and ruining it is much higher in Fortran than in other structured languages, as e.g. SISAL. Developments like SPLIB and the various multitasking libraries, have illustrated where Fortran itself tends to be deficient, and have brought out the concepts of data structures that are not catered for in the language itself. So, looking into the future, the best way ahead is to learn from the lessons of all the multitasking applications and try to press for such facilities to be put into the language. Packages dependent on subroutine calls, on the other hand, may remove some of the clean concepts.

3.2 Layers of Multiprocessing

Over the years, various ways of supporting concurrent programming (multiprocessing) have evolved. But whatever terms are used by the manufacturers to describe these multiprocessing features, it is obvious that multiprocessing can be done at different layers.

The term "Metatasking" has been used throughout this Workshop to define multiprocessing at the layer on top of Fortran. Macrotasking, as provided by Cray, tends to be explicit in creating and managing tasks, and protecting critical regions. Microtasking allows the user to insert directives in the Fortran code to take advantage of the multiprocessor capability. Finally, vectorization and auto-parallelization is usually done by the compiler which identifies and performs FORK and JOIN functions at the loop level, although it still allows the user to insert directives to implement parallel structures. In contrast to macrotasking, both microtasking and auto-parallelization are more implicit in defining the work to be done in parallel, and therefore tend to be self-scheduled.

The multiprocessing software concepts and support are still in evolution. The next years will probably see the emergence of a more coherent user interface and a more friendly programming environment. Thus the question remains, at which level would the users prefer to do multiprocessing. From experience gained at ECMWF with multitasking codes it can be readily said that there is definitely a place for both micro- and macrotasking, and that a combination of both is a very attractive way of exploiting multiprocessing in meteorological models and of removing inefficiencies. So, coming back to the layers of multiprocessing, it seems that one should firstly have microtasking or auto-parallelization at the loop level, and then allow for macro- or metatasking at the top layer.

3.3 Debugging Facilities

There is still a long way to go until one can debug a multiprocessor program as easily as a uniprocessor program. The order of operations being indeterminate in parallel processing, makes it almost impossible to reproduce exactly the same execution behaviour and hence to debug a program easily. Debugging aids are slowly coming along, and the situation is admittedly better now than it was some years ago when ECMWF first started developing a multitaskable version of its model code. But it is still not enough. The vendors are therefore asked to put effort into supplying as many debugging aids as possible, and hopefully also supporting them in the hardware.

Debugging facilities must be something which can be turned on when needed, and turned off otherwise. If different software aids have to be put in, in order to find an error, then - and this happened at ECMWF - the errors may be not reproducable. But reproducability of errors is of crucial importance in debugging multiprocessor programs.

3.4 Reproducability of Results

In changing the underlying programming environment, the concurrent execution of streams in parallel processing also affects the reproducability of results. Since the order of operations is indeterministic and no longer predetermined by the code, there is now an indeterminacy in the results. The differences are of course small and physically insignificant. Nevertheless, they are extremely irritating from the point of view of code maintenance and verification. It is therefore particularly important to have a means of enforcing reproducability of results. Without such a feature it is extremely difficult to check out whether the differences are due to problems at the machine level, or to genuine bugs in the code.

It is probably possible to find ways to force the computation into a certain order (e.g. on the Cray using EVENT logic), but what is really needed is a compiler option which can be switched on/off according to the requirements. The cost of enforcing reproducability can be minimal if the benefits of such a feature are taken into consideration. The vendors should, therefore, make any possible effort to allow the option of enforcing reproducable results in multiprocessing environments.

4. CONCLUSIONS

Meteorological models will continue to have high requirements for large and fast memories. Taking into consideration the possible technological developments, it seems that there is a need for hierarchies of memories in addition to local memory.

The evident deficiencies in memory management should be addressed both by the manufacturers providing memory control structures and by the programming standards bodies providing additional levels of data structures.

A combination of both micro- and macrotasking is an attractive way of exploiting multiprocessing. However, the user-friendliness of multiprocessing programming environments should be improved by providing integrated debugging facilities and options to enforce reproducability of results.

MIX
Papier aus verantwortungsvollen Quellen
Paper from responsible sources
FSC® C105338

If you have any concerns about our products,
you can contact us on
ProductSafety@springernature.com

In case Publisher is established outside the EU,
the EU authorized representative is:
Springer Nature Customer Service Center GmbH
Europaplatz 3, 69115 Heidelberg, Germany

Printed by Libri Plureos GmbH
in Hamburg, Germany